Climate Chaos
Ecofeminism and the Land Question

EDITED BY

ANA ISLA

WITHDRAWN

INANNA Publications & Education Inc.
Toronto, Canada

Published in Canada by
Inanna Publications & Education Inc.
210 Founders College, York University
4700 Keele Street, Toronto, Ontario M3J 1P3
Telephone: (416) 736-5356 Fax (416) 736-5765
Email: inanna.publications@inanna.ca Website: www.inanna.ca

Note from the publisher: Care has been taken to trace the ownership of copyright material used in this book. The author and the publisher welcome any information enabling them to rectify any references or credits in subsequent editions.

Cover design: Val Fullard

Printed and Bound in Canada.

Library and Archives Canada Cataloguing in Publication

Title: Climate chaos : ecofeminism and the land question / edited by Ana Isla.
Names: Isla, Ana, 1948– editor.
Description: Includes bibliographical references.

Identifiers: Canadiana (print) 20190064056 | Canadiana (ebook) 20190064129 | ISBN 9781771335935 (softcover) | ISBN 9781771335942 (EPUB) | ISBN 9781771335966 (PDF)

Subjects: LCSH: Ecofeminism. | LCSH: Climatic changes. | LCSH: Human ecology. | LCSH: Human geography.

Classification: LCC HQ1194 .C55 2019 | DDC 304.2082—dc23

MIX
Paper from
responsible sources
FSC® C004071

To Kai, my three-year-old grandson, and his generation

Table of Contents

Acknowledgements

I would like to dedicate this book to my family—Nelson, Xochitl, Felipe, Alina, and Patrick. Also to my friends at *Canadian Woman Studies/les cahiers de la femme* where my feminism was born and nurtured, and my associates from the journal *Capitalism Nature Socialism* with whom I have participated in different activities, such as publications, debates, and conferences.

I would also like to thank Brock University, where I am a Professor, cross-appointed in the Centre for Women's and Gender Studies and in the Department of Sociology. With students in my "Women and Development" class for the Department of Women's and Gender Studies, we debated the very nature of international economic, social, and political relations at the root of our social and ecological crises. It was in examining the concept of development that we nurtured an ecofeminism as a theory of domination and of liberation that transcends differences of class, age, ethnicity, and speciesism.

With students in my "Environmental Justice" class for the Department of Sociology, we discussed how the social and ecological crises came about, and how they are debated in the institutions of development and environment, as well as those originated outside academia in environmental justice organizations. Our analysis was centred on the scientific consensus that the anthropocentric burning of fossil fuels during the last 180 years is the principal source on the rise of average global temperatures and climate change. During those years, I was pleased to see that year after year, most of the students were eager to delve into the ecological crisis and its potential consequences, notwithstanding that some others were offended and attacked me in academic evaluations and online.

I would also like to thank all the contributors to this important volume: Veronika Bennholdt-Thomsen, Irene Friesen Wolfstone, Patricia E. (Ellie) Perkins, Dorothy Attakora-Gyan, Margaret Kress, Rachel O'Donnell, Leigh Brownhill, Wahu Kaara, Terisa E. Turner, Jennifer Bonato, Reena Shadaan, Klaire Gain, and Ronnie Joy Leah.

Thanks are also due to Luciana Ricciutelli, Editor-in-Chief of Inanna Publications, and to Leigh Brownhill for her skillful copyedits.

In particular, I offer this book to the peasants and Indigenous women and men who allowed me to interview them and to examine their ecological resistance, while they were criminalized by governments, prosecuted by the justice system, and restrained by police.

Introduction

ANA ISLA

THIS SCHOLARSHIP, WITHIN THE TRADITION of ecofeminism, is motivated by the pursuit of justice in the face of climate change, the most urgent problem challenging humanity today. To understand this issue, we need to comprehend the operation of the ecosystem, which runs on physical and chemical laws that are fixed and that we must obey. However, globally, we have ignored the laws of nature.

There are several factors that explain persistent climate change. One of them is the increased amount of extraction and energy used by industrial production in the last two centuries. This has augmented global emissions of CO_2 that have resulted in changes to the natural composition of the atmosphere of the planet.

The atmosphere and the ocean temperatures control the climate. Ninety percent of the extra heat in the atmosphere generated by greenhouse gases has significantly raised the temperature of the planet's oceans, melting the glaciers of the Arctic and Antarctic. If the ice caps and glaciers disappear completely, the atmospheric temperature will rise two or more degrees and sea levels will rise by eighty meters.

These changes in the oceanic and atmospheric circulation patterns have resulted in significant global warming, which could trigger "abrupt" climate change (UCC)—unexpected and dramatic changes in global weather patterns—that we are largely ill-prepared to deal with or adapt to in a time span of decades or even centuries. Further, "As Earth warms, melting of ice caps and glaciers, increased precipitation and other inflows of fresh water to the North Atlantic Ocean may weaken or shut down thermohaline circulation."[1] This change in ocean circulation may shut down the flow of warm water in the North Atlantic creating ice tsunamis. As the concentration of the ocean's salinity increases, the cold water sinks and the warmer water evaporates, in turn changing temperatures in countries bordering the

Atlantic (e.g., colder winters and hotter summers in Europe). This collection examines current ecofeminist debates around the ethical issues that climate change has given rise to, as well as the impact of these changing weather patterns on the lives of women and men around the globe. In the first section, chapters explore the academic field of material ecofeminism; the second section provides an overview of the Land Question; and the third section examines the way in which reigning discourses of "sustainable development" have not only led to a commodification of nature, but have also effaced the multiple visions, uses, and relationships to nature of local human communities. The chapters in this book explore anticapitalist, antipatriarchal, and anticolonial solutions to climate change. Ultimately, we advocate resistance to the colonization of Mother Earth and insist that Pachamama is and will again be supreme.

ON ECOFEMINISM

The discussion in this section starts by presenting the economic contradictions that generate climate change, as identified by material ecofeminism. Ecofeminists have linked the ecological and social crises in a new arrangement of capitalism and patriarchy, the productivism of global capital (Shiva; Salleh "Global Alternatives," "From Eco-sufficiency"; Isla *The "Greening"*). Shiva expands on Veronika Bennholdt-Thompsen and Maria Mies's iceberg model for understanding the economy, and argues that climate change can be explained by the working of three economies in the same space and time: the market economy, the sustenance economy, and the natural economy. Bennholdt-Thompsen and Mies, and Shiva, sustain that each economic act is three-dimensional: *economic* (the market economy is the visible part of the economy), *socio-cultural* (the sustenance economy or social reproduction is the invisible, unpaid, or poorly paid part of the economy that sustains metabolic relations with nature), and *natural* (living organisms reproduced through ecological processes).

The *market economy* has separated the use value from the exchange value of a commodity, while the production and distribution of that commodity is privately owned to create surplus value. This system favours commodity production methods that require the use of techniques and technologies that have exacerbated levels of global pollution. Since WWII, the concentration of global capital has taken control of states, industries, and markets in order to expand commodity production to every corner of the planet. Currently, there is no part of the earth that is not incorporated into the commodity production, which is dominated by big corporations that require increasing levels of fossil fuels to sustain their production outputs. These fossil fuels contain high percentages of carbon that lead to deforestation and the loss of biodiversity of plants, trees, animals, bacteria, and other life forms. Furthermore, the market economy increasingly relies on consumption as "waste." A few examples include: 1) personal expenses, planned obsolescence, packaging; 2) military spending on new weapons and wars, ionosphere weather manipulation; and, 3) financialization of the economy or the growth of banking, asset management, insurance, and venture capital (investment banks, hedge funds). The progression of the market economy depends on the higher entropy of economic growth of global capital-productivism and its military-industrial complex. It expands CO_2 emissions and ozone pollution, in turn overloading the atmosphere and generating global warming. Industrialization, urbanization, and consumerism are, thus, breaking the planet into pieces.

Furthermore, current commodity production ignores the material reality of human existence that is also expressed in a patriarchal manner through the domination of women, feminized and racialized bodies, and nature. It establishes overexploitation of paid and unpaid workers, and manufactures unequal exchange between men and women, white and other ethnicities, and between core and periphery.

In the same space and time, *the sustenance economy* is reproduced by women through the maintenance of the home, bearing children, and reproducing labour power; by peasants working in subsistence farming and horticulture; by Indigenous peoples' cultural survival and lay knowledge; and by colonies or underdeveloped countries that reproduce biological infrastructure for all economic systems. These economies (women, peasants, Indigenous, and colonies) understand the complexity of the everyday ecological world, and bridge human and natural cycles (without changing the CO_2 cycle), because their work is universal, integral, and in touch with the ecosystem. They rarely use up more than is needed for bodily provisioning. All of these groups (women, peasants, Indigenous, and colonized countries) can meet their needs for housing, health care, and nutrition through their resources and knowledge. However, they have been living with violence, reduced to cheap labour or hereditary material (DNA), while their labour and resources are ransacked by military might. Suppressing their voices has had significant and disastrous effects on our ecosystem.

The *natural economy* that conflicts with the market economy is

3

reproduced through ecological processes without producing waste: *the water cycle* is organized around the heat of the sun that causes water in rivers, lakes, and oceans to evaporate. Water precipitates back to the earth's surface in the form of rain. *The oxygen cycle* arises from plants and trees that produce oxygen. Animals breathe the oxygen, exhale carbon dioxide (CO_2), and produce more oxygen. The *nitrogen cycle* is organized by bacteria that help nitrogen change between states— through decomposing biological matter, for example—allowing plants and trees to absorb nitrogen through their roots. We (human and non-human beings) get nitrogen from plants and bacteria; however, since the 1930s, the industrial world has been generating more nitrogen fertilizer (such as ammonium nitrate) than all bacteria on the planet are able to produce.

As capitalism and patriarchy battles the sustenance economy and the natural economy, military and corporate forces also manufacture atmospheric battles with worldwide nuclear radiation hazards (Hiroshima and Nagasaki in 1945, Chernobyl in 1986, Fukushima in 2011) and ionosphere heaters (Bertell). Consequently, today, the disrupted space includes the entire planet, resulting in climate chaos, and leading towards the destruction of the life-supporting systems of the entire planet.

The chapters in this book highlight some of the ecological problems confronting us today on a global level and propose alternative frameworks to debate the issues and search for solutions.

In the first section, Ana Isla's chapter, "Climate Chaos: Mother Earth Under Threat," focuses on climate change as a consequence of the disproportionate amount of carbon dioxide, and other poisons, in the atmosphere. It looks at a cluster of events that point to catastrophic climate change taking place all over the world. Then, it presents the ecofeminist debates on the reproduction crisis, specifically the capitalist commodity production and consumption that attacks human and nature's capacities. Additionally, the chapter highlights ecosocialist arguments on how energy used in commodity production and consumption can affect the biospheric balance that has evolved over hundreds of millions of years. The chapter concludes by pointing to three perspectives that may pave the way out of climate chaos.

Veronika Bennholdt-Thomsen's chapter, "Money or Life? What Makes Us Really Rich," critiques the fact that economic theory is blind to the material basis of our livelihoods. Furthermore, she argues that this blindness has infected our entire modern culture. She explains that the material and services we need to survive are at present only taken seriously when associated with industrial commodities and wage labour. Only products that can be measured, bought, or sold with money have value. Seeing life through this lens, the material and life-giving values of self-sufficiency, self-determination, autonomous work, and unpaid services go unrecognized.

Irene Friesen Wolfstone's chapter, "Deconstructing Necrophilia": Eco/feminist Perspectives on the Perversion of Death and Love," criticizes the Western addiction to self-gratification through consumption, an addiction so intractable that it must be fed even when it clearly contributes to climate change. The author searches for ecocentric philosophies, such as a philosophy of natality, which focuses on the human capacity to bring forth the new, the radical, and the unprecedented into the world.

"The Guardians of Conga Lagoons: Defending Land, Water, and Freedom in Peru," by Ana Isla, is a case study of an environmental justice movement, born in 1999, against the Yanacocha open-pit mining of Newmont Mining Corporation, Buenaventura, and the World Bank. By focusing on women and men subsistence farmers, Indigenous people, and civil society participation, the article brings to the forefront three moments of the struggle: first the defence of the land; second, the need for water protection; and third, the struggle for freedom.

Ellie Perkins's chapter, "Ecofeminism, Commons, and Climate Justice," offers recent work in ecological economics, degrowth, climate justice, and political ecology focusing on the "commons" as an emergent paradigm for sustainable governance institutions to address or rectify the ecological crisis. This chapter summarizes definitions and typologies of the commons, gives some examples of the commons that help to further climate justice, and discusses these ideas from an ecofeminist perspective.

"Finite Disappointments or Infinite Hope? Working Through Tensions within Transnational Feminist Movements," by Dorothy Attakora-Gyan, proposes to reassess food sovereignty in ways that question how power is reinforced and disrupted across various axes of difference. The author debates questions of power, privilege, and representation and how feminists—both academics and activists—can reinforce the very power imbalances feminism seeks to dismantle.

ON THE LAND QUESTION

The second section offers work on women's, peasants', and Indigenous peoples' activism and resistance to ecological exploitation and land dispossession. In the Americas, land is central. First and foremost, it is the homeland of Indigenous people. As capitalism is inseparable from coloniality, the assaults by the productivist economy on the Indigenous sustenance economies of the so-called core and peripheral countries—organized by global capital in association with its military-industrial complex—have resulted in the polarization between rich and poor countries, and in a sexist and racist society. This polarization is responsible for the social and ecological degradation of Indigenous territories and Indigenous bodies. However, as climate chaos expands, various resistance movements claim space to mobilize and analyze how their varied experiences of dispossession are connected, historically and geographically, within patriarchal and global capitalist systems, and its military-industrial complex. In this section, the book looks at the imperial histories and geographies that connect patriarchy, capitalism, and colonialism in the North and the South in the twentieth and twenty-first centuries. To make the connection between confiscation of land and resistance, this section starts with an exploration of South America's history and how the expansion of the fossil economy to increase economic growth was based on policies implemented by the military dictatorships on behalf of commercial banks and financial markets. Then, chapters look at the re-emergence of the environmentalism of the poor and the commons movements to defend territories and lives against global capital's assaults in Canada, the United States, and elsewhere around the world.

Land dispossession has a very high cost among those who are racialized and living on the periphery. In Latin America, the new acceleration of land grabbing and distributional conflicts started during the glory days of the Cold War, when the military-industrial complex—which was created in the name of defending the U.S. from its external enemies—equipped nations with weapons, corrupted governments, and trained militaries to carry out the "modernization project." The School of the Americas (SOA) trained over 64,000 Latin American soldiers in counterinsurgency techniques, including sniper training, commando and psychological warfare, military intelligence, and interrogation tactics. These graduates have consistently used their skills to wage war against their own people. The SOA has left a trail of blood and suffering in every country where its graduates have returned. For this reason, the School of the Americas has been historically dubbed the "School of Assassins" (Gill).

The first military takeover started with the military coups of 1963 in Ecuador, Guatemala, and Honduras in 1963, and then in Brazil and in Bolivia in 1964, in Argentina in 1966, in Peru in 1968, and in Panama in 1968. The second bloody military occupation continued in the 1970s, 1980s, 1990s, and 2000s. This new round of military intervention was deemed necessary to organize neoliberal states based on debt burdens (Roddick; Isla "An Environmental Feminist Analysis," *The "Greening"*) by increasing interest rates in the 1970s. Military dictatorships were fabricated in Bolivia in 1971, in Honduras 1972, in Chile and Uruguay in 1973, in Argentina in 1976, and in El Salvador in 1979. Military dictatorships, which were in place between 1963 and1990, were heavily implicated in disappearances, tortures, and death squads. More than a million Latin Americans were killed during this time (Castañeda).

In Canada, for centuries, Aboriginal people have suffered genocide as a result of state and institutional racism. First, the Truth and Reconciliation Commission has confronted and documented the deliberate genocide of Aboriginal people by the Government of Canada, through Canada's Residential Schools system. Second, the "Sixties Scoop" of Indigenous children from their families of birth, and their subsequent adoption into predominantly non-Indigenous, white middle-class families, resulted in children losing their language, culture, and identity as Aboriginal people. Third, land occupation and enclosure by tar sands production in Alberta, in particular the Athabasca River Basin, has disrupted cultural traditions as well as compromised livelihoods, hunting, and fishing in the homeland of Cree, Dene, and Métis Indigenous peoples (Black, T. et al.). Fourth, Brenda McLeod has examined the lives of Indigenous women within the settler society and concluded that in Canada's *Indian Act*, the female body is not recognized as having rights and an Indigenous woman's status is dependent on the men in her family.

In 2012/2013, three Indigenous women and one settler-descendent launched Idle No More, the largest Indigenous movement in Canada, against resource exploitation on First Nations territories, and demanded respectful consultation on land claims, treaties, water protection, and resource-sharing ("History"). The Idle No More website points out that the treaties between First Nations and the British Crown cannot be altered or broken by either side. The theft of land and resources by the state and corporations has already left many lands and waters poisoned, and has deepened inequality between Indigenous peoples and the settler society. The legacy of the residential school system—racism, extreme poverty, involvement with state child welfare agencies, high rates of addiction, mental health issues, and family violence—make Indigenous women and girls extremely vulnerable to sex trafficking, disappearance, and murder (Cress).

In the United States, Indigenous people are also fighting against 525 years of occupation and enclosure. At the Dakota Access Pipeline that runs through Standing Rock in Sioux Treaty Land, Indigenous people actively demonstrated and protested the expansion of oil pipelines in their territory. Indigenous communities argued that the 1,170-mile-long pipeline under Lake Oahe in North Dakota has the potential to damage sacred cultural sites, contaminate drinking water, and harm tribal members' right to fish, hunt, and gather wild rice in ceded, off-reservation territory. In 2017, police launched an escalating and violent crackdown against Indigenous women, men, youth, and their supporters with dogs, water cannons in subfreezing temperatures, firearms, and arrests ("Standing Rock and the Dakota Access Pipeline").

The authors in the second section of this book examine the ongoing resistance of Indigenous women and men who are claiming back their dignity and fighting against patriarchal capitalist greed and neocolonialism. These anti-colonial, anti-capitalist and anti-patriarchal movements focus on the commons or non-commodified spaces of land, water, and self-determination, and integrate feminist and Indigenous worldviews and experiences in their efforts to achieve justice for all.

In "*Sasipihkeyihtamowin: Niso Nehiyaw iskwewak,*" Margaret Kress argues that in the Cree knowledge systems, place is land, spirit, and body. *Nehiyaw* worldview helps dismantle the dominant narrative, and brings truth to the imparting of colonial histories and the Indigenous resistance that follows. Within the Canadian initiative for reconciliation, settlers must be first to extend the "olive branch" as they recognize and accept that Indigenous peoples are the ones to lead in this renewal; Indigenous peoples are the ones to set the tone in restoring language, ceremony, oral history, culture, and tradition as an ontology of environmental justice.

Ana Isla, in her chapter, "Climate Change and Environmental Racism: What Payments for Ecosystem Services Means for Peasants and Indigenous People," examines The Kyoto Protocol and REDD programs (Reducing Emissions from Deforestation and Forest Degradation) as environmental management perspectives of the United Nations. Both programs propose to resolve climate change through the monetization of nature, meaning the transformation of ecosystem components and processes into commodities or services that can be privately appropriated, assigned exchange values, and traded in markets. This chapter outlines the consequences of these programs on Indigenous and peasant communities in Brazil and Costa Rica.

Rachel O'Donnell's chapter, "Biotechnology and Biopiracy: Plant-based Contraceptives in the Americas and the (Mis)management of Nature," claims that in the present global political economy, the botanical history and contemporary biopiracy of the plant Apacina demonstrates how knowledge and power are intimately linked. Currently, women's knowledge of plants is manipulated to increase their production, even though control violates the ecosystem in the name of progress, science, and development. Biotechnology and biopiracy separate food and medicine, while women in rural Guatemala see food and medicine as one and the same.

In "Building Food Sovereignty in Kenya: From Export to Local Agricultural Value Chains," Leigh Brownhill, Wahu M. Kaara, and Terisa E. Turner discuss the challenges of maintaining a food-first focus in a capitalist market and policy context that is hostile to subsistence agricultural production and trade. In Maragua, Kenya, through direct household food self-provisioning, one group of farmers is tackling hunger and malnutrition by refusing to participate in the corporate market, and instead reviving Indigenous agricultural technologies and practises, and by recreating peoples' markets.

Jennifer Bonato's chapter, "Monsanto and the Patenting of Life: Primitive Accumulation in the 21st Century," draws parallels between the appropriation of women's bodies by the state and church as producers of labour power during the witch-hunt and the current appropriation of the reproductive capacities of seeds' bodies through patenting. She argues that in North America, GMO-label debates and recent developments in the biotechnology industry have proliferated biotechnology's power over legal and regulatory systems.

In "'I Know My Own Body ... They Lied': Race, Knowledge and Environmental Sexism in Institute, WV and Old Bhopal, India," Reena Shadaan argues that the importance of race- and class-based analysis in the context of environmental justice cannot be overstated. By focusing on two experiences, the Kanawha Valley and the Bhopal Gas Disaster, women in her chapter discuss the dismissal of their health experiences, and specifically doctors' refusal to draw connections between their health struggles and their exposure to industrial toxins.

In "Water is Worth More than Gold: Ecofeminism and Gold Mining in the Dominican Republic," Klaire Gain links the exploitation of women and ecology while examining the impacts of Canadian mining corporations within the Dominican Republic. She explores notions of corporate social responsibility and Canadian foreign policy and their influence on the mining sector. A case study of the Pueblo Viejo gold mine in the Dominican Republic reveals the impacts of extraction and exploitation on local ecologies and communities, while also exhibiting the strong women-led resistance to Canadian mining corporations there.

9

In "Indigenous Andoas Uprising: Defending Territorial Integrity and Autonomy in Peru," Ana Isla addresses the social and ecological catastrophes caused by oil extraction activities in Peru by examining a 2009 court case that was part of the aftermath of a 2008 uprising by three Indigenous groups in Andoas town. It argues that Indigenous labour, resources, and territories in the Amazon rainforest are the new sources of accumulation within the global state and "green" capitalism re-launched in the face of ecological crisis.

SUSTAINABLE DEVELOPMENT

Those who have a lot of money to lose, such as the oil corporations and their think tanks, embodied by the United Nations, want to continue to create doubt and confusion about why there is climate change and how to confront it. Instead of confronting the roots of the problem, they have created a new model of capital accumulation, a so-called green economy, based on a key concept—"natural capital"—which is used to refer to the goods and services provided by the planet's stock of water, land, air, and renewable and non-renewable resources. These policies are often driven by large international NGOs and UN agencies that are shaped by global capital. Consequently, "green" politics have been hastening the conditions we critique in this book.

In "The 'Greening' of Costa Rica: A War Against Subsistence," Ana Isla disputes the argument that sustainable development is desirable for the entire world, including and most particularly for the so-called "underdeveloped" world. Isla argues that Costa Rica has been an important "laboratory" for experimentation in innovative environmental governance mechanisms. As a result, new areas of global intervention were opened up and nature entered the domain of neoliberal politics. She has called this process "greening" to indicate how the ecosystems of indebted countries are increasingly becoming destabilized, especially through an ever-growing pressure for resource extraction, at the same time that neoliberal pundits advance the language of more ecologically friendly economic policies and programmes. In this light, "greening" can be understood as a new phase of capital accumulation.

Finally, Ronnie Joy Leah's chapter, "Earth Love: Finding Our Way Back Home," celebrates the power and beauty of the Earth, indeed of all life. The writer carries us forward with an analysis that promotes wholeness, balance, and reconnection in today's fractured world.

In brief, this book assesses specific forms of damage to human nature and non-human nature, and dramatizes the need for change to the logic of policies that value money over lives. It also critiques the approach of the United Nations as well as global capital that is reshaping nature to serve the demand of the world markets, instead of assessing and debating heat waves, forest fires, melting of Artic and Antarctic ice, sea level rise, floods, droughts, famines, loss of genetic diversity, climate-driven refugees, and increased conflicts that have been unfolding. For instance, The Intergovernmental Panel on Climate Change Report (IPCC) 2018, is overly conservative. At the current level of political commitments, the world is on course for a disastrous two or more degrees of warming, economic and climate chaos, and the resulting social upheaval.

ENDNOTES

[1]Thermohaline circulation is mainly driven by the formation of deep-water masses in the North Atlantic and the Southern Ocean caused by differences in temperature and salinity of the water.

REFERENCES

"A History of Idle No More." Idle No More 23 December 2012. Web.

Bertell, Rosalie. *No Immediate Danger: Prognosis for a Radioactive Earth*. Toronto: The Women's Educational Press, 1985.

Bertell, Rosalie. *Planet Earth: The Latest Weapon of War*. Montreal: Black Rose Books, 2001.

Bennholdt-Thomsen, Veronika and Maria Mies. *The Subsistence Perspective: Beyond the Globalized Economy*. London: Zed Books, 1999.

Black, Toban, Stephen D'Arcy, Tony Weis, and Joshua Khan Russell, eds. *A Line in the Tar Sands: Struggles for Environmental Justice*. Toronto: Between the Lines, 2014.

Castaneda, Jorge G. *Utopia Unarmed: The Latin American Left After the Cold War*. New York: Vintage Books, 1994.

Cress, Tori. "Don't you mean, why I still have to march." Idle No More. 27 Jan 2017. Web.

Gill. Lesley. *The School of the Americas. Military* Training and Political Violence in the Americas. Duke University Press, 2004.

Intergovernmental Panel on Climate Change Report (IPCC), 2018. Web.

Isla, Ana. "An Environmental Feminist Analysis of Canada/Costa Rica Debt-for-Nature Investment. A Case Study of Intensifying Commodification." Unpublished dissertation, OISE/University of

Toronto, 2000.

Isla, Ana. *The "Greening" of Costa Rica: Women, Peasants, Indigenous People and the Remaking of Nature.* Toronto: University of Toronto Press, 2015.

McLeod, Brenda. "First Nations Women and Sustainability on the Canadian Prairies." *Canadian Woman Studies/les cahiers de la femme* 23.1 (2003): 47-54.

Ponte, Lowell. *The Cooling: Has the Next Ice Age Already Begun? Can We Survive It?* Upper Saddle River, NJ: Prentice Hall, 1976.

Roddick, Jackie. *The Dance of the Millions: Latin America and the Debt Crisis.* London: Latin America Bureau (Research and Action) Ltd., 1988.

Salleh, Ariel. "Global Alternatives and the Meta-industrial Class." *New Socialisms: Futures Beyond Globalization.* Eds. Robert Albritton, Shannon Bell, John R. Bell and Richard Westra. London: Routledge, 2004. 201-211.

Salleh, Ariel. "From Eco-Sufficiency to Global Justice." *Eco-Sufficiency and Global Justice: Women Write Political Ecology.* Ed. Ariel Salleh. London: Pluto Press, 2009. 291-312.

Shiva, Vandana. *Staying Alive: Women, Ecology and Development.* London: Zed Books, 1989.

"Standing Rock's Fight Against Dakota Pipeline Continues While Tribe Plans for a Fossil-Free Future." *Democracy Now*, 04 July 2017. Web.

"Standing Rock and the Dakota Access Pipeline: Native American Perspectives: Tribal, National, and International Organizations." (n.d.). Berkeley Library, University of California. Web.

Truth and Reconciliation Commission. Canada's Residential Schools, Volumes 1-6. Manitoba: National Centre for Truth and Reconciliation, 2015-2016. Web.

Union of Concerned Scientists (UCC). "Abrupt Climate Change." 09 July 2004. Web.

On Ecofeminism

1.
Climate Chaos

Mother Earth Under Threat

ANA ISLA

CLIMATE CHANGE IS ALREADY UNDER WAY with unpredictable consequences. Evidence of changes to the earth's physical, chemical, and biological processes is obvious everywhere. Greenhouse gas emissions have increased the carbon dioxide concentration in the atmosphere. In the past, half of this carbon was stored in forests, while the other half was removed by oceans. Currently, as a result of global deforestation and warming oceans, oxygen is at its lowest breathable point. According to NASA and the National Oceanic and Atmospheric Administration ("NASA, NOAA Data"), "The planet's average surface temperature has risen about 2.0 degrees Fahrenheit (1.1 degree Celsius) since the late nineteenth century." For example, 2018 was the warmest year on record, smashing records set in, 2017, 2015, and 2014. In January 2017, Badlands National Park tweeted facts about climate change, despite gag orders from President Trump.

- "The pre-industrial concentration of carbon dioxide in the atmosphere was 280 parts per million (ppm). As of December 2016, 404.93 ppm."
- "Today, the amount of carbon dioxide in the atmosphere is higher than at any time in the last 650,000 years."
- "Ocean acidity has increased 30% since the Industrial Revolution."

Ecological degradation is devastating the planet and the earth is becoming increasingly inhospitable with unprecedented weather events. Changing temperatures alter the balance of communities and further degrade vulnerable ecosystems. For instance, as a result of temperatures dropping to –20°C, in Puno, Peru, between July and August 2015, seventeen children died; in addition, around 171,000 alpacas died from hypothermia, diarrhea, pneumonia, and starvation. The Peruvian government is ill prepared for major crises; consequently, support was late and inadequate (Sanchez). Meanwhile in Canada, as a result of a drier winter combined with an unusually hot, dry air mass over Northern Alberta, the temperature climbed to 32.8°C (91°F) ("Daily Data Report"). Consequently, in May 2016, there were 49 simultaenous wildfires, with one fire in Fort McMurray that ran out of control and destroyed a significant portion of the twon. The wildfire covered an estimated 522,892 hectares, including 2,496 hectares in Saskatchewan. Mandatory evacuation orders were put in place for several towns. Battling fires across the province were 2,794 firefighters and support staff, 147 helicopters, 16 tankers and 233 pieces of heavy equipment ("Final Update"). During the subsequent summer of 2017, hundreds of wildfires spread through the provinces of British Columbia, Manitoba, and Saskatchewan.

Destructive events due to warmer ocean surface temperatures are also taking place. "Researchers were able to document how the dramatic shift in oxygen concentrations was characterized by disturbances in the global carbon cycle" (David). Warmer oceans hold less dissolved gases, including oxygen, which affects marine organisms, particularly mammals. In January 2014, in Peru's Pacific, more than 400 dead dolphins washed ashore (Foley); similarly, in New Zealand, in February 2017, more than 400 whales beached themselves to die ("Farewell Spit"). El Niño, which is a cold, low-salinity ocean current that runs along Ecuador, Peru, and Chile, has been heating and altering weather in all Pacific Rim countries. Each El Niño and La Niña cycle in the past 20 years has occurred with increased frequency and violence. In March 2017, Peru's northern coast's sea surface temperatures rose from 6° to 27° Celsius resulting in intense rain that produced historic flooding and mass destruction in four of the Northern regions (Climate Prediction Centre).

"The largest overall volume of oxygen was lost in the largest ocean—the Pacific—but as a percentage, the decline was sharpest in the Arctic Ocean, a region facing earth's starkest climate change" (Mooney). In the name of progress, a schizophrenic madness fueled by the primacy of industry, is trying to "save" the Arctic with geoengineering. As the Arctic ice is melting, a group of researchers from the University of Arizona are planning to refreeze it using around ten million wind-powered water pumps installed around the region (Martins). Meanwhile, the Canadian High Arctic Research Station (CHARS), a federal research organization, is doing an inventory of the resources contained in the Arctic to provide permits for mining and oil projects, and thus legalize harm. Glaciers in Antarctica are also being destabilized. The ice shelves in Larcen A, B, and C collapsed in 1995, 2002, and July 2017. "When an ice shelf collapses, the glaciers behind it can accelerate toward the ocean." This has changed the landscape of the Antarctic Peninsula and the level of seawater, promoting the collapse of other vulnerable ice-shelves (Patel). Furthermore, as ice collapses, methane frozen deep underneath the ice is being released into the atmosphere, producing a powerful greenhouse gas with the potential to double global warming. Professor Guy McPherson, from the University of Arizona, sustains that the Artic ice will disappear in 2026, resulting in the end of the grain production as the basis of the "civilized society" ("Methane Unveiled").

In fact, the erosion of sea ice strikes at the very root of the Arctic ecosystem, for it provides a surface on which algae – the basic material on which the entire food chain in the region depends—can grow. "Algae lingers on the underside of sea ice and as spring begins there is a major increase in its growth," said Brown. "It is then eaten by tiny creatures called zooplankton, and they in turn are eaten by fish that are in turn eaten by seals, which are in turn consumed by polar bears. But if algae levels drop the whole food chain is disrupted." (McKie)

In sum, scientific evidence shows that carbon dioxide emissions in December 2017 surpassed 406.82 particles per million (ppm) in global carbon concentration (the maximum should be 300 ppm), and total methane CH4 (a major hothouse gas) has reached 1865 particles per billion (ppb) (the maximum should be 800) (NOAA). The Mauna Loa Observatory tells us that we are approaching climate catastrophe: the global average temperature is rising, and if another decade of business-as-usual fossil fuel emissions continues, we can reach +2°C, a dangerous warming threshold. Further, in the Arctic and Antarctica high temperatures are snowballing at rates twice the global average. The Artic and Antarctica are critical for climate stability because of the albedo effect, which is the ability of a surface to reflect sunlight. Ice- and snow-covered areas have high albedo, and an ice-covered Arctic reflects solar radiation that would otherwise be absorbed by the oceans and cause the Earth's surface to heat up. As global warming causes these ice- and snow-covered surfaces to melt, they become low-albedo areas, more heat is absorbed and sea levels rise exponentially, threatening island communities and low-lying coastal areas. Global warming is also destabilizing the polar jet streams, meaning more extreme weather events. Some scientists argue that the northern hemisphere's jet stream had crossed the equator into the southern hemisphere, threatening global food supplies and an "end of winter" (Irfan, Barclay and Sukumar; Adams).

By this time, we all understand that the sun supplies energy that keeps the earth "alive"; at the same time, solar energy is reflected back to outer space, which helps in maintaining a global equilibrium. If this did not occur, the planet would keep getting hotter, destroying all life. Leading scientists have developed an analysis of nine "planetary boundaries" that are crucial to maintaining life on earth. Global warming is one of these. Others include ocean acidification, stratospheric ozone depletion, the nitrogen and the phosphorus cycles, global fresh water use, change in land use, biodiversity loss, atmospheric aerosol loading, and chemical pollution.

In general, climate change is a modification of global climate patterns. Many things can cause the climate to change, such as the earth's changing distance from the sun and oceanic changes when volcanoes erupts. However, most scientists agree that humans can change the climate too (Steffen, Crutzen and McNeill), because civilization is a heat engine, whether we use solar panels or fossil fuels. Currently, there is an overwhelming scientific consensus that global warming and climate change is due to our use of and dependence on fossil fuels. "Burning these things puts gases into the air. The gases cause the air to heat up. This can change the climate of a place. It also can change earth's climate" ("What is Climate Change?").

The response to climate change has been varied. The establishment, through the United Nations, has articulated its response through its publication, *Our Common Future* (1987), which identified three agents of global warming: "poverty, uneven development, and population growth" (Chapter 1). In fact, all of these are products of "economic growth" as it is currently defined and pursued. Yet, the Report offered "sustainable economic growth" as the solution, which means extending markets to Nature! Since the UN adopted this approach at the Earth Summit in 1992, the "survival of the planet" has become the justification for a new wave of intervention, where people and nature become a domain of politics (Sachs). My study, *The "Greening" of Costa Rica: Women, Peasants, Indigenous People and the Remaking of Nature* examines Costa Rica's experimentation in "innovative environmental governance" and reveals the United Nations' initiatives that support "green capitalism." Theirs is not a

18

war against climate change, but against subsistence in order to expand capital accumulation.

In this chapter, my goal is to understand why climate change is occurring so suddenly that it is shifting the state of the earth's systems. I focus on global warming as a consequence of disproportionate amounts of carbon dioxide (CO_2) in the atmosphere, as well as the economics and military developed by science and technology. First, I discus the reproduction crisis where capitalist commodity production and consumption based on science and technology are attacking humans' and nature's capacities to the point of distorting organic bodies. I explore the materialist ecofeminist perspective on the interlocking oppressions of diverse forms of exploitation of bodies—human and ecosystems. Second, I highlight ecosocialist debates on the fallacies, contradictions, and problems with economic growth-dependence, pointing out how energy used in commodity production (surplus) and consumption (waste) can affect the biospheric balance that has evolved over hundreds of millions of years. Third, I look at the military-industrial complex and its biophysical ramifications. Fourth, I present several perspectives for the way out of climate chaos. I conclude with the demand that the way out must include the equitable participation of women, peasants, and Indigenous people in local and international collective political responses and actions.

ECOFEMINISM ON THE REPRODUCTION CRISIS: ATTACKING PEOPLE'S AND ECOSYSTEMS' ABILITY TO FUNCTION AS ORGANISMS

Ecofeminist theorists in particular have contributed to the analysis of male domination of women and nature. Reproduction became a political category in the 1970s during the rise of the Women's Liberation Movement. In their book, *The Power of Women and the Subversion of the Community,* Maria Rosa Dalla Costa and Selma James argued that "housework" was a precondition to capitalist production since it reproduced labour power and thus enabled the very process of capital accumulation. This reproductive labour was made invisible as work. Unpaid and performed in isolation, reproductive work has undermined women's physical integrity. The wages for housework political program thus demanded state remuneration.

Ecofeminists broadened the scope of feminist theory to include feminized bodies. They illuminated that it has been possible to sustain the illusion that economic growth is a positive and benign process because the costs have been borne primarily by what has been devalued: women, peasants, Indigenous people, colonized countries, and nature. Maria Mies, Veronika Bennholdt-Thomsen, and Claudia von Werlhof, in *Women: The Last Colony,* describe women's labour as the archetype of global non-wage labour. They also point out how many Third World workers are engaged in non-waged forms of production that are based on family labour. They describe Third World men and women as the "world's housewives," whereby the relationship between husband and wife is repeated in the relationship between the First and the Third Worlds. In this way, they link First and Third World women's labour to patriarchy and imperialism. Silvia Federici, in *Caliban and the Witch: Women, the Body and Primitive Accumulation,* draws a long historical line from the enclosure of the commons and the witch trials in Europe to modern day prostitution and the monetization of nature globally. She theorizes the systematic subjugation and appropriation of women, nature, bodies, and labour. She demonstrates how capitalist social relations are enabled by patriarchal re-arrangements that separate production from reproduction, that use wages to command the labour of both waged and unwaged workers, and that devalue women's status. The criminalization of women's bodies was transposed to the colonies and generalized to all "non-whites." Federici revalorizes the work of reproduction and reconnects our relation with nature, with others, and with our bodies to regain a sense of wholeness in our lives.

Carolyn Merchant's *The Death of Nature* offers an interpretation of the male domination of Western science and industrial progress. In examining the scientific revolution, she made an essential argument that the rise of modern science, technology, and the economies they enabled, was the driving force behind the oppression of women and nature. She also argued that the masculinist takeover of reproduction during the witch-hunts resulted in the triumph of mechanism over organism and the relegation of caring labour to a devalued domestic sphere. In the same line of thinking, Vandana Shiva and Ingunn Moser, in their book *Biopolitics,* demonstrate that for the last 70 years the market economy has been taking control of the sustenance economy through science and technology, in turn attacking people's ability to function ecologically and as organisms embedded in ecosystems. They highlight the fact that the mechanistic paradigm of genetic engineering has reduced all behaviour of biological organisms, including humans, to the level of genes. As a result of this reductionist thinking, genetic engineering exchanges genes between different species and designs new genetic material that have not previously existed. Life on earth is altered by these techniques, and the novelty contains negative ecological consequences, as knowledge developed under controlled conditions has limited validity in open systems.

Vandana Shiva also argues that food security comes from seed sovereignty. She illustrates further the assault on people and nature by genetic engineering. Shiva demonstrates that after World War II, particularly since the Green Revolution in the 1960s, chemicals used for warfare are now being used in agriculture as pesticides and herbicides. The industrial paradigm of agriculture that uses genetic engineering is centralized in large monoculture farms that exacerbate the rise of CO_2, displacing diversity to produce commodities oriented toward corporate control. The objective is to make the farmers pay for seed, soil, and climate data. Shiva exposes several myths of genetically modified organisms (GMO) including the myths that:

• "GMO are an 'invention' of corporations, and therefore can be patented and owned." In this way, it denies the innate self-organizing and self-reproducing capacity of living organisms, and it makes saving and sharing seeds among farmers illegal. As a result, the "industry is appropriating millions of years of 'nature's evolution', and thousands of years of farmers breeding." (xi-xii)

• "Genetic engineering is more accurate and precise than conventional breeding." Genetic engineering introduces genes from unrelated species into plants. Because it is not known that it will work, "GMO uses antibiotic-resistance genes that can mix with bacteria in the human gut and aggravate the crisis of antibiotic resistance we are currently facing." (xii)

Discussing the consequences of genetic engineering for humankind, Stephanie Seneff argues that chemicals used by the American agrochemical and agricultural biotechnology corporation, Monsanto, such as the herbicide Roundup, are implicated in new epidemics that are destroying people's bodies. Seneff shows that the autism epidemic today is a result of exposure to environmental toxicants, in particular to glyphosate (an active ingredient in Roundup). She explains that:

the incidence of autism has risen alarmingly over the past two decades in the United States, exactly in step with the increased use of Roundup as an herbicide on corn and soy crops. While the autism rate was estimated to be 1 in 10,000 in 1970, the most recent number from the Centers for Disease Control and Prevention (CDC) was 1 in 68 for twelve-year-olds in 2014. This number reflects the age group who were born in 2002—the number would surely be much worse for children born today. (77)

Seneff concludes that glyphosate used on corn and soy:

chelate the mineral manganese ... [and] other minerals, including iron, zinc, cobalt and magnesium, making them unavailable to the plants, and therefore making the plants deficient in these minerals. People who eat the plants would also become deficient.... Mineral deficiencies are the key source of many of the ailments facing humans in the modern world. (81)

Considering the results of genetic engineering in nature, my study on the Kyoto Protocol shows that genetically-modified trees, planted to sell carbon credits, are implicated in forest ecosystem destruction, in particular in its inability to regenerate secondary forests, which conserve biodiversity and regulate hydrology. In the carbon credit framework, forests are recategorized as "natural capital" and communities that used to live in the forest are declared enemies of the rainforest. The Costa Rican government, through its Ministry of Environment and Energy (MINAE), assesses the ability of large-scale agriculture entrepreneurs, in association with international capital, to sell carbon credits. They promote genetically-modified forest species of high yield and great market value, such as gmelina (*Gmelina arborea*), eucalyptus (*Eucalyptus deglupta*) and teak (*Tectona grand*), trees that are native to South Asia and Australia. Mono-arboriculture has been defined in this system as "reforestation" even though these plantations constitute artificial ecosystems that use massive amounts of chemicals to growth and increase the levels of CO_2 emitted into the atmosphere. As a result, these trees are susceptible to fire, pests, and disease. Further exposing the myth that these projects are about "reforestation," corporations are allowed to cut the trees down after fifteen years of growth and transform them into wood for floors or paper, boxes for fruit export, and for making furniture.

The monoculture of tree species has become a time bomb for biodiversity in Costa Rica. For instance, teak plantations have resulted in the erosion of flatlands, since the teak root system grows deep into the soil, but in the rainforest, the systems of nutrient and water absorption are at the surface. In general, nutrients and water are concentrated at a depth of between 70 and 100 centimetres. Consequently, teak trees are encircled by flaked soil. In addition, when it rains, the large-sized leaf accumulates great amounts of water that then pours violently onto the soil below. Water descending onto the soft surface destroys the soil. The far-reaching spread of the roots and the shade produced by the teak leaves obstruct the vegetative growth on the lower layers of the forest, which could otherwise prevent soil damage from the violent cascades. Under these trees, food for healthy bacteria, viruses, and people has disappeared. In contrast, the trees that are indigenous to the area are connected directly to each other through the multitude of creatures that relate to them as food, shelter, and nesting places; through their shared access to water, air, and sunlight; and through an underground system of fungi that links all the trees as a super-organism. By planting genetically-modified homogeneous forests, the Costa Rican rain forest ecosystems have been converted into sterile monocultures.

In summary, ecofeminists (Isla; Federici; Merchant; Mies and Bennholdt-Thomsen; Seneff; Shiva and Moser) have for years pointed out that the political economy of science and technology applied to commodity production and consumption is distorting the human and natural bodies of the planet. Next, this chapter discusses the political economy of the ecological crisis to explain climate change as a result of CO_2 increase. It looks at how the practices of commodity production and consumption combine with the physics, chemistry, and biology of our world. Ecosocialists (Altvater; Deleage; Malm) argue that in bioeconomic terms, every economic process must be seen from the point of the principles of thermodynamics, insisting that these principles apply both to natural systems as well as to social systems. Economic activity does not escape the laws of physics, and living beings are also subject to entropy.

ECOSOCIALISTS ON THE ECOLOGICAL CRISES ENGENDERED BY COMMODITY PRODUCTION (SURPLUS) AND CONSUMPTION (WASTE)

Industrial civilization in itself is a heat engine. Empirical analysis as well as historical data confirms the existence of a close connection between the quantities of fossil energy consumed, the state of wage work activity, and overall economic output. Growing primary material consumption affects the climate mainly because of the large amounts of energy involved in extraction, use, transport, and disposal, all of which create pollution.

Jean-Paul Deleage argues that the planet is being destroyed by the functioning of production and consumption begun a long time ago, accelerated by the capitalist system, and in particular by capital globalization and its military-industrial complex. He argues that capitalism has always treated nature as unlimited, therefore it has accelerated the entropic irreversibility of the planet. He sustains that in "bioeconomic terms, the extraction of raw materials and their elimination in the form of waste constitute the first and last phase of all economic activity" (39). Consequently, social and ecological problems are twofold: 1) they arise at the beginning of production, the first phase of all economic activity, when the raw materials are extracted and consumed; and 2) at the end when they are unloaded in the environment in the form of waste, the last phase of all economic activity.

Andreas Malm argues that the historical homeland of global warming is England. The British access to colonies and the textile factory system based on water power produced unprecedented quantities of commodities. However, everything changed when the handicraft worker was replaced by an engine as the central power of the factory. The engine required constant attachment to energy. Coal as a source of energy, with high carbon content as the mechanical prime mover, fostered the industrial revolution. The release of methane from fossil fuels increased entropy to threatening levels. Andreas Malm argues that "[W]ith coal placed right under the driving fire of capital accumulation, as the fuel transmitting motion to the labour process, a spiral of growing fossil fuel combustion was, for the first time, integrated into the spiraling growth of commodity production" (292). He associates "fossil capital" with the burning of fossil fuels (the main source of human-induced climate change) and capitalism's development of industrial production.

Following Marx's general formula of capital, Malm writes that the general formula of fossil capital is as follows:

$M - C$ (labour + Means of Production (F))... P (CO_2)...$C' - M'$

M = Money as capital;

C = Capital buys commodities (labour + raw material + machinery) including F (energy or fossil fuel);

P = Production burns fossil energy and dissipates carbon dioxide;

C' = Final Commodity;

M' = Increased money as capital; i.e., profit.

Malm argues that:

...fossil fuels are now subject to productive consumption...the combustion of fossil fuels in their solid form and the consequent release of CO_2 do not in themselves generate any value for the capitalist, but they are material requirements for value creation (profits).... But fossil capital is also a process. It is an endless flow of successive valorizations of value, at every stage claiming a larger body of fossil energy to burn. (289, 290)

The growth of fossil fuels in commodity production has fed on a succession of disruptions of ancient local and regional ecological balances. For instance, by 1952, London was affected by severe air-pollution arising from the use of coal (Davis, Bell and Fletcher). Elmer Altvater found major sources of contradiction in the functioning of capitalism through identifying the natural process of time and space, described as "ecological modality," as well as the attempt by capital to codify and control time and space with the aim of speeding up the process of capital accumulation, defined as "economic modality." He argues that two different modalities of space and time conflict upon a territorial-social reality, as biological time and reproduction are slower than economic time or commodity production. As capitalism is a system that only understands value in terms of money capital, the perpetual drive toward short-time, or "t" (economic time) accumulation, is in direct conflict with the ecological limits of "T" (historical time) that allow and provide for life on this planet. By applying the principles of entropy in the use of energy in production and consumption where no transformation of energy or matter is perfectly efficient, Altvater concludes that recycling is thermodynamically impossible, thus energy and raw materials are used only once. As a result of this ecological and economic conflict, the extinction rate has accelerated since the escalation and spread of industrial capitalist development, and has created the sixth mass extinction of biodiversity event currently under way that eliminates at least 150 species a day. Gerardo Ceballos, Paul Ehrlich, and Rodolfo Dirzo "indicate that beyond global species extinctions Earth is experiencing a huge episode of population declines and extirpations, which will have negative cascading consequences on ecosystem functioning and services vital to sustaining civilization.... [W]e emphasize that the sixth mass extinction is already here" (7).

The next section argues that climate change could not have increased as fast as it did in the last 70 years without the military-industrial complex's dominance, which has been undertaking counter-insurgency warfare (Pursell), terrorizing populations (Castaneda), repressing internal dissent (Dahi), and destroying the atmospheric layer (Bertell *Planet Earth*).

THE MILITARY-INDUSTRIAL COMPLEX AND
ITS BIOPHYSICAL RAMIFICATIONS

Indicating contempt for the value of life, the ruthless worldwide expansion of neo-colonial and neo-liberal capitalist politics was organized with the support of the military-industrial complex. It is defined by Carroll Pursell as:

an informal and changing coalition of groups with vested psychological, moral, and material interest in the continuous development and maintenance of high levels of weaponry, in preservation of colonial markets and in military-strategic conceptions of international affairs. The groups include ... congressmen, labor leaders, corporation executives, church spokesmen, university professors, and professional soldiers.... At the center of the complex ... is the unique relationship between the Pentagon as buyer and the war industries as seller. (ix)

Pursell shows that after WWII, a decision was made to fit the war program into the existing economic structure to support trade. "A peaceful world secure for trade was henceforth to be guaranteed by a standing military establishment backed up by a war economy prepared to offer massive support to foreign interventions" (7). Since the 1960s, the military-industrial complex has been key to the new imperialism as it enhances the war-making capability of the U.S. and it dominates people and nature. "This increases the likelihood of recourse to 'solutions' based upon military power" (Melman 290). Currently, global capital, led by the United States, and through war, has seized oil from Iraq and Libya in the Middle East, and minerals from Afghanistan.

The biophysical ramifications of capitalist military[1] alienation is discussed by Rosalie Bertell (*Planet Earth; No Immediate Danger*). Bertell's *No Immediate Danger: Prognosis for a Radioactive Earth* critiques war and the nuclear industry of the military-industrial complex that starts in the research lab but ends in the atmosphere. She maintains that during the Cold War, the UK, the U.S., France, and the Soviets explored the atmosphere with war intentions. Bertell provides evidence of the U.S. nuclear atmospheric testing in the Pacific Ocean since 1946. In ten years, the U.S. set off more than 86 nuclear bombs, and since 1983 have also fired missiles in the Marshall Islands of Micronesia (Pacific Ocean). She also details that between 1966 and 1974, France's Centre d'Expérimentation du Pacifique detonated seven nuclear tests in the Pacific Ocean. She argues that as a result of these nuclear bombs, the magnetic field of the earth, the Van Allen Belts,[2] which protect the earth from destructive solar wind charged particles, has been damaged. Disturbance in the electromagnetic field may also explain the current contamination and reduced oxygen levels of the Pacific Ocean that now is disgorging dead mammals (see my introduction). She describes that in 1974, Peru denounced the radiation fallout at the United Nations and broke diplomatic relations with France because the tests poisoned its fisheries and citizens. Since then, "El Niño" or the Humboldt Current, which is a cold, low-salinity ocean current that runs along Ecuador, Peru and Chile, has been heating and altering weather in all Pacific Rim countries, affecting Australia, New Zealand, and Fiji (Bertell 102).

Furthermore, in *Planet Earth: The Latest Weapon of War,* Bertell argues that the "U.S. military's aim of dominance extends to atmospheric heating" (125) in the ionosphere, and that it is experimenting with weather and climate manipulation that distorts the organic bodies. After WWII, the military researched and applied atmospheric modifications with chemicals introduced in the atmosphere, and its manipulation could create rain, and direct hurricanes, tornados, and monsoons. Lowell Ponte, in *The Cooling: Has the Next Ice Age Already Begun,* provides evidence that weather modification started during the Vietnam War, when the U.S. investigated ways of using artificial lightning as a weapon in Project Skyfire and hurricanes in Project Stormfury. The military also investigated the possibility of destroying the ozone layer over North Vietnam with lasers or chemicals, causing damage to crops and humans. Bertell adds that "[t]wenty million gallons (of Agent Orange) were sprayed over 10 per cent of Vietnam, reducing dense jungles and mangrove forests to barren wasteland. Many children in the polluted areas were born with learning difficulties or severe impairments" (Bertell *Planet Earth* 158).

Bertell also demonstrates that the military researched and applied atmospheric modifications with waves, heat, or electromagnetic force introduced in the upper atmosphere, to interrupt or distort the normal wave motions. Bertell uncovers a military network of ionospheric heaters, The High-Frequency Active Auroral Research Program (HAARP), that fires electromagnetic waves that can seriously damage the ecosystem, particularly the Artic. HAARP modifies the ionosphere for a military dominance agenda. Its prime goal is the manipulation of the atmospheric electrojet as a weapon, capable of redirecting significant amounts of electromagnetic energy toward the enemy; its additional usage is for the generation of low-frequency waves for communication, such as radio, Internet, cell phones, satellite television, and submarines. HAARP's massive transmitters produce incisions in the ionosphere that are repaired naturally by the sun. However, Bertell states that these man-made incisions "may destabilize a system that has maintained its own cycle for millions of years" (*Planet Earth* 123). She presents the Federal Environmental Impact Statement filed by the air force for HAARP that its transmissions "can raise the internal body temperature of nearby people, ignite road flares in the trunks of cars, detonate aerial munitions used in electronic fuses and scramble aircraft communications, navigation and flight control systems" (*Planet Earth* 126). HAARP low-frequency waves, which are reflected back to the earth by the ionosphere, create vibrations at great distances through the earth. This could be used to induce effects such as earthquakes (Bertell *Planet Earth* 130) as well as generate direct weather effects. Around HAARP there are other military networks, such as SuperDARNS that monitors ionospheric heaters in the lower atmosphere; and GWEN (Ground Wave Emergency Network) a storm making system and also weapon and drone guiding system.

Furthermore, Bertell found that since the 1990s, the U.S. and its allies deliberately polluted the environment as a strategy of war in Kosovo and Iraq to test the newest military weapons based on depleted uranium (DU) ordnance (*Planet Earth* 28). In the Kosovo war, NATO used DU bombs and missiles. When fired, DU burns at over 3000 degrees Celsius and becomes uranium aerosol containing microscopic radioactive particles that contaminate all life (land, air and water) and continue to harm long after the end of the war. The dependence of the powers on Gulf oil caused NATO to drop 88,000 tons of explosives (seven times the force of the Hiroshima bomb) on Iraq. The U.S. used "heat weapons" with solar energy to burn the country. Between the U.S. military attacks and Iraqi forces, tankers, pipelines, oil wells, refineries, and storage tanks were destroyed. Bertell reported that enormous clouds of toxic smoke from millions of barrels going up in flames each day, and at least 40 tons of radioactive dust on the battlefield of Kuwait and Iraq (*Planet Earth* 36) impacted the atmosphere, and produced fierce storms, typhoons, flooding, and acid rain all over that part of the world. Bertell's prognosis is that the military is preparing for a weather war that will lead to warlike conditions on our planet and militarization on a global scale. Further, "weaponization" of all our life conditions through weather manipulation has also been confirmed by the Action Group on Erosion, Technology and Concentration (ETC Group).

The productivist economy of the colonizers and settlers proposes to take control of the global thermostat, this time through "civil" geoengineering. According to ETC group "civil" geoengineering is a technological approach involving intentional large-scale intervention in the earth's oceans, soils and/or atmosphere to "reduce CO_2." It can refer to a wide range of schemes that are already a reality without public notification, including: blasting sulphate particles into the stratosphere to reflect the sun's rays; dumping iron particles in the oceans to nurture CO_2 absorbing plankton; firing silver iodide into clouds to produce rain and genetically engineering crops so their foliage can better reflect sunlight (ETC Group). For example,

[O]n April 15, 2017, the largest mainstream geoengineering program to date officially kicked off at Harvard University. The program has raised over $7 million so far, aiming for $20 million over 7 years. It will include a field test phase in which scientists will spray tiny, highly reflective particles at altitudes of greater than 7 miles above the Arizona desert, in order to test the efficacy of one of the most widely-known geoengineering methods, called solar radiation management. (Freedman)

Actors leading the advance geoengineering are the military, universities, the United Kingdom's Royal Society, and the U.S. National Academy of Sciences, joined by Canada, Germany and Russia (Freedman). Geoengineering continues despite the fact that the U.S. National Academy of Sciences has said "Geoengineering is not a viable response" (Hamilton).

What an organism (human animals and non-human animals) needs to survive is a habitat that nurtures and protects life. In the next section, I present a way out of climate chaos that demands and includes the voices of those who have been silenced.

THE WAY OUT OF CLIMATE CHAOS FROM THREE PERSPECTIVES: ECOCENTRIC, SUBSISTENCE, AND PACHAMAMA

The claim of the three perspectives presented here is that "we need to move out of the commodity production frame as well as from patriarchy" (Mies and Bennholdt-Thomsen; Salleh *Global Alternatives,*

"Eco-Sufficiency; Basadre; Marañon-Pimentel; Kovel) to see the way out of climate chaos in order to ensure the survival of the planet and all its living systems. Joel Kovel, in *The Enemy of Nature*, suggests a restructuring of society based on an ecocentric mode of production that includes a new value theory, which he calls intrinsic values. He argues on the need to free use-value from exchange-value and open up the appropriation of intrinsic value to build ecosystem integrity; that is, to change property relations so that the earth, viewed as the source of all use-values and all ecosystems, is appropriated by "associated producers." He says that *prefiguration*, understood as a way to organize society, "is intrinsic to ecological production, rendering the *provisioning* of ecofeminism as the *previsioning* of a Utopian moment" (218). He leans on Indigenous cosmovision and ecofeminist provisioning to direct the way out of the ecological and social crises.

The ecofeminist subsistence perspective sees local subsistence economies as alternative models of social, economic, gender, and ecological justice. For example, Maria Mies and Veronika Bennholdt-Thomsen propose a subsistence perspective that is based on production that includes all work expended in the creation and maintenance of immediate life and which has no other purpose. They present this subsistence perspective as a form of resistance against the effects of global capitalism and colonialism. Subsistence producers all over the world—the majority of whom are women and/or feminized—should be seen as experts who can lead the way to recovering autonomous ways of living, starting from territorial and bodily autonomy, food security in small farms and energy efficiency. Ariel Salleh (*Global Alternatives*) developed the concept of meta-industrial workers—understood as the unpaid or poorly paid members of households, peasants, and Indigenous peoples—with the unique capacity for *provisioning*, meaning that their labour satisfies their ecological and cultural needs. The notion of meta-industrial labour as a regenerative and metabolic fit provides the critical mass for change that sustain metabolic relations with nature, and is interconnected and politically inclusive. The meta-industrial workers reproduce labour power, biological infrastructure for all economic systems and lay knowledge. Moreover, Salleh ("Eco-Sufficiency") maintains that the socio-cultural economy (the sustenance economy) is an eco-sufficient sphere of regenerative labour and use value, and is more important for economics than the sphere of production for exchange (300).

Within the anticolonial perspective, Pachamama is the Andean way of unifying terrestrial ecosystems and the living things that inhabit them (Basadre). Pacha is universe, world, time, place, while Mama is mother; in this way, the term translates as "Mother Earth." During 500 years of oppression, communities in the Andes resisted and survived by their precise and detailed knowledge of the history, language, culture, vegetation, mountains, and other physical and socio-cultural attributes of the region. In times of heightened risk and uncertainty, they are falling back on tested and familiar methods of mutual aid, where community denotes a sense of shared ownership, and production is for satisfaction and pleasure. Boris Marañon-Pimentel argues that in the Andean peoples' cosmovision, everything is related and everything depends on a fair balance, "good living," or "*Buen Vivir.*" Their perception of solidarity, reciprocity, and collective labour is so strong that Indigenous peoples in Peru, Bolivia, and Ecuador return to the commons as *Buen Vivir* every time they are able to do so. For instance, the government of Peru has re-established in many places the institutions of the commons, such as Ronderos in Cajamarca. The commons idea has been established in the Constitution of Ecuador (2008), which is expressed as *Buen Vivir* or *Sumak Kawsay* in Kichwa, while in the Constitution of Bolivia (2009), it is invoked as *Vivir Bien, Sum Qamaña* in Aymara, and *Ñande Reko* in Guaraní. In the construction of *Buen Vivir,* the errors and limitations of so-called development are visualized, its fundamental bases anchored in the idea of progress are questioned, and at the same time the doors are opened to alternative options. *El Buen Vivir* disseminates a spirit oriented to decolonize our own thoughts and to break the chains that have for so long subordinated Indigenous knowledge. Currently, Indigenous peoples are able to foresee possibilities to reverse dispossession and enclosure, having been empowered by the rebellions of Movimento dos Trabalhadores Rurais Sem Terra in Brazil; the Zapatistas in Mexico; Los Derechos de la Madre Tierra in Ecuador; Vivir Bien in Bolivia; Conga No Va! in Peru; Berta Caceres Vive in Honduras; Idle No More in Canada; and the Water Protectors in the U.S. These movements have been creating spaces in which the voices and histories of people deeply impacted by climate chaos can been heard.

CONCLUSION

The crises described in this chapter are fruits of patriarchal-capital's economic rationality that has concentrated on increasing wealth and political power while ignoring the finiteness of resources and the decline of Planet Earth, and the oppression and exploitation of feminized labour and bodies. As capital's accumulation system has now expanded its operations to embrace the entire planet, it has disrupted the biogeochemical processes of the Earth itself, expressed most dramatically in the form of climate chaos. The way out must include all the voices that have traditionally been silenced: women, peasants, Indigenous people, and people living in colonized countries. The only solution climate chaos today is a broad-based social and ecological revolution in which we all mobilize together to protect the future of humanity and our home, the Earth.

ENDNOTES

[1]The "invisible hand" of the market has been expanded on a global scale by tU.S. militarization. David Vine describes the U.S. as a military-base nation that extends militarization around the world. According to the Pentagon, there are now around 800 U.S. "base sites" in around 80 countries: 174 in Germany, 113 in Japan, and 83 in South Korea. Bases are also in Aruba and Australia, Bahrain and Bulgaria, Colombia, Kenya, and Qatar, among many other places. In addition, "there are U.S. troops or other military personnel in about 160 foreign countries and territories, including small numbers of marines guarding embassies and larger deployments of trainers and advisors like the roughly 3,500 now working with the Iraqi Army. And don't forget the Navy's 11 aircraft carriers. Each should be considered a kind of floating base, or as the Navy tellingly refers to them, "four and a half acres of sovereign U.S. territory." Finally, above the seas, one finds a growing military presence in space" (Vine).

[2]The Van Allen Belts are two energy fields that surround the earth: The interior one, between 1,000 and 5,000 kilometres of land surface, and the exterior one, between 15,000 and 20,000 kilometres of surface. They begin about 1,000 kilometres above the Earth's surface and are subject to sun radiation that influences their expansion and contraction.

REFERENCES

Adams, John Luther. "The End of Winter." *The New Yorke* 27 March, 2018. Web.

Altvater, Elmer. "Ecological and Economic Modalities of Time and Space. *Is Capitalism Sustainable? Political Economy and the Politics of Ecology.* Ed. Martin O'Connor. New York: The Guilford Press, 1994. 76-90.

Basadre, David Roca. "Introduccion al Ideario de Tierra y Libertad."

Academia.edu (2017): 1-15. Web.

Bertell, Rosalie. *Planet Earth: The Latest Weapon of War.* Canada: Black Rose Books, 2001.

Bertell, Rosalie. *No Immediate Danger? Prognosis for a Radioactive Earth.* Summertown, TN: Book Publishing Co., 2000.

Burns, Sean. *Revolts and the Military in the Arab Spring: Popular Uprising and the Politics of Repression.* London: I.B. Tauris & Co. Ltd., 2018.

Castaneda, Jorge G. *Utopia Unarmed: The Latin American Left After the Cold War.* New York: Vintage Books, 1994.

Ceballos, Gerardo, Paul R. Ehrlich, and Rodolfo Dirzo. "Biological Annihilation Via the Ongoing Sixth Mass Extinction Signaled by Vertebrate Population Losses and Declines." *Proceedings of National Academy of Sciences of the United States of America* 114.30 (July 2017): n.p. Web.

Climate Prediction Centre / NCEP. "ENSO Alert System Status: El Niño Watch." *ENSO: Recent Evolution, Current Status and Predictions.* 11 February 2019. Web.

"Daily Data Report for May 2016." Environment Canada. 6 May 2016. Web.

Dalla Costa, Maria and Selma James. *The Power of Women and the Subversion of the Community.* Bristol, UK: Falling Wall, 1975.

Dahi, Omar S. "The Political Economy of the Egyptian and Arab Revolt." *Institute of Development Studies (IDS) Bulletin* 43.1 (January 2012): 47-53.

Davis, Devra L., Michelle L. Bell, and Tony Fletcher. "A Look Back at the London Smog of 1952 and the Half Century Since." *Environmental Health Perspectives* 110.12 (December 2002): A734-5.

David, Josh. "Low Oxygen Levels in the Ocean Could Lead to More Wildfires in the Future." *IFLSCIENCE!* 15 May 2017. Web.

Deleage, Jean-Paul. "Eco-Marxist Critique of Political Economy." *Is Capitalism Sustainable? Political Economy and the Politics of Ecology.* Ed. Martin O'Connor. New York: The Guilford Press, 1994. 37-52.

ETC Group. "Geopiracy: The Case Against Geoengineering." *Communiqué # 103* October 2010. Web.

"Farewell Spit: More than 400 whales stranded on New Zealand beach, majority reported dead." *ABC News* 9 Feb. 2017. Web.

Federici, Silvia. *Caliban and the Witch: Women, the Body and Primitive Accumulation.* Brooklyn, NY: Automedia, 2004.

"Final Update 39: 2016 Wildfires (June 10 at 4:30 p.m.)." Alberta.ca

10 June 2016. Web.

Freedman, Andrew. "We're about to test out hacking the Earth's climate. That should scare or inspire you." *Yahoo.com news* 19 April 2017. Web.

Foley, James A. "More than 400 dolphins washed ashore dead in Peru in January." *Nature World News* 03 Feb. 2014. Web.

Hamilton, Clive. "Geoengineering Is Not a Solution to Climate Change. Using Technofixes to Tinker with Global Climate Systems Is an Excuse to Avoid Unpopular but Necessary Measures to Reduce Carbon Emissions." *Scientific American* 10 March 2015. Web.

Irfan Umair, Eliza Barclay and Kavya Sukumar. "America is warming fast." Weather 2050. 26 November 2018. Web.

Isla, Ana. "The Kyoto Protocol." *Eco-Sufficiency and Global Justice: Women Write Political Ecology.* Ed. Ariel Salleh. London: Pluto Press, 2009. 199-217.

Isla, Ana. The *"Greening" of Costa Rica: Women, Peasants, Indigenous People and the Remaking of Nature.* Toronto: University of Toronto Press, 2015.

Kovel, Joel. *The Enemy of Nature: the End of Capitalism or the End of the World?* Blackpoint, NS: Fernwood Publishing, 2007.

"Huaicos en Perú: Panamericana Norte permanece interrumpida por desborde del río La Leche [VIDEOS]." *La República* 25 March 2017. Web.

Malm, Andreas. *Fossil Capital: The Rise of Steam Power and the Roots of Global Warming.* London: Verso, 2016.

Marañon-Pimentel, Boris. "Crisis Global y Descolonialidad del Poder: La emergencia de una recionalidad liberadora y solidaria." *Buen Vivir y descolonialidad. Critica al desarrollo y la racionalidad instrumentales.* Ed. Boris Marañon-Pimentel. México: Universidad Nacional Autónoma de México and Instituto de Investigaciones Economicas, 2014. 20-59.

Martins, Daniel. "Thep to re-freeze the Artic with millions of wind pumps." *Digital Reporter* 19 February 2017. Web.

Mauna Loa, Hawaii, Observatory. National Oceanic and Atmospheric Administration (NOAA). Earth System Research Laboratory, Global Monitoring Division (n.d.). Web.

Melman, Seymour. "Pentagon Capitalism." *The Military-Industrial Complex.* Ed. Carroll W. Pursell. New York: Harper & Row, 1972. 286-296.

Merchant, Caroline. *The Death of Nature: Women, Ecology and the Scientific Revolution.* New York: Harper and Row, 1990.

"Methane Unveiled: Q&A on Abrupt Climate Change with Dr. Guy McPherson." Margo's Healing Corner 31 May 2018. YouTube.

Mies, Maria and Veronika Bennholdt-Thomsen. *The Subsistence Perspective: Beyond the Globalized Economy.* London: Zed Books, 1999.

Mies, Maria, Veronika Bennholdt-Thomsen and Claudia von Werlhof. *Women: The Last Colony.* London: Zed Books, 1988.

Morton, Timothy. *Hyperobjects: Philosophy and Ecology after the End of the World.* Minneapolis: University of Minnesota Press, 2013.

Mooney, Chris (2017) "Scientists have just detected a major change to the Earth's oceans linked to a warming climate." *The Washington Post* 15 February 2017. Web.

National Oceanic and Atmospheric Administration (NOAA). Earth System Research Laboratory, Global Monitoring Division. "Trends in Atmospheric Carbon Dioxide." NOAA 18 Jan. 2018. Web.

"NASA, NOAA Data Show 2016 Warmest Year on Record Globally." *NASA Latest* 18 Jan. 2017. Web.

Patel, Jugal K. "A Crack in an Artic Ice Shelf Grew 17 Miles in the Last Two Months." *The New York Times* 07 Feb. 2017. Web.

Ponte, Lowell. *The Cooling: Has the Next Ice Age Already Begun? Can We Survive It?* Upper Saddle River, NJ: Prentice Hall, 1976.

Pursell, Carroll W. *The Military-Industrial Complex.* New York: Harper & Row, 1972.

Sachs, Wolfgang. "Environment." *The Development Dictionary: A Guide to Knowledge as Power.* 2nd Edition. Ed. Wolfgang Sachs. London: Zed Books, 2009. 24-37.

Salleh, Ariel. *Global Alternatives and the Meta-Industrial Class Futures Beyond Globalization* Abingdon, UK: Taylor & Francis, 2004.

Salleh, Ariel. "From Eco-Sufficiency to Global Justice." *Eco-Sufficiency and Global Justice: Women Write Political Ecology.* Ed. Ariel Salleh. London: Pluto Press, 2009. 291-312.

Sanchez, Kleber. "Ola de frío en región Puno deja 21 fallecidos y el deceso de 171 mil alpacas." *La República* 11 July 2015. Web.

Seneff Stephanie. "Autism and Gliphosate: Connecting the Dots." *Seed Sovereingty, Food Security. Women in the Vanguard of the Fight Against GMOs and Corporate Agriculture.* Ed. Vandana Shiva. Berkeley, CA: North Atlantic Books, 2016. 77-103

Shiva, Vandana. "Seed Sovereignty, Food Security." *Seed Sovereignty, Food Security: Women in the Vanguard of the Fight Against GMOs and Corporate Agriculture.* Ed. Vandana Shiva. Berkeley, CA: North Atlantic Books, 2016. vii-xxi.

Shiva, Vandana and Ingunn Moser, eds. *Biopolitics. A Feminist and Ecological Reader on Biotechnology*. London: Zed Books, 1995.

Steffen Will, Paul J. Crutzen, and John R. McNeill. "The Anthropocene: Are Humans Now Overwhelming the Great Forces of Nature?" *Ambio* 36.8 (2007): 614-621.

Mckie, Robin. "How disappearing sea ice has put Artic ecosystem under threat." *The Guardian* 4 March, 2017. Web

Vine, David. "Our Base Nation." TomDispatch.com, 13 September 2015. Web.

World Commission on Environment and Development, ed. *Our Common Future*. Oxford, UK: Oxford University Press, 1987.

"What is Climate Change?" *NASA Latest* 14 May 2014. Web.

2.
Money or Life?

What Makes Us Really Rich

VERONIKA BENNHOLDT-THOMSEN

SUBSISTENCE, NOT CRISIS: WE CAN'T EAT MONEY!

ON THE ONE HAND WE SAY "it can't go on like this!" But on the other hand we have no idea how it *should* go. Against our better judgment—and consciences!—we accept measures most of us see as harmful in many ways. We remain trapped in the straightjacket of the current capitalist money and commodity system. But I am convinced that we can free ourselves if we focus on subsistence. "Subsistence" means having what we really need for our lives. Yet the term "subsistence economy"—an economy focused on life's necessities—is met with resistance and frequent comments like: "This means going back to the Stone Age."

Does that mean that an economy organized to provide all of the necessities of life is not considered desirable? As a question, this can shed light on the ideological prejudices leading to the rejection of a subsistence-based economy. These stem from a firmly established perception of a modern economy of abundance where all products on earth are available as commodities, where everyone can live in comfortable prosperity, and no one has to worry about the basic necessities.

From the viewpoint of this fantasy, subsistence evokes visions of poverty, insecure and primitive living conditions. But what if all of a sudden it becomes clear that this fantasy of a modern age of abundance for all is just that, a fantasy? Such a moment of clarity seemed possible with the financial crisis.

But then something like the subsidy for junking older cars is offered as a countermeasure, and gets an overwhelming response. "Germany addicted to the junking mania," reported *Spiegel* magazine in 2009 (issue no. 14). Although everyone knows that cars contribute to global warming and oil reserves are dwindling. Still, we are glad that this way thousands of jobs may be saved. What nonsense!

Or take the so-called bailouts. The banks that caused the crisis receive billions of Euros in subsidies and guarantees from the national treasury. We all know that in this way money belonging to all citizens jointly is handed over to private profit-making interests. We know that in the future these funds will be lacking for community projects and that all of us, especially ordinary people—and thus the majority of the population—will have to pay the price. The process that began a long time ago continues: the poor get poorer and their number grows, and the rich get ever richer.

But there are no massive protests. Evidently the majority has the impression that governments have no choice, that without these bailouts the entire economic system would collapse and things would get even worse than during the Great Depression of 1929. And most so-called alternative proposals stay within the narrow frame of the chosen direction: Save certain banks and firms—but not others, increase state control here, relax it there. But no one questions that the system has to be subsidized with public money for banks and companies.

Intuitively, however, many people see that for quite some time something has gone fundamentally wrong with our kind of economy. But the longer they have organized their lives and expectations to fit a growth economy, the less they know what an economy organized around providing necessities could look like. For decades we believed that nothing was more important than *making lots of money*. And the experience of the post- war economic "miracle" and subsequent decades of prosperity seemed to prove us right. Everything, every handshake, was aimed at making money. And for a long time there never appeared to be a problem in transforming that money into tangibles like food, clothing, the roof over our heads—what we need to live.

But what happens if this kind of transformation breaks down? What if we no longer get anything for the handshake? Then we realize that although money can evaporate into thin air, as has happened since the beginning of the financial crisis, it cannot fill us up. In short, we cannot eat money.

We realize that we don't know how to grow food, build a roof or mend old clothes. Because in our highly specialized world with its ever increasing division of labor, only some have these skills. We no longer know, other than through the exchange of money, how to come together and share this knowledge. On a very basic level, namely the level of subsistence, we lack communication and block it from our consciousness.

In order to be able to recognize the specific, material, life-sustaining value of things and services instead of purely their cash value, we need a new term:

> Subsistence production—or production of life—includes all work that is expended in the creation, re-creation and maintenance of immediate life and which has no other purpose. Subsistence production therefore stands in direct contrast to commodity and surplus value production. For subsistence production the aim is "life." For commodity production it is "money," which "produces" ever more money, or the accumulation of capital. For this mode of production life is, so to speak, only a coincidental side-effect. (Mies and Bennholdt-Thomsen 20)

THE ECONOMY—WHAT IS IT REALLY?

"What is good for the economy, is good for us all." This sentiment seemingly overcomes any misgivings about the purpose of economic stimulus packages and subsidies for bailing out banks and corporations. But it's wrong. Because the economy is more than just bankers, businessmen and trade union bosses, it includes all of us, and is in fact a social process. We all determine its course, and it molds us too, as well as our culture and outlook on life. That is why the present financial and economic crisis is also a crisis of our society and the values that define us. Self-interest and fear of scarcity buttress a worldview based on industrialization and growth.

In this belief system the understanding that every person lives and thus is acting economically, does not exist. Instead, there is only *the* economy, with employees who must be happy if they can sell their labor. Capital, not people, is credited with economic know-how and maybe even the exclusive skill of acting economically. There exists, at best, a vague notion of an economically acting community where people work together and depend on each other. In the present crisis however, this is changing. Not so long ago, the Deutsche Bank decided to downsize its savings and current account business, indeed, to phase it out by offering poor service to its smaller customers, concentrating instead on large-scale financial operations. Now everyone, as well as the Deutsche Bank, fears that these customers will withdraw their savings, threatening to result in more bank failures.

A Separation Between the Financial and the Real Economy Is Pure Fiction

The economy appears to have as little to do with useful goods as it does with people, at least in recent decades. Instead, it seems to be limited to number games with derivatives, futures, certificates and funds. But the financial jugglers, those glamorous heroes, have juggled so poorly that - just as with the Chinese jugglers porcelain plates in the circus—their offerings have come crashing down. And now they remember that, to continue the metaphor, the plates they juggled are actually real, made from clay and soil. Now we hear that there is "concern that the financial crisis could affect the real economy." This sentence was read and heard everywhere, coming from the German chancellor Angela Merkel as well as many commentators.

Let's think this over. Were they actually admitting that the financial economy has disengaged from the real economy? In fact, ordinary citizens have been under this impression for quite some time, as they attempt to make sense of hedge funds, futures and derivatives or to understand why managers receive severance packages worth millions despite (or because of?) the fact that they fail to prevent risky investments or the take-over of their firm. Bonus payments to managers of ailing companies in the middle of the financial crisis follow the same principle. But why, if separating monetary gains from economic performance is considered legal and acceptable, is the collapse of the financial economy suddenly such a threat to the real economy that all forces must be mobilized to prop it up?

The confusion stems from a classic neoliberal idea. The separation of the financial and the real economy is pure fiction. Yet policies have been pursued as if this fictional separation were real. Economic courses have been set as if the financial sector alone were the actual economy. However, it is easy to see that the financial economy is not separate from the real economy. Capital market gains may be used to purchase real goods at any time. And yet up to now we have been told that the financial economy was *the* supporting pillar of the economy, while the rest was merely a supplier of means, ensuring that financial operations could run like clockwork. For instance, German cities sold their sewage systems and streetcars, sewage treatment and waste incineration plants to a so-called investor in the U.S. Then they leased back their former property and miraculously, and ostensibly without any problems, money poured into city treasuries. This financial game was called cross-border leasing (Rügemer).

Most of the German railroad was privatized and—although the

actual date has been postponed due to the financial crisis—its initial public stock offering remains on the agenda and is expected to raise lots of money. The question is, for whom, and who will benefit? Certainly not those who until now have been the actual owners of the railway (or the postal services or the telephone network or the waterworks, etc.), namely the citizens of the state who funded these common goods or who inherited them from their grandparents.

On the one hand, it is good that the smoke and mirror illusions of the financial market have finally been recognized, in particular by the politicians in charge. On the other hand, it seems as if they and the entire audience are complaining that they were not cheated well enough and that the magic tricks are still recognizable as such. Yet they voted for this hokus-pocus in the first place. And now, they are angry that the magical illusion turns out to be a real illusion after all. But when impresarios like the sacked boss of Deutsche Bahn, who after all had been employed for the staging, are reprimanded for doing everything possible to let the show go on and then ask for their share of white rabbits from the top hat, this bodes ill: Individuals are made into culprits and the theater of illusions is maintained.

However, the real economy is already infected by the illusionary financial sector's greedy profit expectations. City centres are deserted because large department stores shut down branch stores,   even though business is doing well. They are taken over, merged or closed down because profits no longer meet the expectations of managers and shareholders. These excessive profit expectations, also in the real economy, are especially obvious in the case of so-called jobless growth. Despite good profits, employees are fired en masse to accommodate shareholders expectations for higher dividends through "lean production" methods.

The most severe consequences for all of us come from the introduction of financial market mechanisms and profit expectations into agriculture. Because the amount of arable land does not increase. And if there is an equally good harvest year after year, why not be content? After all, there will be enough bread for the coming year. However because of the financial economic pressure for profit maximization, new and risky technologies have been developed for agriculture. But the natural limits of agriculture, which is the true real economy, cannot be overcome. The consequences of this mixing of fictitious and real value are the loss of real family farms and the increase in real hunger. And this happens everywhere, not only in the countries of the southern hemisphere.

Dependent Wage Labuor as a Condition for Citizenship

Because of confusion about what the economy really is and the mixing-up of money and the true value of life-sustaining resources, it has become the seemingly irrevocable duty of any normal person in industrial society to sell his or her labour as a commodity.

To ensure that this is internalized once and for all by everyone, the German government introduced its Agenda 2010[1] and Hartz IV[2] "reforms in particular. It seems that those who do not "earn" money, have earned the privilege of being treated without dignity. The employment office appoints a guardian for the wage-less person and if he or she fails to obey the orders of this so-called case manager, the already measly basic social assistance payment may be cut in stages down to zero. The housing allowance may also be reduced, or the person may be forced to move if the size and cost of their rented apartment exceed locally permitted limits. Most striking in the large array of possible sanctions is the obligation of having to work for an expense allowance of only one Euro per hour. This obligation to work for a wage of one Euro demonstrates more than all other disciplinary measures that in this so-called welfare state, the right to eat, live, be clothed and have a sense of belonging is reserved for those who work for a wage, in whatever form.

Apparently, a majority of Germans agree. At the very least, protest against this legal atrocity has failed to materialize. It was silently approved because many people think that social welfare recipients and the so-called long-term unemployed are living at their expense. So, rhetorically speaking, does this imply that the unemployed steal food, take up the living space, and ruin the clothes of others? While it is not without irony that now, five years later, a conservative Christian Democrat-Liberal government grants at least a little bit more money than the previous government of Social Democrats and Greens, the basic lack of solidarity remains.

How did this momentous confusion between money and the real value of life-sustaining resources come about? How can it shape an entire culture to such an extent that many lose their feeling for human dignity and human solidarity? A glance back into the history of modern economic thought helps answer this question.

The Invention of the "Invisible Hand": A Theory from the Nineteenth Century

Adam Smith is reputed to be the father of modern economics. His famous thesis remains the central principle of economics to date: If everyone pursues their own self-interest, they will also be maximizing the economic well-being of society. Accordingly, human activities are directed by some sort of magic economic authority, like an invisible hand. If everyone busily focuses on increasing their own profit, wealth will increase overall and all will do well. In other words, if enough money flows around, all will have enough to eat and a roof over their heads. At the present time, we can see more clearly than ever that this is not true.

The "invisible hand" and David Ricardo's principle of comparative advantage are theories from the nineteenth century.[3] In those days, national economic circulation prevailed and even when this was enormously expanded through colonial trade, we were, at the end of the eighteenth and in the early nineteenth century, still at the beginning of international trade and industrialization. The looting of the colonies was just preparing the basis for it, as was the expropriation of peasants' land through enclosures.

Smith and Ricardo had no idea how extensively all necessities of life would change into commodities of global trade during the next two centuries, although they were laying the theoretical foundations for this very process. For most people, food and the supply of life's necessities still had very little to do with money. Self-supply, and use of common land as well as rural and neighbourly reciprocity were the foundations of subsistence. While money was used (as at fairgrounds, cattle markets, for pre-industrial paid labour, tithes paid to land owners, etc.), it only played a marginal role in the day-to-day economy and above all was handled and valued according to the rules securing subsistence (moral economy). Smith and Ricardo thought about national economy and international trade against the background of locally and regionally organized subsistence.

In their theoretical work, they only considered and paid attention to an abstract exchange or cash value, and not to the value of sustenance or usefulness. One could say that their intention was to "free" economic policy thinking from its attachment to material, specific use in order to leverage the exchange value economy. They were the preeminent theoretical exponents of the emerging bourgeois class, which was in the process of breaking away from the absolutist state and its mercantile theories. According to mercantilism, the state controls the national economy and improves its own trading position towards other states through war and conquest. The trading volume is seen as naturally limited and trade is handled almost like feudal sinecures. Smith and Ricardo, however, with their concepts of the invisible hand and of comparative advantage, argue for a *bourgeois freedom of trade* claiming that in this allegedly peaceful way, the wealth of nations will increase.

With their arguments they substantially contributed to the fact that dependence on nature and peoples' work and production for subsistence are "forgotten" in modern economics. The substance or physical reality of food, clothing and shelter is ignored and replaced instead with its trade value. In the late eighteenth and early nineteenth centuries, this omission may be seen as bourgeois-political and liberal-theoretical pragmatism and not necessarily as ignorance. Today it is different.

In the twenty-first century, the majority of the world's population depends on wage labour, peasant and artisanal production are in danger of disappearing, and the basic goods necessary for subsistence have become internationally traded commodities. Thus most of humanity has become directly tied to the globally organized capitalist process of accumulation. This is the case even with those not in formal wage labour, who work in the so-called informal sector or as peasant farmers, and with self-employed tradespeople, as hardly any local or regional economic circulation remains that strengthens subsistence. Not only is Coca Cola, the symbol of the internationalization of consumer goods, found everywhere, but a limitless number of products for everyday needs are almost exclusively supplied by the international supermarket economy, replacing local goods, produce and marketing.

Via Campesina, the international peasant movement, demands access to their own local markets. Meanwhile, even the most essential basic foodstuffs are affected. Take Mexico, for example, was self-sufficient in maize (corn) for their daily tortillas until the 1960s. Now maize is mostly imported from the U.S. The so-called food riots that erupted in 2008 first in Mexico and then in various other countries were the direct result of the uprooting of local staples. Due to speculation with foodstuffs, such as maize used to make biofuel, this basic food has become much too expensive for many households. Profit maximization for some directly threatens the availability of food for others, because no or only minimal regional and local structures remain to secure subsistence. Conclusion: The "invisible hand" does not give, it takes!

From the "Invisible Hand" to the Invisibility Of Subsistence

The fact that economic theory is blind to the material basis of our livelihoods is one problem. Another is that this blindness has infected our entire modern culture. Since industrialization began in the nineteenth century, our awareness of the economic importance of subsistence has faded, although—or maybe because—industrialization and commercialization have taken over ever more areas of human needs, culminating in the mass consumption of the twentieth century. The things and services we need to survive are at present only taken seriously when associated with industrial commodities and wage labour. Only things that can be measured, bought or sold with money have value. Seeing life through these lenses, the material and life-giving values of self-sufficiency, self- determined, autonomous work and unpaid services go unrecognized.

Five elements in particular impede perception of subsistence today:

1. Disregard for women's work in the modern sexual division of labour

With industrialization, regard for women's work did not increase, but decreased. This was systematized by the model of the male breadwinner and female housewife. Women's crucially important caretaking is not recognized as work and is thus seen as having no economic value. The negative evaluation of women's subsistence work has put its cultural mark on all of women's activities. That is the reason women are on average paid less than men, and why they seem predestined for temporary and short-term employment in regards to wage work. We have called this phenomenon the "housewife-ization" of wage labour (Bennholdt-Thomsen "Marginalität"; Mies; Werlhof). It is one of the primary ways that the miserable conditions of wage labor—for both women and men—could proliferate unhampered during this crisis.

2. Disregard for peasant farming

For decades it has been difficult to integrate peasant farming, as opposed to the landed nobility[iv] and colonial plantations, into the maximization economy. Defense mechanisms against industrialization and profit orientation come from the relative frugality of peasant farmers, their attachment to the land and their communities, as well as their concept of growth in line with nature. In the opinion of the non-peasant majority, this outlook was and is still considered to be uneconomic, as peasant farming has reputedly no value for the overall economy. The dominantly negative connotations still associated with the concept of subsistence stem from this discussion. For instance, that economic activity for life's necessities—instead of for profit—is backward, in fact not an economic activity at all and thus must be overcome. This subsistence-destroying viewpoint has shaped both the World Bank's policies and European Union's agricultural policies.

3. Disregard for nature

Since the modern era, Western thought has regarded nature as something to be freely used and exploited, apparently without consequence. Nature has value only when it becomes private property, thus acquiring monetary exchange value. The concept of "the commons," commonly shared access to nature's bounty, is quickly disappearing. The modern concept of nature is ultimately the backdrop against which women's work, colonial and neo-colonial regions, and peasant production are considered economically irrelevant, if not completely invisible.

4. Colonialist looting

The colonies were seen as both a seemingly inexhaustible reservoir of natural resources and of Indigenous labour. Both were appropriated by force, in fact plundered. In the twentieth century, the economics of colonial rule were maintained through development policies, now continued with the policies of globalization.

5. Fear of scarcity

The collective and neurotic fear of scarcity prevents us from recognizing the substance or physical reality of subsistence.

THE INVENTION OF SCARCITY

One of the most important advocates of the ideology of scarcity was Adam Smith. He contributed to its broad acceptance by presenting a consistent theoretical framework showing why the economy must continually grow. Since then, economics has been seen as the science of how unlimited needs can be met with limited resources.

Adam Smith invented the modern fear of scarcity by conceptualizing prosperity in a completely abstract, non-material way. A reminder: for Smith, economic well-being for all is achieved when each person acts according to their own self-interest, seeking the greatest return possible. And the reverse is that someone who does not strive for the greatest possible return bears the blame both for his own and general shortages.

Smith made the fear of scarcity seem virtuous, by presenting other peoples and those from earlier times, who lived well on their given resources, as "savage and barbarous." The threat is obvious: If people in the "civilized" and "thriving nations" fail to make an effort to conquer the inevitable scarcity through economic growth, they could become as "miserably poor" as those who "from mere want ... are frequently reduced, or at least think themselves reduced, to the necessity sometimes of directly destroying, and sometimes of abandoning their infants, their old people, and those afflicted with lingering diseases, to perish with hunger, or to be devoured by wild beasts" (8-9).

With this racially-coloured horror scenario, the colonization of other countries is not only legitimized, but, as far as economic principles of growth are followed, even considered a "good deed" for the colonized peoples. The main outlines of international development policy after WWII were thus already defined. Looking back on early colonization, we see how the Christian message of salvation and missionary work among the 'heathens', which legitimated the colonization of Latin America (Conquista), was replaced by profit maximization and free trade. It remains to be mentioned that a good half-century after Adam Smith's death, conditions much like his horrific example actually took place, namely the great Irish potato famine (under English rule with free trade ideology).

To sum up: Fear of scarcity, as it is in all our minds, spread widely through Smith's definition of prosperity. The real riches—healthy food, protective homes and social ties—can no longer be seen through the euro-centric, greedy glasses of the bourgeoisie. For them, only money and commodities are real. The subsistence economy of caring work without money, as in the family, and the appropriation of natural goods without money, as in the peasant commons economy, are de-economized and delegitimated. With all that is life-sustaining defined away, subsistence can now be withheld without moralistic or legal doubts. This creates a modern paradox: scarcity leads to growth and growth leads to scarcity.

Capital's Worldwide Mission and the Colonization of Subsistence

With international development policy after the Second World War, development, meaning the spread of the same selfish profit-oriented economic goals, was pursued worldwide. In his inaugural address in 1949, U.S. President Truman called large areas of the world "underdeveloped." Where "economic life is primitive and stagnant," the U.S. should help with "its scientific advances and its industrial progress," as well as "its capital" to achieve economic growth.

Development policy and development aid were created. The responsible international bodies were the recently founded Bretton Woods institutions: the World Bank and the International Monetary Fund (IMF). The newly created United Nations then approved both the gross national product of a country and the average per capita income in U.S. dollars as the internationally binding criteria for defining development and under-development.

How people really live and work is not of interest, only how much money is in circulation. For example, under its president McNamara, the World Bank promoted credits for projects fighting "absolute and relative poverty," which should "draw farmers from subsistence to commercial agriculture" (World Bank 20). Much later, looking at the growing number of poor in the world, McNamara called this policy a failure. But by then his development projects had left a trail of destruction.

The clear failure of the economic model of development policy has not prevented the International Monetary Fund's imposition, since the 1980s, of a neoliberal policy in ever more countries that are now no longer called "underdeveloped" but "developing" countries. Structural Adjustment Programs (SAPs) force national markets to align with the world market of powerful corporations, ignoring borders and economic customs. The SAPs, namely the suspension of protectionist measures for national local production, as well as rigid cutting of suspected protectionist subsidies for staples or for health and educational systems, can be imposed because the affected countries have fallen in the credit debt trap. This, in turn, was the result of the former development policy. Money for growth (development credits) creates scarcity. The SAPs have led to the impoverishment of millions, especially in Africa, and the collapse of entire national economies, for example in Argentina. Ultimately, the World Trade Organization founded in 1995 included agriculture—and thus peoples' basis of existence—in its maelstrom of limitless corporate dominated international markets.

As a result, today one large, globalized circulation of commodities has developed. The non-consumerized subsistence background of the nineteenth century food-producing and care-taking work, which still partly existed in Europe until well into the twentieth century, and in some regions of the Third World still exists today, has been successively absorbed by the internationalized circulation of commodities. The autonomous or self-determined supply of life's necessities has shrunk so far and become so much part of the international commodities system that it has taken on totalitarian traits.

What has developed with the "invisible hand" (Smith), "behind the back" of the producer (Marx), is the absorption of subsistence, the "transformation of all things" (Wallerstein) and services into commodities, the "colonization of the lifeworld" ("Lebenswelt," Habermas), and finally a totalitarian commodity system.

Globalization: Totalitarianism of the Money and Commodity System
As with every totalitarian system, there are two aspects constituting the totalitarianism of the money and commodity system:

* •first, *the ideological blindness* of the overwhelming majority;
* •secondly, both the structural and direct use of *violence*, from which, in view of the all-encompassing power structure, there is almost no escape.

The real problem of our time and our present civilization is the moral failure of society, which has raised the standards of capital to ethical human values. An example is the violence currently inflicted on refugees at the borders of Europe in the name of these values. Also wars that—in the name of these values—are defined as "peace missions." And what about the millions of people forced to go hungry? All of these realities are part of what Jean Ziegler calls the *"Empire of Shame."*

In a globalized world and giant throw-away culture, the excessive corporate profits, and the even more excessive profits from financial speculation, cannot be had without more than a billion people going hungry. In October 2009, the World Food Organization FAO estimated the number of hungry people at 1.02 billion. This is twelve times as many people as live in Germany. The fact that we have allowed this to happen should be our most terrible problem.

The propaganda machinery of international development policy has been especially effective in Germany (and the U.S.), where many people share the ideology of "developed" and "underdeveloped" economies and societies. The judgment of "underdeveloped" recalls the debates of the "Conquista" over the worth or worthlessness of Indigenous peoples, and whether or not they could actually be considered as human beings. Today, acting in a "developed" way for maximum profit seems to be a human value set in stone. And yet, the number of people on this planet who are currently suffering from hunger is probably much higher than it ever was in the supposedly underdeveloped subsistence economies. The ideology of development policy perpetuates the colonial claim of the master race's power. This "master race" abrogates to itself the one true faith and imposes its "correct" economic strategy everywhere, thus further enriching itself in a colonial way. A concept of the diversity of cultures and respect for other economic values and criteria is completely lacking from this development ideology.

But like every totalitarian system, this one also eats its children. It is an illusion to think that the violence-charged mechanisms such as greed (self-interest), theft (of land and other basics of subsistence) and conquest (colonialism, developmentalism, globalization), which are produced by our economic social system, will stop at specific boundaries.

Rosa Luxemburg thought that at some point this inhuman spiral would have to come to an end. She was the only Marxist theoretician of the period to recognize the fundamental importance of the subsistence economy—her "natural economy"—for the functioning of the capitalist production method, namely profit realization. Once the external sinecures are exhausted, with all production sites and all markets integrated into the capitalist system of money and commodities, then, Rosa Luxemburg believed, the limits of continuous profit-making would be reached, and the system would fall apart.

But for all her perceptiveness, Rosa Luxemburg overlooked the inexhaustible aspects of subsistence, both the unpaid caring work done mostly by women, as well as nature's enormous riches, which could still be colonized. The colonization affects the wealth of the natural world, as well as life itself. Who would have believed, one hundred years ago, that on the threshold of the twenty-first century living organisms could be patented and thus transformed into capital? Or that even water could become a commodity? Or the air, in that CO_2 emissions would be traded as carbon certificates?

Helpful souls in science and politics develop ever new means of profitable commercialization. What is necessary for life is dissected into ever smaller parts, as with DNA from genes, which can then be manipulated, privatized and turned into money. Or by venturing into ever larger dimensions of the atmosphere, as when climate change provides grounds for new areas of financial speculation.

The mechanisms used to commercialize the basics of life are everywhere, affecting everyone, colonizing people even in the so-called developed North. The noose around our necks is continuously tightened, as almost everything we really need is turned into a corporate-capitalist commodity. So we become commodity consumers in ever broader basic aspects of our lives, from our health and the food we eat, to our leisure time and communication, to our care for the sick and elderly, and more. This means that ever more independent knowledge is lost, and ever more areas of our human and ethical self- determination are taken from us.

COMPLICITY AND MONEY

It is a shock to realize that—and how—we in our society are ideologically and culturally entangled in this mania of maximization. However it is worthwhile to take a closer look at this truth, as it also holds the comforting notion that if we are the problem, we are also the solution. We can free ourselves from totalitarian constraints. Acting economically, how people live their lives in this world, has turned into a war. We have become accomplices because we believe that participation in this war is inevitable, even natural. Without being aware of it, we use the weapon of this war every day—money. Once we accept money as the basis of subsistence, we are recruited.

But we can make peace if we reclaim sovereignty over life's necessities. This can be achieved in several ways. On the one hand, mentally, ideologically, in relation to awareness, we must free subsistence from its invisibility, as it has been banished to our subconscious by money. On the other hand, we must actively take ever more of what is necessary to live into our own hands.

Supplying ourselves and doing manual labour are much more than the mere production of necessary goods. They are in themselves processes of understanding. We discover that we have skills, we experience ourselves empowered and able to give something of ourselves. We learn once again to make room for feelings of closeness with our natural environment.

As long as money flows, environmental destruction does not seem to be a loss since what has been destroyed is replaced by its supposed equivalent in money. But in reality, essential resources melt like the polar ice caps once they are taken over by money. The diversity of plants and animals disappears, species die out. Soil erosion and desertification advance, while our understanding and compassion for the world and the other beings that share it correspondingly decrease. Eventually we must recognize what avarice and greed mean: that my consumerism harms others. However money conceals the fact that my cheap purchase deprives others of their livelihoods, causing land expulsions, migratory labour, low-paid jobs and temporary employment elsewhere.

In the North as well as the South, money has become the most effective mechanism for destroying subsistence. It replaces the immediate force and direct violence of the former feudal and colonial tributary systems, but remains no less powerful. Indeed, it greatly surpasses them. In particular, the use of force as conveyed through the money system is far more subtle, and the respective power structures are significantly more blurred.

Unlike older, openly violent systems of power, the separation between victim and offender—and thus the responsibility for violence—is now much harder to determine.

The social boundary between those who suffer and those who

benefit from the present money and commodity economy has become permeable. While the leap from shoeshine boy to millionaire is still a fairy tale, many people continue to believe in it and for some, it indeed does come true. Compared to rigid historical estate structures, the caste system or slave-owning societies, the racial and class barriers of current societies are surprisingly low. The majority believes in the legitimacy of the money system. Therefore, its mechanism of rule is more important than today's racial and class limitations. Heads of states and corporate bosses are right: the central social issue of our time is money. Yet in a completely different way than what they think. The issue is about our freeing ourselves from the constraints of the all-encompassing system of money and commodities.

The fundamental solution required in our time is thus some form of collective *self-liberation*. Such a new vision can also result in a new understanding of politics, one appropriate to the twenty-first century and its economic and social organization, but one for which we doubtlessly first must develop specific tools. This much is clear: The old concepts of victim versus offender, of rulers versus subordinates, of war versus liberation struggles, these concepts are no longer suitable motives for determining our actions to achieve another economy and society.

"HOMO OECONOMICUS" IN QUESTION

Economic reason as *the* criterion for all political and personal decisions is one of the pillars of modern complicity. It indiscriminately supports both capitalist and socialist ideological concepts. André Gorz, an eminent advocate of a more just society, especially for wage workers, published his "Critique of Economic Reason" in 1989, with the subtitle: "Search for Meaning." Here he contradicts the socialist vision that injustice will end once workers own their factory:

If, from the outset, the development of the means of industrial production had been in the hand of ... worker's cooperatives, enterprises might have been managed and controlled by the people working in them, but *industrialization would not have taken place....* only ... the separation of the worker from the means of producing ... made it possible to rationalize and economize labour....

People want to realize their potential through their own activities! Into the nineteenth century, work was experienced as meaningful activity with immediately obvious usefulness, as opposed to repeating the same action for hours at a time and at a pace set by a machine. But things have developed in exactly this direction, so that today it is almost obscene to ask about the meaning of salaried work, especially in times of crisis and rising unemployment. What matters is that one is doing paid work, and that the economy thus continues to grow. While as late as the beginning of the twentieth century, it was noted that "Henry Ford had to employ 900 workers if he needed 200, because 700 could not tolerate the monotonous work under time pressure and soon stopped coming" (Scherhorn and Meyer-Abich 171 [my translation]). It should be added that they could obviously afford to stay away. In other words, alternatives to meaningless wage labour existed, namely the independent generation of subsistence.

But acceptance of the growth economy is in question, not only regarding production, but also exchange. This was underlined by a team of social scientists who in 1981 launched the Mouvement Anti-Utilitariste dans les Sciences Sociales (M.A.U.S.S.), an anti- utilitarian movement opposed to the dictate of economic reason in the social sciences. Their message was that modern men and women and modern society do not function primarily according to cost benefit analyses, as postulated almost across the board in the social sciences. Rather the process of give and take is individually as well as socially very differently motivated, primarily by a longing for attachment. Humans, as expressed through their actions, including their economic actions, want to interact with each other instead of having to compete as opposing interests, or in fact not interacting at all.

The work of the anti-utilitarian movement builds on the academic work of French sociologist and ethnologist Marcel Mauss (1872-1950), especially his essay "The Gift." Here, based on research studies on Indigenous populations, he analyzed the phenomenon of giving and taking and its importance for social cohesion. Mauss concludes that the longing for mutual acceptance is a basic human need, the basis of social bonds expressed in the exchange of gifts as a form of mutual indebtedness. This exchange, the process of giving, receiving and reciprocating, is not solely self-interested, according to Mauss, because then society would no longer exist.

Alain Caillé, who contributed significantly to reintroducing Marcel Mauss's anti-utilitarian version of work to the discussion, stresses that this viewpoint has nothing to do with moral judgments. The non-economic, non-self-interested aspect of the gift exchange is not reduced to donations or charity, but is simply part of the activity of exchange, which ultimately is, and will remain, a process of human socializing. "There is no more urgent task today, on a theoretical as well as an ethical and political level, than to break with the prevailing economistic world view, according to which human motivation is exclusively economically determined—whatever this term may mean—and that ultimately the world is ruled by economic considerations and forces alone" (213 [my translation]).

But what about the very real force exerted by the mechanisms of the ruling economic system on individuals? Is it really possible to act differently?

THE EMPOWERMENT OF THE INDIVIDUAL

How can the conflict between the decision to act in a different economic way and the ruling system be resolved? In response to Attac[5] scientific advisory board member Ulrich Brand's claim that the proposed economic reform known as the "Green New Deal" would only reinforce the ruling economy, Sven Giegold, European Parliament member from Alliance '90/ The Greens,[6,] stated:

In such a crisis, we have to take advantage of the situation to achieve at least what's possible—even if it means a pact with the devil. To reject the ecological and social taming of capitalism because it doesn't fundamentally question the economic system, seems to me cynical, in view of the climate crisis. We don't have that much time any more. (taz[7] [my translation])

How social change can be achieved is an old subject, at least for the left. For a long time there was something like a personified idea of *the* system, one that could be exchanged for another through a revolutionary coup. This idea is no longer considered appropriate, not least because of negative experiences with revolutions. But the question behind it, as to how individuals and society are connected and how individuals as a whole form a society, remains fundamental to social science. What is clear is that this cannot be answered generally, but only for specific historic epochs and cultures. If currently the need for a new social contract is discussed everywhere, then it is in this context, and addresses the changed conditions of our era.

We have been working with the subsistence theory for more than three decades, specifically on the bridge between the individual and the modern society, between close relationships and social structure. We have seen that a focus on daily sustenance, and researching it as an integrated part of the economy as a whole, challenges the monolithic concept of economy that blocks any revision of the outdated social contract (Mies, Bennholdt-Thomsen and von Werlhof). With a subsistence perspective, empowerment of the individual against the "diabolic" power of the "system" becomes tangible. Feelings of powerlessness against the authorities over us are unnecessary. Rather we must stop seeing ourselves as victims. And this is possible because we ourselves are taking charge of our lives. The call to the individual and the civil society (as a term for associated individuals instead of class, people, or nation) to campaign for an ecologically and socially just world, as raised today by a broad spectrum of movements, arises from this view from below.

The isolation of individuals is the historic result of the development of capitalism into a global commodity market system of goods and finance. This is also the historically new power structure specific to our era. The individual is more or less directly subjected to the totalitarianism of the commodity system according to the motto of the wage earning society: that each of us is *individually* responsible for our own survival. The state plays an ever smaller role as an intermediating authority, whether between wage work and capital, or as provider of basic social needs. Instead of mediating, the state has taken over the role of guarantor for the functioning of the economic system of maximization—in the west and east, north and south—using outdated institutions of enforcement.

In view of the social power structures of our era, the old quandary of whether "we" should abolish the plundering capitalist system or whether "we"—since there is no time for systemic issues—should concentrate on reforms, has become obsolete. What matters now is that *we*—all sovereign individuals capable of acting responsibly—withdraw from the forced maximization economy by refusing to participate. The individualization through the globalized commodity economy, along with the isolation it creates, is probably the biggest problem of our time. But it is also our biggest opportunity, because the need for our own subsistence necessities is the source for political, social and economic empowerment of the individual.

One of the most important insights of subsistence theory is the widely researched fact that subsistence production did not disappear in the twentieth century and will not disappear in the twenty-first, even under the conditions of the generalized wage/money/ commodity society. But it changes its appearance. While the appreciation of subsistence production and work declines, direct and self-sufficient caring activity is not disappearing and cannot disappear. For without nurturing and being nurtured, without caring and being cared for, without giving and receiving gifts, we could not exist. Empathy and caring attention cannot be turned into commodities.

For caring is physical and tangible: providing good food, a warm blanket and those vegetables passed over the garden fence. The existential requirements and needs for subsistence remain out of reach for *homo oeconomicus,* mister money. The culture of subsistence, the subsistence knowledge that still exists, do-it-yourself and self- sufficiency, are the basis for a civil society of liberation from the straightjacket of the growth economy. While money separates individuals from each other, the immediacy of subsistence brings us together.

FROM *"HOMO OECONOMICUS"* TO *"HOMO DONANS"*

How does money separate people from each other? Isn't money a wonderful invention of civilization, easing the exchange between people? In this complex society is it possible to manage without money?

In her reply to these questions, the Texan-Italian author Genevieve Vaughan (*For-Giving; The Gift*) picks up from where the anti-utilitarian deliberations of M.A.U.S.S. left off. In her view the anti- utilitarian insight is correct and important, in that economic activity in modern society is not mainly motivated by the so-called economically rational calculation of wanting more, but rather by the desire for social attachment. Yet the authors of M.A.U.S.S. adhere to a basically utilitarian proposition, namely that human interaction is always bound to a cycle of the "triple obligation" of giving, receiving and reciprocating, i.e. returning a gift—with special emphasis on the return. According to Caillé, this is universally valid, and is, in fact, an anthropological constant.

Vaughan rejects this idea of humankind. She criticizes its patriarchal narrow- mindedness, which ignores that the primary human experience develops in the attachment to the mother, which has nothing to do with this triple obligation. The infant is nurtured and cared for because this is essential for the child's survival. The mother or another motherly, caring person gives without expecting a corresponding return. The fact that this is so is due to far simpler reasons than an essential goodness of the mother. If infants are not looked after, they die; there would be no society. Vaughan also emphasizes the anthropological constant that is valid for all epochs of humanity. And also for our modern age. Therefore, Vaughan contrasts the dominant idea of humankind, the

"*Homo oeconomicus,*" with that of "*Homo donans,*" the giving person (*For-Giving; The Gift/*).

"If the fathers of capitalist theory," with Adam Smith leading the way, "had chosen a mother rather than a single bourgeois male as the smallest economic unit for their theoretical constructions, they would not have been able to formulate the axiom of the selfish nature of human beings in the way they did," remark Ronnie Lessem and Alexander Schieffer ironically (124). The decisive criticism of these nineteenth century theses finally came in the course of the new women's movement and women's studies from the 1970s on. Women's work (care economy), the history of women (Her-Story) and socialization according to motherly principles (matriarchal studies) were freed from invisibility. Women philosophers in the Italian group "Diotima" show how patriarchal language considerably hampers women's ability to think of their own person in female categories. The subject/individual in this language ("homo," "man," "mankind," etc.) is not intended to be a woman. "She" is the other, the one who is notoriously absent in the philosophical discourse (Cavarero). Luisa Muraro shows how the "symbolic order of the mother" was ousted from the dominant symbolic order and how it can be reappropriated.

The term "*Homo donans*" is one such reappropriation. It reflects the motherly giving, nurturing side of humans. The fact that Genevieve Vaughan (*For-Giving; The Gift*) adheres to the word "homo" sends two signals. On the one hand, it is that male giving is not excluded from the image of motherly giving. In fact, she considers it a traumatic experience for the small boy that during his socialization into manhood he is basically forced to distance himself from his own gift-giving self because of its motherly-female connotations. And this is a tragedy for all of us. In our patriarchal society, masculinity is connected with an attitude of demanding and taking, which as a whole has become the general economic attitude. On the other hand, the term "*Homo donans*" signifies that not every woman must bear children to be a gift-giving person. Rather, it emphasizes the simple fact that everyone's primary experiences are deeply influenced by gift-giving because we all, men and women, have been nurtured in a motherly way.

Doubting the Exchange Paradigm of Taking, Instead of Doubting The Paradigm of Giving

But how can a complex society function according to the principle of unconditional giving? As a matter of fact, society did function this way for thousands of years. Genevieve Vaughan calls this principle "transitive": I give, you give, the next person gives and so on. In this case, giving is not ego-oriented as in: "I give so that you will give to me" but follows the needs of others: I give because I see, feel, perceive, know your needs. Thus, a chain begins and finally a circle of givers, community and society develops. Giving is the basic pattern of communication, of material and immaterial com-muni-cation (*munus* = lat. "gift"), neither taking nor with the obligation to reciprocate (Vaughan *For-Giving*).

Socialization along the lines of the gift principle is fundamentally different from the market society that follows the exchange principle, emphasizing wanting and taking. The structure of exchange itself already contains the fear of scarcity, which ultimately becomes the starting point of the modern economy. Exchange is always accompanied by the concern: "Will I get enough in return?" However, societies where material communication follows the gift principle assume and experience abundance. Nature's gifts are equally available to all. Members of the society make use of them according to their diverse and varied needs. No abstract measurement is required; equality must not first be (re) created. It is a premise of the society that all humans are of the same equal human status because all were born from a mother. Whereas the exchange principle reflects the existence of a higher authority, a "justice ... from above [that] prescribes to each his share, distributing or retaliating" (Bloch [my translation]). Giving and expecting to receive in return, the exchange paradigm, is tied to a measure defining whether the gift and return gift are actually equal. Because the focus of the exchange is not the other, but the ego and its needs. In order to establish equality, a measure is required.

The myth of equivalence, that a commodity, such as a loaf of bread or a piece of land, is actually represented by a specific sum of money, i.e. that equality has been established, is the bait for our complicity. It is the background for the concepts of fair prices and fair wages. The sum of money, not need or necessity, becomes the measure. But is not the value of bread actually quite different for those who are hungry than for the well-fed? And yet, it costs the same for both, and the hungry can only eat if they raise the money required. The equalizing justice lies supposedly in the fact that the hungry are given money, perhaps as charity, instead of enabling them to have enough to eat in dignity. People who need food and who know how to grow food should have access to the land for themselves and their children. Instead, those with money can buy up the land. They buy it at an allegedly equivalent value, and then sell the food at a high price: this is called development and economic growth.

The equation, which is already a part of the concept of exchange, is a purely mathematical abstraction removed from any sensual experience. Human warmth can quickly chill. That explains statements like that uttered by Hubert Markl, former president of the German Research Foundation and later Max Planck Society, who argued that "the human masses" of soon "ten or even twelve billion people ... in the future will probably require up to half of the total net biomass of the earth," and thus can only be fed with genetically engineered mass-produced food (*Der Spiegel* 48/1995).

By ignoring the manifold, different subsistence needs of people, human dignity is disregarded. But this is not a problem for Mr. Markl. Rather, the modern iscourse of justice overflows with measurements, UN Millennium Goal No. 1, *Eradicate extreme poverty and hunger*: "Halve, between 1990 and 2015, the proportion of people who suffer from hunger." No comment.

Modern humans deprive themselves of their sovereignty and with it their human dignity by daily allowing the laws of the money system to rule their lives. By believing in the equation—*Money = Existence = Food = Livelihood*—they subordinate their lives to a superior power. The collective fiction that money really has an essential value for survival is only possible if everyone believes that the value of a sum of money is actually the claimed value. Hence, Hans Christoph Binswanger talks about "Money and Magic." This magic only works as long as everyone accepts the rule of the superior power that guarantees the equation, i.e., the value of money. This fact was more obvious in former times when pieces of money still carried the ruler's portrait. By virtue of his power he guaranteed that the assumed value of the piece of money was backed by tangible valuables. In other words, power attributed, rule accepted.

Therefore, the most important aim of the governments in the current financial crisis is to stabilize confidence in the value of money. The fact that this not only stabilizes the governments, but also the entire oligarchic power of finance is in line with the power structure of our era, where international corporations and financial institutions are the "new rulers of the world" (Ramonet). And it reflects our era's subordination as consumers: the majority pays for the measures taken to stabilize their confidence.

DECOMMERCIALIZATION: THERE IS ENOUGH FOR EVERYONE!

We, *our* society, *our* economy must decommercialize our heads, our hearts and relationships. "But it's impossible to live without money," you say. Correct, at least for the time being. But a lot is possible without money! Decommercialization is first and foremost an attitude of mind. Orienting ourselves not to money but to what we actually need puts all decisions in a new light. Modern insatiability can be replaced by the satisfaction of having a need fulfilled. Decommercialization is a guideline for individual action aimed at social change.

Decommercialization of heads and hearts means no longer counting on money and capital as the instruments that allegedly organize life in a global society. The daily politics of decommercialization aim for a new form of horizontal relationships between free, self-responsible individuals who can meet on the basis of their shared essential needs. Instead of a central abstract standard, their communication is determined by their respective specific, diverse and vital necessities of life. There is enough for everyone!

We need a collective self-liberation from the psychotic fear of scarcity. By always asking: "What does it cost?" and "How much money will it make?" this fear is continuously invoked, like a mantra. Decommercialization is a means of self-defence against the totalitarian mechanisms of the commodity system, and is an act of self-empowerment. Decommercialization cannot be achieved overnight, but rather through a process of collective learning—and that has already started.

Free Communication

Worldwide, more and more mostly young people are getting together to campaign for freely accessible knowledge and freedom from state control of data communication. The movement emerges where the internet is widely used: in Argentina, Chile and Brazil; in the U.S. and Canada; in South Africa, Australia and New Zealand, in Europe and in Russia. In seven European countries these groups are officially registered as the so-called Pirate Party. In the 2009 German parliamentary elections, they immediately gained two percent of the votes.

As far as information available via the internet, and the technology required for accessing this information, another morality distinct from the one of exchange has been evident for some time. Free software and free downloads of texts, images and music are considered a form of community property to be shared throughout the internet community. The viewpoint that knowledge and information are common property has found its expression since 2001 in Wikipedia, the online encyclopedia where knowledge is collated and made freely available through voluntary work, free from copyright or similar restrictions.

Helping others, for hours on end and unselfishly, has been a part of dealing with information technology right from the start. Today the internet is teeming with forums where people who don't know each other take the time to answer all kinds of questions, and pass on advice and help. This seems to have resulted in a new culture of helpfulness among young people, which was not at all expected in a society where time equals money. This movement for free software and free knowledge coincides with—and may even have been initiated by—the development of a new type of youth culture. Especially in the cities, it is characterized by a lifestyle of frugality and sharing. It appears that immaterial communication has rubbed off on material communication.

In line with Vaughan's concept, communication is a quasi-archaic property of subsistence. Language, the material that knowledge transfer and exchange of life experiences are made of, is a gift, which, among many other things, is given to us in the same way that we are given life. The neoliberal attempt to subject ever more immaterial areas to the exchange logic is met with resistance from a deeply rooted feeling that is not so easily corrupted by the ruling logic of instrumental reason. Consequently, movements for free communication attack an important building block of the World Trade Organization (WTO) structure, that of TRIPS (Trade Related Intellectual Property Rights). With the help of TRIPS, ancient common knowledge is turned into law and can be declared the private property of a company, which, for example, turns a plant extract known for as long as anyone can remember for its healing properties into an international commodity. Local users can now be charged with a license fee.

Decommercialization in the City

When we in the urbanized society of the twenty-first century, one entirely dictated by consumption and drip-fed by supermarkets, want to strive for an economy where the necessities of life are available for everyone, what are the socially transformative measures available to us here and now?

Some answers are provided in the comprehensive study by Friederike Haberman on networks in Germany and Austria, "A new way of everyday life and economy" [my translation]. In many cities initiatives have been set up with the declared aim of undermining the money economy. These include free stores where clothing and other goods are brought and taken away without payment. There is rent-free housing, not only in squats but also in city communes and other common property, such as intergenerational housing. Resource and skill centres share tools and know-how. Free food is available in "Volxküchen" (peoples' kitchens), and by collecting discards from supermarket dumpsters. The motivation behind these initiatives is not so much for personal use, or need. Rather it is about a new form of communication not mediated by money. "You are now leaving the capitalist sector," reads a sign at the entrance of a free store in Berlin. To date (2018) nearly one hundred free stores have been set up in Germany. Food from dumpsters is quickly redistributed through informal networks, free food shelves in basements, and one group in Berlin is currently setting up a dumpster-diver co-op. The goal of this and the other give-away campaigns is to build community (Habermann).

What a contrast to the aim and intention of German charity food banks! These are supported by the McKinsey consulting firm with a guideline, based on experiences in the U.S. where low standards of state welfare are much older. McKinsey also formed part of the commission that contributed to the dismantling of the German social assistance system in 2001. Since then, people in need regularly line up at the food banks and are often, as in Bielefeld for example, poorly treated. Meanwhile, the supermarkets that donate food receive tax rebates for their donations, save disposal costs, are seen as charitable sponsors—and the wheel of the undignified dependency on money and commodities keeps turning.

A society that believes that supply depends on money and that money depends on selling one's labour, is increasingly moving towards polarization and an end to solidarity. At present the income of about one quarter of the total population of Germany is at the level of social welfare or Hartz IV, as Olaf Scholz, then German minister for social affairs stated in October 2009. And poverty will certainly increase, since extensive lay-offs have been announced as a consequence of the financial crisis. What are the prospects for these people?

Under the Garbage, the Field is the translated title of Elisabeth Meyer-Renschhausen's German book on community gardens in New York City. This excellent report with many interviews reads like a novel and encourages self-empowerment. During the 1990s, people in U.S. cities transformed wasteland into community gardens for vegetables, salad and fruit. Where once rubbish was dumped, plants now blossom. In 2003, the American Community Garden Association counted more than 6,000 such neighbourhood gardens. They bring people together; they provide fresh food to some for the first time in their lives; they take young people off the street and away from gangs and drugs; they build bridges across ethnic boundaries; and they change ravaged districts into pleasant green living areas.

Two thirds of the community gardeners are women. It is noticeable that subsistence oriented supply is apparently mainly of concern to women, and especially to feminists, practically and theoretically. And no wonder, since the prevailing disrespect for subsistence as a whole coincides with the disrespect for the non-monetary supply activities of women. It was the proud and self-confident action of the women's movement, claiming that these unpaid contributions are in fact socially necessary, that brought about a civil society movement for freedom from the constraints of the growth economy. The studies by Elisabeth Meyer-Renschhausen and Friederike Habermann were supported and published by the (German) Women's Initiative Foundation (Stiftung Fraueninitiative) under the auspices of Carola Möller.

Christa Müllerset up the foundation "Stiftung Interkultur," which is mainly dedicated to intercultural gardens, as a project of the foundation "anstiftung & ertomis" in Munich. These special neighborhood gardens enable migrants in Germany to 'gain new ground' and, similar to a plant, to "sink new roots." In their gardens, refugees and migrants from different countries grow fruits and vegetables, often varieties from their far away homelands. They feel more at home, have a supply of fresh food, make new contacts and, last but not least, have something to give their German neighbours over the fence. The first intercultural garden was created as an initiative of Bosnian women refugees in 1996 in Göttingen, Germany. In 2018, there are more than six hundred intercultural gardens across Germany, and another fifty are being developed ("Stiftung Interkultur"; Müller).

There Is More than Wage Labour! The Decommercialization of Work
The spirit and feeling of refusal to be completely taken over by the industrial exploitation of labour lives on. In our society, labour has long been separated from self-realization through work, and people have adapted to dependent, mostly meaning-less jobs. However, there are indications that independence, creativity and community spirit are alive and well. In Germany, evidence for this can be seen in the diverse ways that craftspeople, health experts, artists, teachers and writers have developed to generate income at a time of dwindling long-term contractual employment.

The 1980s and 1990s were increasingly shaped by information technology and the globalization of markets with accompanying "jobless growth" and privatization of public services. During this period, many workers in Germany found ways to organize their own incomes following the guiding principle of meaningful work.

Combining paid and unpaid labour, they developed a "patchwork" of income-creating activities. With the help of short- term job creation programs, state subsidies for non-profit institutions, unemployment benefits earned through previous formal employment, and all types of self-employed work, many were still able to more or less do their work of choice.

Apparently, it was exactly this independence from wage labour that proved to be a thorn in the flesh of the commission that devised Hartz IV and of the members of parliament who passed the bill. Along with Peter Hartz (at the time, board member of Volkswagen AG and now a convicted lawbreaker), the commission included a manager from Deutsche Bank, a member of the Roland Berger management consulting firm, a director from McKinsey's and other business heads. What really incited this aggression towards the "patchworkers" that such drastic disciplinary measures had to be drawn up to force them back into the coercive system? After all, they did not amass any wealth and at times had to fear for their livelihoods due to lack of security. Yet they were able to defy the situation and to mobilize primeval energies. It was possible to focus on goals such as being creative or serving the common good instead of having to do wage work in a job only endured for the paycheck. Day care centres and women's shelters, fair-trade and whole-food shops were set up, academics lived from one lectureship to another, did free-lance writing, others organized free theatres and artist collectives, developed woodworking and sewing workshops or managed to complete a university degree thanks to social welfare support for young mothers. People got by, some better, others not so well, but the necessities of life could be generated: food, clothing, accommodation, education for the children, social life, travel. Above all, one was at peace with oneself.

Adrienne Goehler[8] identifies this large group of people—in line with a U.S. study—as the German equivalent to the "cultural creatives" who are said to account for a quarter of the population (Ray and Anderson). Goehler defines her concept of culture similar to the US authors. "To be culturally active is a concept that not only refers to the manageable group of those making a living from culture, but also those who understand culture as matrix for creativity and thus a fundamental human ability." In Germany, a trend can be noted towards a cultural change, away from the identification with economic growth and increasing profits and consumption. Goehler makes the case for a policy paving the way for fragmented livelihoods, one that is open for new types of occupations (*Kurskontakte* Dec. 08/ Jan. 09).

The Decommercialization of Money

It sounds like a paradox, but it isn't. Money has two fundamental functions: as a medium of exchange and as a commodity. As a neutral exchange medium it can facilitate exchange by simplifying, for a large number of individuals, transactions with a wide variety of goods. As a commodity, money itself has a price, namely the interest, which carries the increased risk that money takes on a life of its own as opposed to its function as a medium. Money may be held back if the expected interest to be earned appears too low, as was the case in 2010. Banks are accused of hesitating to extend credit to the real economy, despite the fact that they are receiving government funds for exactly that purpose. Indeed, the entire crisis results from the detachment of money. The trading of money has spiraled without any relation to the goods for which it was meant to act as a medium.

The decommercialization of money first of all involves reducing the function of money as a commodity. This can happen on many levels and is most urgently required at present. The money involved in speculative transactions—the commodity that has taken on a life of its own—must be separated from money used as an exchange medium for necessary goods. The current so-called rescue packages continue to suffer from the construction faults of the monetary system, namely that the real economy and the financial economy operate with the same money. What some need to live on can be "play money" for others. Therefore, separate currencies are necessary. Anyone wanting to go to the casino must first change money from the real economy into play money. A proposal of the governor of the Bank of England and the former chairman of the U.S. Federal Reserve and later presidential advisor deals with at least part of the problem. They argue that, "normal banking, i.e., current and savings accounts as well as corporate and private loans, should be separated from the considerably more risky investment banking" (*Süddeutsche Zeitung* 22/10/2009).

However, there are probably only a few people authorized to make decisions on financial policies. But many others are by no means unable to act in financial matters. It is important to make ourselves aware of this possibility for action. Especially when it comes to matters of money and currencies, people tend to believe that only a higher authority can make the rules. This belief leads us to project ourselves into the director's chair, instead of addressing our own situation in the here and now. However, we can in fact take action on the monetary system. For decades, various initiatives have made possible the decommercialization of money for every man and every woman in their daily lives. Local and regional alternative currencies and local exchange trading systems are a first step towards the decommercialization of money.

Two basic motives can be distinguished among the diverse approaches to alternative money. The older school of thought focuses on the question of interest or no interest. This goes back to Silvio Gesell (1862-1930) and his theory of "free money." The range to which the "free economics" applies is relatively broad, including the complete national economic system (INWO). The new school of thought around local exchange trading systems starts, however, with an explicit view from below, from everyday life. Members join forces locally, form face-to-face relationships and find quite different practical solutions to existing problems. Both schools of thought come together in the local currency movement, as documented by the German social economy journal, *Zeitschrift für Sozialökonomie* (edited since 1963 by Werner Onken). This journal also documents the close relationship between the search for alternative money and the social movements for economic and ecological justice since 1968.

In England, exchange systems flourished in reaction to Margaret Thatcher's neoliberal gutting of social welfare. These Local Exchange Trading Systems, LETS, contained the central idea of current alternative exchange systems: "I may not have a job or many possessions, but I am skilled and can offer something that others need. In return I can find enough for my own needs in the exchange circle." Generally, the alternative exchange systems create their own currencies; some call them "gems," or "bees." The unit of exchange can be based on time, such as fifteen minutes of work. Or it can be based on an exchange rate, such as one Euro = one "Waldviertler" as used in this region of Austria. According to Friederike Habermann, there are currently 271 exchange systems in Germany. She also reports on the *"Gib & Nimm"* (Give & Take) initiative in Wuppertal, where some forty members exchange services like ironing, computer servicing, childcare etc. with no central exchange unit or bookkeeping involved.

Right from the beginning, members of "Give & Take" consciously opted against any form of measuring and counting and they are now experiencing how difficult it is to let go of the exchange concept. "You first have to break down barriers in your mind," says Marie, one of the members interviewed, and concludes why this is so: "Normal alternative exchange systems basically mirror our society." Regardless of the currency unit and even when money is reduced to its function as simple mediator, it is hard not to remain stuck in the prevailing discourse of self-interest. The emancipation of individuals, and finally society, from the super-ego of "*Homo oeconomicus*" is a process in which values have to be reconsidered and changed. A process that takes time. Meanwhile, stories of the courageous actions of individual initiatives will eventually make history. Habermann's book presents many of these stories.

The Microcredit Bluff and Lessons to Be Learned from It

With microcredits, commercialization invades the last corners of minds, hearts and communities. In 2006, the Bangladeshi banker Muhammad Yunus was awarded the Nobel Peace Prize for this. With the granting of credits to the "destitutes," these were able to purchase "a calf or a sewing machine. Millions of poor people have become entrepreneurs in this way." The journalist quoted here however also allows her doubts on the peaceful intentions of this profiteering to shine through. "The Grameen Bank charges 20 percent interest and yet, Muhammad Yunus received the Nobel Peace Prize for establishing this" (*Spiegel* online 13/10/06).

Farida Akhter, who has for decades denounced the violence against women caused by development aid, for example through forced sterilization, clarifies: "Money circulated through the poor communities expanded often to 130 percent, appropriating the remaining resources of the poor in the form of interest. Indebting the poor has become the new game of development and swept the development discourse and the former practice away" (112). Akhter also sees a parallel between the programs to control population growth and the microcredit programs. Ninety-one percent of those sterilized were women and 82 percent of microcredit borrowers are women, because they are more easily pressured: by the fieldworkers of the cooperating NGOs, who earn their own income this way, and from their own husbands, who want to use the money for their own purposes. "Violence has not diminished, but has rather increased with the microcredits.... She even has to sell her chicken or get another personal loan." The repayment rate of microcredits granted to women is 95 to 100 percent. In the past, development aid workers were involved in setting up support for economic activities. Today, what matters is the granting of credits and securing repayment (Akhter).

The biggest problem is the fact that the credits undermine local mutual solidarity, also concerning the way credits were commonly originated. The rural organization of credit ("grameen" means "village") used to be similar to our historic credit and savings co- operatives. Today the local economic solidarity is being usurped by distant banks and transformed into the opposite, namely instruments of mutual control and liability.

In this way village women are directly connected to the national and international financial system. Akhter reported how BRAC, an NGO involved in the granting of credits, joined forces with Monsanto in order to foster the distribution of genetically engineered seeds with the help of microcredits. This example shows what effects microcredits can have, namely further curtailment of subsistence sovereignty.

In November 2009, the third "Vision Summit" took place in Berlin with the theme "Social Business—Another Wall to Fall." Muhammad Yunus, who created the concept of social business, was the main speaker. The intention of 'social business' was revealed at the award ceremony of the Vision Summit. The award was given to the Grameen Danone Community for the first Global Social Joint Venture: Grameen Danone Foods.

> Grameen Danone Foods Ltd. will produce a special yogurt from pure full cream milk which contains protein, vitamins, iron, calcium, zinc and other micronutrients to fulfill the nutritional requirements of children [and] adults. The price of each ... yogurt will be ... cheaper than other available yoghurts in the market and in line with what low-income people can afford. Grameen Danone Foods will reduce poverty by creating business and employment opportunities for local people since raw materials including milk needed for production will be sourced locally. The companies that make up Grameen Danone Foods Ltd. have agreed not to take out any of the profits of the company. ("Zidane Inaugurates")

What is the reality beyond these dazzling slogans? Bangladesh is a country of peasant farmers. Almost sixty percent of the population work in agriculture. Owning a cow for self-sufficiency and the sale of dairy products is an important goal for the peasant women. It is here, in this context, that local production happens. Here is the local market and the jobs that prevent malnutrition. The Danone-Grameen project in fact threatens this local subsistence economy, as it takes over and destroys local and regional markets.

What do these examples in Bangladesh tell us about the situation in the metropolitan north? How long will it take for German banks to discover social welfare recipients as a target group for microcredits? Already, they are publicly referred to as "customers" of social welfare. The next step seems to be mapped out. As is well known, the model of wage labour as a means of profit maximization is reaching its limits and some have already—wrongly and mistakenly—predicted the "end of the working society." Where thus far the employment contract served as a purchase agreement for the worker who sold his or her work as a commodity, money or a credit agreement will take its place. The ties that bind the worker may now be more impersonal, more mechanical so to speak—obeying the mechanism of money—but no less compulsory.

The way out is when we stop supporting the money mechanisms, and instead focus on subsistence. Communities should create a body of regulations beyond the exchange logic of money.

Food Comes from the Earth! The Decommercialization of Agriculture

What about our food? Don't we need money as mediator between city and rural areas? Is a decommercialized agriculture possible?

Experience shows us that focusing on subsistence leads to a decommercialization of agriculture and vice versa. In agriculture, focusing on subsistence means that crops are cultivated and animals are kept for the purpose of food creation and not for profit. This is exactly what rural logic and economy are about. Farmers feed themselves with their harvest, milk and meat, and sell the surplus. In industrialized agriculture, however, self- supply is the exception, as for example when hog breeders raise a pig for their personal consumption on straw instead of concrete crevices, with feed free of standard pharmaceutical additives. Apparently many breeders prefer to avoid the meat of animals from their own factory farms, which are fattening 1,000 to 10,000 animals at a time. (Baier, Bennholdt-Thomsen and Holzer).

Decommercialization means that agriculture is no longer ruled by money. Seed can be selected and reproduced by the farmers. The soil recovers when turbo fertilizers, growth regulators, herbicides, fungicides and insecticides are no longer used to increase yields. Animals are bred naturally without artificial insemination or hormonal treatment; a cow will live for twenty-five years and not just five. Decommercialization of agriculture ends the paradox that in rural areas many people are starving; that the farmer is skinny and the tractor is fat; that in India, thousands of peasant farmers are committing suicide because they cannot repay the money they borrowed to purchase seed from Monsanto. Decommercialization means a reversal of the World Bank's 1975 motto: "To draw farmers from subsistence to commercial agriculture" (20).

In the cities of the North, a subsistence orientation means that people must stop plundering the forests, the earth, and its inhabitants in other parts of the world with "their" money by consuming cheap agricultural produce that destroys the subsistence agriculture in those parts of the world. Even here, low food prices are causing farmers to lose their livelihood: ever more farmers are giving up because they can no longer compete with increases in livestock numbers and tons of grain per hectare and want to avoid ruinous debts. This is the same process that is happening in India. Decommercialization brings an end to the paradox that rich harvests or bountiful milk production are considered a disaster called "overproduction." Decommercialization and a subsistence orientation mean careful handling of humans, animals, plants, and the landscape.

Subsistence agriculture does not imply that peasants are producing only for their own consumption, eating alone in self-sufficiency, as is often insinuated. That has never been the case. This myth was in fact created in line with colonial development policy, which established the market for international looting and destruction of local and regional economies as well as of independent national economic cycles. In Europe, peasant women's markets (almost everywhere women ran the market stalls) existed up to the twentieth century. These markets, of which only a few have survived, offered the surplus from self-sufficiency as well as goods and items produced especially for sale.

For centuries, the rural economy in Europe was a market economy; in parts of Central America, Africa, the Middle East and South-East Asia, this has been the case for thousands of years. Market economy is simply not the same market economy everywhere and forever.

The network of peasant markets that covered all continents for thousands of years with supply lines—obviously with great cultural diversity—was plainly ignored by economists from Adam Smith to Robert McNamara and was as a consequence destroyed. When a Swiss organic farmer from Möschberg—one of the cradles of organic farming—almost desperately declared at the Möschberg meeting of organic growers in 2009: "I want to remain a nurturer," it became clear what rural markets are about, and how much the dimension of real provisioning has been destroyed even in organic farming by the commercialization of the supermarket era.

The human measure of manageable size has been pushed aside by the megalomania of economic growth. This was the analysis of the working group "A different way of acting economically" at the above-mentioned fifteenth Möschberg talk organized by Bioforum Switzerland. This working group developed the vision that economic activity can recover, provided the measures of value of the peasant economy, especially the principle of manageable size, are applied in all economic sectors.

"Profit for the peasant farmer is not to be confused with the profit an entrepreneur must aim for under the thumb of capitalism." (Heindl).

RE-RURALIZATION: FOR A NEW URBAN-RURAL RELATIONSHIP

An economic system that is not conceived of as a war for growth, and that follows a social contract based on the equality of all, would see many more people working on the land. Because if we want an economy that is not at other peoples' expense, in a peaceful world, we must try to live from the resources in that part of the earth we call home.

In the twenty-first century, re-ruralization obviously means something different than it did at the end of the Roman Empire. It includes people in the city and the development of the city itself. The social division of labour, in which the countryside provides food for the population centers, and the extreme separation between the rural and urban we know today, first came to a head with industrialization. This undesirable development can be changed. United Nations data show that for the first time in human history more people live in cities than in the countryside and this seems to indicate that urbanization and further development of megacities with mega-slums is unavoidable. However, this development, as the continuation of the growth economy, is not inevitable. In fact, for some time there have been indications that megacities are shrinking, as the economic patterns creating them are in crisis, and that counter-movements exist.

Re-ruralization in the twenty-first century includes three aspects:

- re-ruralization of the city
- re-ruralization of the urban-rural relationship
- re-ruralization of agriculture.

Re-ruralization of the City

"Agriculture is moving back into the city," stated organizers of the "Urban Agriculture" conference in 2009. This meeting focused into the city states the foundation "anstiftung," focusing especially on examples of "neighbourhood gardens; intercultural gardens; small, herbal and school gardens; farms for children; and community gardens" in German-speaking regions.

In the U.S., Detroit, Michigan has become famous for its urban agriculture movement in a deindustrializing city. There, members of the civil rights movement sought new forms of organization in reaction to the city's disintegration. In her autobiography, Grace Lee

Boggs describes their challenge, to create local, more self-sufficient and sustainable economies, ones that would redefine our relationship with the Earth, and that could rebuild communities destroyed by consumer oriented monocultures and the transformation of relationships into commodities. The new goal was to bring the countryside to the city and create a new culture there. Thus began the "Detroit Summer" initiative, which for twenty years has tried to create a new economy and lifestyle in a city that is shrinking and falling apart (Boggs).

A similar kind of re-ruralization of the city is being promoted through the relatively new Transition Town Initiative (TT). Starting in England and Ireland, hundreds of TT groups have also formed in continental Europe. A main theme is that the era of big oil is coming to an end ("peak oil'), and a move into the post-oil era must be found. Transition *Town* and not *City* reflects the goal of finding a middle way between cities and villages, the places where most people in Europe live. One important initiator is Rob Hopkins, an Irish expert on permaculture, an alternative integration of fields, gardens, and housing.

The transition to the new era should happen through local or regional integration of supplies of energy and food (with self-sufficiency through neighbourhood gardens and an urban-rural connection through farmers' markets) and of other basic resources needed. Materials will be recycled and free stores will be supported by independent artistic production, a local currency and much more, all of which strengthen public spirit. Above all, the TT philosophy reveals a great openness towards the diversity of possible paths to a different, ecologically adapted city. Since 2006, the town of Totnes in south-eastern England has been especially successful with this initiative.

Re-ruralization of the Urban-Rural Relationship

For some time, city dwellers have sought closer connections to the land without having to move there. "Community Supported Agriculture" (CSA) has been spreading increasingly in North America and Europe since the 1980s. Farm owners plan the next planting season together with a group of city dwellers, deciding what and how much will be planted, and how to organize distribution in the city. The people from the city finance the crop and thus their own harvest. Usually they can also participate in basic work, where many hands are needed, like weeding or harvesting, for example potatoes. Vandana Shiva reported that in the U.S. some three thousand CSA groups had formed in the past seventeen years. In western Switzerland alone, there are more than forty groupings of this kind. In Germany, there are fewer than ten projects.

However, among them is the well-known Buschberghof, which has successfully functioned this way for twenty years.

In Germany, the system of regular delivery from farms is widespread. Households that receive their weekly "green box," find an assortment of vegetables, fruit and salad that varies according to the season and weather. The customers thus adapt to real natural conditions, far different than shopping in the globalized world supermarket.

A very successful and common city-country relationship is the formation of communal groups of city people in the country. Since the 1970s, rural communes have been founded in Germany, some now with up to seventy adult members. Members work in agriculture or in workshops, organize kindergartens and seminars, pooling their income. In the same spirit and following the growing environmental awareness, eco-villages were formed somewhat later. Still developing, such a village is seen as home for up to three hundred people. The most important principles are self-sufficiency and self-organization. In addition, there are rural communities founded in a Christian spirit of community, some dating from the late 1950s. From the 1980s, communities with a Buddhist spirituality formed. Then there are groups that are mainly focused on organic agriculture, organized as cooperatives, which produce on one hundred hectares (two hundred and fifty acres) of land or more. The diverse projects established in eastern Germany after 1989 are also relatively well-endowed with land. A look at the documentary, *Gemeinschaften und Ökodörfer in Europa* ("Communities and Eco-villages in Europe") shows that Europe is covered with rural communes, especially Germany, which has more than one hundred such projects.

The communities outlined above remain part of an urban culture. Although they have moved to the countryside, they remain closely connected with similarly oriented groups in the city. The contrast to the traditional rural culture around them is clear, and can sometimes lead to friction. But the new communities' members generally make a real effort to establish bridges between the cultures of city and countryside. This approach is successful, especially when people know each other longer and because the outlines of rural culture are also increasingly blurred. One could say it is becoming citified.

Re-ruralization of Agriculture

In the second half of the twentieth century, agriculture became industrialized. The degree of industrialization differs according to the world region and is also different within these regions. But the international development organizations and the WTO are actively working to remove the last, also cultural and psychological, barriers to the commercial take-over of rural areas. Agriculture should be treated in international agreements like every other economic activity. A result of this development is the massive buying out of peasant farmland and corresponding expropriation of the resident population ("land grabs"). The buyers of these lands include: "Countries that since the food supply crisis fear for their own food security—especially Arab states with huge oilfields but little productive agricultural land.... In addition, there are the financial investors such as Morgan Stanley or BlackRock." (*Süddeutsche Zeitung*, 11/02/2009 [my translation]). The U.S. investor Philippe Heilberg bought 400,000 hectares (onr million acres) of the best land in southern Sudan (!) from war lords there. George Soros acquired land in Argentina for production of liquid fuels. The Italian clothing company Benetton controls more than 900,000 hectares (2.25 million acres) in Argentina for wool production, while the Russian hedge fund "Renaissance Capital" controls 300,000 hectares (750,000 acres) in Ukraine. The South Korean "Chaebol" Daewoo Logistics has a 99-year lease for 1.3 million hectares (3.25 million acres) in Madagascar, half of the island's fertile land. Meanwhile, German investors buy land in Romania through the investment firm Agrarius Ackerland, which went on the German stock exchange in November 2008 (*Süddeutsche Zeitung*, 31/03/2009).

A re-ruralization of the countryside means bringing local peasant farmers back into agriculture. This is part of a process through which the "Earth Democracy" or "Democracy of all Life," as Vandana Shiva puts it, can be achieved. Everywhere worldwide, the displacement of small farmers, men and women, through the industrial, monocultural, capital-intensive plantation economy must be stopped, and the farming methods of small farmers, as well as regional marketing structures and opportunities must be strengthened. This is not just necessary for social justice, but, as the 2008 World Agricultural Report[9] makes clear, is also required for the ecological recovery of the earth. The protection of biodiversity, the reduction of greenhouse gases and carbon dioxide emissions, the production of healthy food within the regions and the protection of the earth's water supplies can only be achieved through the diverse economy of small-scale farming.

Food sovereignty is unfailingly bound to the small-scale farm perspective. The term and idea came from a 1996 meeting of the "Via Campesina" international movement of farm workers and peasant farmers in Tlaxcala, Mexico. The term 'food sovereignty' includes both food production—access of the local population to land, seeds, forests and water—as well as the consumption of food: its qualities of freshness, taste, nutrition, and special ways of preparation. These can satisfy the eater and are filling, as opposed to "*mal bouffe*" or junk food— the unnatural industrial supermarket offerings condemned by

French farmer-activist José Bové and others. Diverse, local and regional autonomous food supplies must replace the monopolized, corporate-led monocultures of the globalized food industry which has seized control of the entire food chain from seeds to meal. Food sovereignty is at the heart of the movement to reclaim the independence of the individual and civil society as opposed to the totalitarian system of money and commodities.

FOR A POLITICS OF DAILY LIFE

Not everyone wants to work the land, live communally, get around by bike or wear other peoples' clothes. And we can't forget about money or self-interest overnight. Just as we cannot replace wage work with self-determined community work, even if many would like to do that right away. Even so, there is a lot we can change in our daily lives. The politics of daily life is the decisive means of self-empowerment for individuals and civil society, against the disabling moloch of the money and commodity system. Another world is possible, and we can do something to achieve it.

Starting points for another economy exist. All the studies about the practice of subsistence, the work done every day without paying or being paid for, the things that are necessary for life and are done "so that life goes on," come to the same conclusion. People actually do much more for immediate subsistence than they realize, *and* they are filled with pride and satisfaction when they realize the extent of their contributions. Mainly, these involve activities they are happy to do anyway, without admitting that this is so. Because the dominant discourse that disciplines us and directs the socialization process sends a completely different message: that only work for wages or, in other words what brings in money—the more the better—belongs to a socially accepted lifestyle.

We should consciously distance ourselves from this money ethic. Only in this way can a new social contract emerge, beyond the culture of multiplying money. Because the goal is not to get through the next crises, but to stop the mechanisms of impoverishment and destruction. It is about achieving another relationship with our fellow humans, our fellow creatures and the world as a whole. To achieve this Earth

Democracy, another mindset and behaviour is demanded from us all. By now it should be clear enough that it is ethically-morally impossible to delegate our own human responsibility given the framework of current global structures of power and governance. Take, for example, the following facts. The U.S. president Barack Obama received the Nobel Peace Prize yet continued the politics of war, sending more battalions of soldiers to Afghanistan. At the world climate summit in Copenhagen (December 2009), national leaders from 192 countries haggled over ways to prevent pollution of the atmosphere, without meaningful results. The same happened with the climate summit in Paris in 2015. And in 2018, the U.S. government under Donald Trump even withdrew from the Paris climate agreement.

In our era, the dominating power is solely legitimated by economic growth. This legitimation has to be denied, and by all of us. The responsibility for policies of an Earth Democracy lies with each individual, and can be realized in our daily actions. Will money drive our decision-making, or will it be replaced by an orientation on subsistence, on what is really necessary for a good life for all?

As we seek a guiding policy for our daily lives, it makes sense to ask:

- What do I do without money?
- Which of my relationships have nothing to do with money?
- Which communication, be it material (goods) or immaterial (social relations) has no calculated, hence no exchange bias?
- and finally: What makes us really rich?

Money leads to misanthropic separation and the atomization of society. The abstraction of money and the anonymity of exchanges through money lead to our estrangement from each other and weaken feelings of community. That is why money is also the central building block in the "Empire of Shame" (Ziegler). Daily actions against this imperialism can only happen if we reduce anonymity and abstraction as much as possible. Also when money is involved we can think about another attitude towards money. Fair trade, environmental certification and organic certification appease our consciences, but are by no means sufficient. They may prevent the worst excesses of combined, plundering self-interest. But they do not change the basic fatal attitude, that similar to a supposed colonial master's right everybody can lay claim to all the goods of this earth just because they have money.

We have to ask ourselves how Christian or humanitarian it is at all, when fertile land in East Africa is used to produce roses for Europeans, even if certified as fair trade. Or when ever scarcer water supplies in Africa, Latin America, and Asia are used for sugar cane, cotton, soya and wine for export, instead of food crops for the local population. We should provide for ourselves with what our land, climate and the farmers of our region can produce, instead of easing our consciences by paying a few cents more for certificates. They don't help anyone, because you can't eat money!

This article was originally published in German with the title, "Geld oder Leben: Was uns wirklich reich macht" (Munich: Oekom Press, 2010). The article was translated into English by Sabine Dentler and Anna Gyorgy. Second English edition: August 2011. © Women and Life on Earth e.V., Bonn, Germany, info@wloe.org.

The English translation was made possible by the generous support of Genevieve Vaughan, whose valuable work on the gift economy also contributed to the development of "Money or Life." (See: www.gift-economy.com/forgiving.html).

The author's thanks go to the German foundation community anstiftung & ertomis for supporting the study "From an Economy of Taking to an Economy of Giving," which provided the framework for this essay.

Women and Life on Earth: Women in International Cooperation for Peace, Ecology and Social Justice (WLOE e.V.) is a German non-profit association based in Bonn. WLOE e.V. offers and supports the work of women especially, in connected areas of ecology, peace and global justice, often through translation and editing of original texts. The website is active in English, German and Spanish at www.wloe.org.

ENDNOTES

[1] A series of laws aimed at reforming the German social system and labour market. The declared aim of Agenda 2010 was to improve economic growth and thus reduce unemployment.

[2] The Hartz IV reform merged the previous unemployment insurance payments and social welfare benefits, leaving both at approximately the lower level of the former welfare benefits and ending insurance benefits for the unemployed after only one year. For German, it is a clear break-up of the century old insurance standards and social welfare net.

[3] "David Ricardo ... was an English political economist, often credited

with systematising economics, and was one of the most influential of the classical economists, along with Thomas Malthus, Adam Smith, and John Stuart Mill.... Perhaps his most important contribution was the law of comparative advantage, a fundamental argument in favour of free trade among countries and of specialization among individuals" (http://en.wikipedia.org/wiki/David_Ricardo).
[4]Landed aristocrats, or "Junkers" of the nineteenth century owned most of the arable land in the Prussian and eastern German states, giving them virtual monopoly on all agriculture in the German states east of the Elbe river.
[5]Attac is an international movement working towards social, environmental and democratic alternatives in the globalization process.
[6]Green political party in Germany, formed in 1993 from the merger of existing West German Green Party (founded 1980) and the eastern Germany Alliance '90.
[7]The abbreviation for *die tageszeitung*, Germany's seventh largest national daily newspaper, cooperatively-owned and administrated through workers' self-management.
[8]Former active Green Party politician.
[9]See Romig for comments on this report.

REFERENCES

Akhter, Farida: *Seeds of Movements: On Women's Issues in Bangladesh*. Dhaka: Narigrantha Prabartana, 2007.
Baier, Andrea and Veronika Bennholdt-Thomsen. "The 'Stuff' of Which Social Proximity Is Made." *Regional Approaches to Sustainable Economy: Potentials and Limits, Experiences from German Case-Studies*. Eds . Thomas Kluge and Engelbert Schramm. Frankfurt: ISOE-Materialien Soziale Ökologie, 2002. 10-17.
Baier, Andrea, Veronika Bennholdt-Thomsen and Brigitte Holzer. *Ohne Menschen keine Wirtschaft. Oder: Wie gesellschaftlicher Reichtum entsteht. Berichte aus einer ländlichen Region in Ostwestfalen.* Munich: oekom, 2005.
Bennholdt-Thomsen, Veronika. "Marginalidad en América Latina. Una crítica de la teoría." *Revista Mexicana de Sociología* 43.4 (1981): 1505-1546.
Bennholdt-Thomsen, Veronika. "Marginalität in Lateinamerika— Eine Theoriekritik." *Lateinamerika: Analysen und Berichte V.* Eds. Bennholdt-Thomsen. T. Evers and K. Meschkat Berlin: Olle & Wolter, 1979. 45-85.

Binswanger, Hans Christoph. *Money and Magic: A Critique of the Modern Economy in Light of Goethe's Faust.* Chicago: University of Chicago Press, 1994.

Bloch, Ernst. *Natural Law and Human Dignity.* 1961. Cambridge, MA; MIT Press, 1986.

Boggs, Grace Lee. *Living for Change: An Autobiography.* Minneapolis: University of Minnesota Press, 1998.

Caillé, Alain. *Anthropologie der Gabe.* Frankfurt: Campus, 2008.

Cavarero, Adriana. *For More Than One Voice.* Palo Alto: Stanford University Press, 2005.

Der Spiegel 48/ 1995.

Die Tageszeitung (taz) 5/6/10/2009.

Einfach Gut Leben e.V. ed. *Eurotopia, Gemeinschaften und Ökodörfer in Europa.* Poppau, 2009.

Gemeinschaften und Ökodörfer in Europa (Communities and Eco-villages in Europe). L.O.V.E. Productions, Documentary, 2009. Web.

Gesell, Silvio. *The Natural Economic Order.* London: Peter Owen, 1958.

Goehler, Adrienne. *Verflüssigungen. Wege und Umwege vom Sozialstaat zur Kulturgesellschaft.* Frankfurt: Campus, 2006.

Gorz, André. *Critique of Economic Reason.* London: Verso, 1989.

Habermann, Friederike. *Halbinseln gegen den Strom. Anders leben und wirtschaften im Alltag.* Königstein/Taunus: Ulrike Helmer, 2009.

Habermas, Jürgen. *Theorie des kommunikativen Handelns, Band 1: Handlungsrationalität und gesellschaftliche Rationalisierung; Band 2: Zur Kritik der funktionalistischen Vernunft.* Frankfurt am Main: Suhrkamp, 1981.

Heindl, Bernhard. "Hoffnung vom Hof? Plädoyer für eine neue Übersichtlichkeit des Wirtschaftens." *Kultur und Politik* 2 (2009): 5-6. Bioforum Schweiz, Möschber.

Initiative für Natürliche Wirtschaftsordnung (INWO). Fairconomy. n.d. Web.

Kurskontakte Dec. 08/Jan. 09.

Lessem, Ronnie and Alexander Schieffer. *Integral Economics.* Farnham: Ashgate/Gower, 2010.

Luxemburg, Rosa. *The Accumulation of Capital.* London: Routledge, 2003.

Mauss, Marcel. *The Gift: The Form and Reason for Exchange in Archaic Societies.* London: Routledge, 2002.

Meyer-Renschhausen, Elisabeth. *Unter dem Müll der Acker (Community Gardens in New York City).* Königstein: Ulrike Hellmer, 2004.

Mies, Maria. "Subsistenzproduktion, Hausfrauisierung, Kolonisierung." *Beiträge zur feministischen theorie und praxis* 9/10 (1983): 115-124S.

Mies, Maria and Veronika Bennholdt-Thomsen. *The Subsistence Perspective: Beyond the Globalised Economy.* London: Zed Books, 1999.

Mies, Maria, Veronika Bennholdt-Thomsen, and Claudia von Werlhof. *Women: The Last Colony,* London: Zed Books, 1988.

Müller, Christa. *Wurzeln schlagen in der Fremde. Die Internationalen Gärten und ihre Bedeutung für Integrationsprozesse.* Munich: oekom, 2002.

Muraro, Luisa. *L'ordine simbolico della madre.* Rome: Editori Riuniti, 1991.

Ramonet, Ignacio. *Nouveaux pouvoirs, nouveaux maîtres du monde.* Montréal: Fides, 1996.

Ray, Paul H., and Sherry Ruth Anderson. *The Cultural Creatives: How 50 Million People Are Changing the World.* New York: Three Rivers Press, 2000.

Ricardo, David. *Über die Grundsätze der politischen Ökonomie und der Besteuerung.* Marburg: Metropolis, 2006.

Romig, von Friedrich. "Warum Rückkehr zur eigenen Währung? Einige grundsätzliche Erwägungen." *Zeit-Fragan* 29 (2012): np. Web.

Rügemer, Werner. *Cross Border Leasing. Ein Lehrstück zur globalen Enteignung der Städte.* Münster: Westfälisches Dampfboot, 2004.

Scherhorn, Gerhard and Klaus Michael Meyer-Abich. "Suffizienz in Konsum und Produktion." *Umwälzung der Erde: Konflikte um Ressourcen, Jahrbuch Ökologie 2010.* Ed. Günter Altner. Stuttgart: Hirzel, 2009. 171-179.

Shiva, Vandana. *Earth Democracy, Justice, Sustainability, and Peace.* Cambridge, MA: South End Press, 2005.

Stiftung Fraueninitiative (Women's Initiative Foundation) (n.d.). Web.

"Stiftung Interkultur." Anstiftung.de (n.d.). Web.

Smith, Adam. *An Inquiry into the Nature and Causes of the Wealth of Nations.* 1776. Chicago: University of Chicago Press, 1977.

Spiegel Magazine No. 14 (2009).

Spiegel online 13/10/06.

Süddeutsche Zeitung 11/02/2009.

Süddeutsche Zeitung 31/03/2009.

Süddeutsche Zeitung 22/10/2009.

Vaughan, Genevieve. *For-Giving: A Feminist Critique of Exchange.* Austin: Plain View Press, 1997.

Vaughan, Genevieve, ed. *The Gift, Il Dono: A Feminist Analysis.* Rome:

Metelmi, 2004.

Werlhof, Claudia von. "Lohn ist ein Wert, Leben nicht? Eeine Replik auf Ursula Beer." *Prokla* 13.50 (March 1983): 38-58.

World Bank. Agricultural Credit. *Sector Policy Paper.* 1975.

"Zidane Inaugurates First Grameen Danone Dairy Plant." Grameen Bank 17 July 2017. Web.

Ziegler, Jean. *L'empire de la honte (The Empire of Shame).* Paris: Fayard, 2005.

3.
Deconstructing Necrophilia

Eco/feminist Perspectives on the Perversion of Death and Love

IRENE FRIESEN WOLFSTONE

HANNAH ARENDT'S PHILOSOPHY OF NATALITY asserts that we are born to live, to create, and to begin. Before we can begin to imagine a paradigm of natality, we must first understand how necrophilia pervades the dominant Western paradigm. With a clear understanding of necrophilia, we can join Grace M. Jantzen and Adriana Cavarero in the emancipatory project of envisioning a paradigm of natality that builds on two assumptions: all humans are born of a mother, and we live an embodied existence within a living landscape.

If anthropogenic climate change represents the apex of the Western culture of death, then we must accept that Western civilization is catapulting toward mass extinction. However, if you, like me, hang on to shreds of hope that there is life after the collapse of Western values, then you join me in the search for ecocentric philosophies. I am convinced that the "way out" of the climate crisis must be different that the "way in"; therefore, a framework for climate change adaptation involves a radical change in how we think and how we are in this world that we call home. A philosophy of natality may be a radical "way out"—a new way to think and be.

Natality is under-theorized, yet offers potential for an emancipatory politic. Natality (n. from Latin *natalis*; adj. natal) means "pertaining to one's birth" or "native" in reference to a place (Barber). Arendt's philosophy of natality focuses on the distinctly human capacity to bring forth the new, the radical, and the unprecedented into the world (Arendt 178). She insists that, despite the fact of death, humans do not live in order to die, but to begin. Arendt offers three definitions of natality. First, natality refers to the fact of birth. The activity of birth is linked with nature and organic matter, which is governed by cyclical time of life, death, and regeneration (96). Every human enters the world as a baby, born of a mother. Secondly, natality refers to belonging to a world characterized by plurality, where each ensouled and embodied natal is different from anyone else who ever lived, lives, or will live (8). The acceptance of difference is necessary for harmonious communities. Our second birth occurs in the public realm where natals differentiate and take responsibility for their own creative initiatives. Thirdly, natality is a political philosophy in that each revolution is a new beginning enacted through collective agency (9). Initiative is the greatest political activity; thus, natality, not mortality, is the central category of Arendt's political thought.

Necrophilia is a word with many layers of meaning. Necrophilia (n. from Greek *nekro* meaning corpse; -*philia* meaning love) means morbid and erotic attraction to death or corpses (Barber). Necrophilia implies sexual assault on an inert female body, illustrated by the classical narrative of the battle of Troy in which Achilles kills Penthesilea, queen of the Amazon warriors, and then violently rapes her corpse. Mary Daly defines necrophilia, "not in the sense of love for actual corpses, but of love for those victimized into a state of living death" (Daly *Websters* 59). Following Grace M. Jantzen, I interpret necrophilia as obsession with death, where obsession means a state of disordered thinking in which death is confused as love—the perversion of *eros* and *thanatos*—a perversion that robs death of dignity (Jantzen 135). The use of the date-rape drug is necrophilic in that it silences the victim and perverted in that it robs the victim and the sexual act of dignity, and indicates how necrophilia makes perversion normative by silencing resistance and outrage. In the following compressed overview, I identify only seven of the many ways in which death is valorized and perverted in the dominant Western paradigm.

DEATH OF NATURE

Western science favours mechanism and reductionism—two theories that separate humans from nature and support a worldview which holds that nature is an inert and mindless compilation of parts which have no inherent meaning. Francis Bacon, the so-called father of modern science, turned science into a gendered activity in which men exercise hegemony over nature and "others" (Sardar 2). Bacon was Attorney General for King James VI, who reigned during the worst of England's witch-hunts, and this fact provides context for his misogynist language in which nature is no longer a wise, venerated Mother Earth, but a wanton female to be conquered by a male aggression. Using the language of the Inquisition, Bacon urged the domination of nature for human use:

He compared miners and smiths whose technologies extracted ores for the new commercial activities to scientists and technologists penetrating the earth and shaping "her" on the anvil. The new man of science, he wrote, must not think that the "inquisition of nature is in any part interdicted or forbidden." Nature must be "bound into service" and made a "slave," put "in constraint," and "molded" by the mechanical arts. The "searchers and spies of nature" were to discover her plots and secrets.... "Nature placed in bondage through technology would serve human beings." (Merchant *Radical* 45)

The science of mechanistic reductionism reduces Land to a machine that has value only insofar as it has utility for humans and can be converted into a commodity that supports capitalist economics. Newtonianism posits that the cosmos is "like an immense clock, a mechanism whose basic components and principles could be revealed and examined through science. According to a Newtonian worldview, nature is a machine and is no more than the sum of its parts," meaningless in itself and subject to control by humans (Suzuki 15).

The transformation of Earth from a living, nurturing mother to inert matter enabled capitalism to expand its exploitation of nature (Merchant *Death* 182). "The removal of animistic, organic assumptions about the cosmos constituted the death of nature—the most far-reaching effect of the Scientific Revolution" (Merchant *Radical* 48). Today, mechanistic science is the ideology that legitimates extractivist capitalism and its domination of land (58), feeding a culture of greed that is emotionally disconnected from the earth. Necrophilia is indicated by Western addiction to self-gratification through consumption, an addiction so intractable that it must be fed even when it clearly contributes to climate change. The perversion is manifest as pervasive disavowal— the state of disordered thinking which observes our culture's slow death due to climate change but denies that it is so and makes no effort to reduce carbon emissions. A philosophy of natality contributes to the project of disrupting mechanistic reductionism by drawing on the sciences of relationality that understand organic nature as an amalgamation of creative, self-organizing systems which are active, intelligent, communicative, and intentional. Donna Haraway, in *Staying with the Trouble*, writes that it matters how we relate to the material

84

world: "Materialist, experimental animism is not a New Age wish nor a neocolonial fantasy, but a powerful proposition for rethinking relationality, perspective, process, and reality" (165). She calls on us to make kin with land and with the human kin and other-than-human kin with whom we co-inhabit the land.

MATRICIDE

Male appropriation of birthing was a way of erasing the matriculture that existed prior to Western patriarchy. In the Olympian myths, Zeus swallows pregnant Metis, mother of Athena, and later gives birth to Athena from his head. Zeus' matricide in the Olympian myths over-writes an earlier matricultural mythology (Cavarero 108). The male appropriation of birthing is linked to male desire to become divine by claiming the ability to create life (Jantzen 141; Daly *Gyn/Ecology* 65). The Abrahamic religions disavow the mother and establish a jealous male god in the sky who creates the world with his *logos*, not by gestation.

Francis Bacon argued that nature's womb harboured secrets that could be wrested from her using technology. Modern medicine executed Bacon's science by taking over the work of midwives and medicalizing women's reproductive functions from conception to parturition. Cultural knowledge embedded in birthing myths and rituals fell into dis-use as Western medicine took control of birthing away from mothers. Birthing became technologically-oriented and detached from natural phenomena. Many women lost their awareness of natural regeneration cycles and birthing processes, acquiring a type of nature blindness (Haarmann 259). Necrophilic perversion is indicated when drugs are administered to reduce a mother's consciousness of the birthing process, making her the abject object, not the active and conscious subject of birthing. Necrophilic matricide is condoned whenever the obscenity *motherf–r* elicits no outrage.

Adriana Cavarero asserts that the lack of attention paid to the fact that we are born from woman has given Western philosophy a preoccupation with death rather than birth. Western philosophy juxtaposes life and death in a way that disavows culture's dependence on women's generative maternal force. She critiques the academy for spurning the abundant "documented evidence of the existence of an original matriarchy" by claiming it "does not add up to the kind of proof accepted by every scholar" (5). Cavarero investigates the "traces of the original act of erasure" contained in patriarchal records, exposing Zeus' crime of matricide and interpreting that act as symbolic of patriarchy's erasure of the Great Mother (7f). She emphasizes that cultural continuity depends on the maternal power to generate. Continuity is assured only when the mother/daughter relationship is visible to human eyes. Earth's beings flourish only when females give birth to daughters. When the maternal no longer has power to generate, we approach "the threat of nothingness" (61).

It is ironic that the Canadian Métis culture carries the same name as Metis, Athena's Titan mother who was swallowed by Zeus. Metis' fate mirrors the fate of the aboriginal mothers whose identity was swallowed by colonialist fur traders who claimed the mothers' children as their own by giving them Scottish and French surnames. Indigenous women are reclaiming matriculture by decolonizing their bodies. Leanne Simpson links the material, the political, and the spiritual when she declares that Indigenous women are reclaiming their responsibility to serve their communities as carriers of culture (28):

If more of our babies were born into the hands of Indigenous midwives using Indigenous birthing knowledge, on our own land, surrounded by our support systems, and following our traditions and traditional teachings, more of our women would be empowered by the birth process and better able to assume their responsibilities as mothers and nation-builders (29).

Natality and matriculture are linked in the common value of mothering and birthing, not as an essentialist impulse, but as a cultural system embraced by both men and women, mothers and not mothers, for their contribution to cultural continuity. Indigenous cultures appear to be positioned to midwife the rebirth of matriculture, first in their own cultures and then in Western cultures. The Sedna myth integrates a core principle of indigeneity: living in balance and harmony with the cosmos is impossible if the culture does not venerate generative power. Matriculture refers to cultural traditions that valorize natality, in its literal and metaphoric meanings, and elevate The Maternal for its creative, spiritual, affective, educational, and judicial contributions to cultural continuity. Matriculture does not presume the subordination of men, but rather an egalitarian partnership between the sexes, and the expected division of labour determined by gender (Passman 85).

DOMINATION

The ideology of dualism and human separation can be traced to Greek philosophers, and is embedded in the Abrahamic religions; however, René Descartes is considered the father of modern dualism. Descartes' philosophy consolidated and augmented Bacon's reductionism and formed the intellectual context for the current ecological crisis (Plumwood *Nature*). Descartes held that there are two kinds of existing things: physical and mental. He argued that self-conscious awareness is a unique human achievement that elevates humans above all other species (Suzuki 15). Cartesian dualism seeks to master the body in order to reside in purely rational, intellectual states. Dualistic thinking categorizes phenomena into binary opposites in which one part of the binary is valuated as superior while the categorical "other" is devalued as inferior or primitive; thus, Cartesian dualism bastions hierarchical systems of domination: anthropocentrism, sexism, racism, androcentrism, colonialism, ableism, and classism.

Human-centeredness, or anthropocentrism, is the hyper/separation of humans as a special species; it weaves a dangerous set of illusions about the human condition into the logic of our basic conceptual structures. Human-centeredness is a complex syndrome which rationalizes the "delusions of being *ecologically invulnerable,* beyond animality, and 'outside nature,'" and thus beyond the reach of the sixth mass extinction event (Plumwood *Nature* 115). Human/nature dualism conceives humans as not only superior to but different in kind from other-than-human beings, which are relegated to a lower non-conscious and non-communicative physical sphere (Jantzen 32f). Cartesian thinking is necrophilic in that its goal is to control the mind by transcending the body in order to achieve immortality and divinity in death. Christian and Islamic Fundamentalisms are similarly preoccupied with an external world and yearn to escape embodied life in this world for a heavenly home. The secular obsession with transcending the body manifests in celebrating war and building elaborate war memorials that beautify youth who die in battle.

Natality is situated in the continuum of ecocentric philosophies which include deep ecology, social ecology, ecofeminism, New Materialism, and indigeneity (Wolfstone 195). Freya Mathews's ecofeminist philosophy moves beyond deep ecology to explore ecological interconnectedness or "oneness" to describe personhood as the embodied relation of self to the self-realizing universe in the extended region of spacetime (149). An ecocentric philosophy recognizes that all beings are equal and interdependent in earth's living systems that are agentic and animated. As relationality with earth deepens, we acknowledge our ecological vulnerability and our animality. Interdependence is linked to the principle of sufficiency (enoughness) which frees humans from the drive to acquire and consume in accordance with the competitive ideology of capitalism (Plumwood *Feminism* 5).

LOSS OF COSMOLOGY

Western culture's scientific heritage from Bacon and Newton has bankrupted its cultural imagination; consequently, Western culture manifests symptoms of cosmological destitution such as anxiety, alienation, anomie, and a massive confusion over values (Mathews 134). The loss of cultural cosmology embedded in narratives and accessible to the entire community is accompanied by reduced a capacity for symbolic thinking, big-picture thinking, and ethical thinking. Modernity remains stuck in Newton's atomistic social order, which regards humans as atoms, individuals keen on survival, and competitively self-interested while adrift in a meaningless universe (30). Instead of becoming rugged individuals, Arendt is concerned about homelessness and rootlessness linked to an ungrounded worldview and suggests that the Copernican paradigm shift toward a heliocentric and astrophysical worldview made it difficult for natals to trust their senses (Arendt 261). The Copernican paradigm shift took only decades, yet modernity appears to be incapable of making another rapid paradigm shift to a relational worldview in spite of the urgency of multiple ecological crises. Popular film culture manifests necrophilia by attempting to create beauty out of terror and horror and to mimic male dominance, capitalism and warfare in space, thus perpetuating the Cartesian myth of transcendence by escaping an over-heated planet in order to colonize other planets.

A philosophy of natality recognizes that "we stand in need of cosmological healing" (Mathews 47). Cosmology contributes stability and groundedness by evoking imagination and a shared symbology to express the sacredness of the whole and of ethical relationships. Indigenous cultures perceive Land as the source, the ground, and the womb, of life. Land is sacred under the aspect of Mother Earth, the great body that sustains all natals (156). Reclaiming cosmology restores the sacred. Some postmodern critics may level the accusation of essentialism regarding the Mother Earth metaphor that is so frequently used not only by ecofeminists, but also by many who hold to ecocentric and Indigenous philosophies. In this study, "Mother Earth" is a cosmological term regarding relationality to Land and timespace; it recognizes Earth's agency, intelligence, and communication.

SILENCING THE "OTHER"

Bacon silenced nature so that she could be exploited. Similarly, continental psychoanalysts silenced women by denying them a subjective voice and European colonizers silenced other cultures by denying them subjectivity. Silencing the "other" is a colonialist strategy: "the de-mothering of nature through modern science and the marriage of knowledge with power was a source of subjugating women as well as non-european people" (Shiva *Staying* 18). Freud and Lacan projected their male morphology onto the entire female gender, presenting their phallocentric imaginary and symbology as a universal truth:

> According to Lacan, there can be no women subjects. Subjectivity requires language, and language is masculine, grounded in the Phallus as universal signifier. Women qua women, therefore, cannot speak. When women speak, when women take up subject positions, it is not as women, but as imitation males, men in drag. (Jantzen 43)

Grace M. Jantzen diagnoses Lacan's disordered thinking as masculinist repression and suggests a therapy by which the "material and maternal basis must be brought to consciousness" (97).

Gayatri Spivak uses the term "othering" for the process by which colonialist discourse creates "others"—those that are homogenized and marginalized by mastering them. When Spivak asks, "Can the subaltern speak?" she does not mean that it is impossible for the subaltern to reclaim a voice; she posits that when the subaltern speak, they create a voice consciousness that may not be perceivable by dominators because it is not pertinent or useful to the dominator (Spivak 80).

Necrophilic perversion violates the "other" after the "other" has been silenced and rendered incapable of giving consent; it is indicated by the ethical void of globalization in which multi-nationals appropriate the homeland of Indigenous cultures without their consent, degrade their landscapes, pollute their environments and impoverish their people in order to feed the insatiable addiction of Western consumers. In contrast, a feminist philosophy of natality embraces difference and plurality.

MONOCULTURE

Multinationals promote monoculture by marketing seeds that are genetically modified so that they cannot self-propagate, compelling farmers to purchase seeds annually instead of seed-saving. In India, more than 280,000 farmers have committed suicide after being forced into bankruptcy after investing in expensive, unreliable patented seeds. Vandana Shiva is a critic of multinationals who claim Indigenous farmers' collective knowledge as their invention through biopiracy patents—a type of enclosure of the intellectual and the biological commons. Shiva advocates for farmers' right to save and exchange seeds in order to preserve biodiversity. Necrophilia is indicated by the complicity of governments and multinationals to make seed-saving illegal.

Indigenous farmers protect their biocultural heritage by actively resisting Monsanto. On the face of it, their protests may appear to be conservative resistance to modernity, but at the heart of their active resistance is a radical reclamation of the traditional knowledge that sustained biocultural diversity in the past. Like the Roman Janus, they look into the distant past in order to look deeper into the future, while Monsanto takes a short-term view of future profits by promoting a culture of death in "Roundup-Ready" seeds. Monoculture is three dimensional; it manifests as loss of biodiversity, loss of languages, and loss of cultures. The emerging field of biocultural studies has collected data that indicates the rates of culture loss and language loss parallel the rate of loss of biodiversity (Maffi 412). According to Luisa Maffi, an anthropogenic extinction crisis is indicated by the massive loss of biodiversity in Earth's plant and animal species and in the health of the ecosystems that sustain them. Cultures and languages are vanishing under the rising tide of global monoculture, and Maffi worries that we are rapidly losing critical life-support systems and the human knowledge that can teach us how to live in balance with our planet (414).

Natality celebrates plurality and difference and recognizes that biocultural diversity is critical to cultural continuity. Vandana Shiva's work reflects the intersection of biocultural diversity, matriculture, and political revolution: "When nature is a teacher, we co-create with her—we recognize her agency and her rights" (Shiva *Everything* n.p.).

CHOKING DEMOCRACY

A healthy democracy is participatory and consensual, but Western democracies are collapsing into a post-political condition that makes democracy unrecognizable. The post-political condition has also been theorized as post-democracy, depoliticization, and de-democratization.

Governments, controlled by corporations, induce passivity in citizens by feeding an addiction to consumption and by discouraging activism. A passive citizenry is indicated by declining voter turnout. The lack of overt citizen consent and participation makes Western nations increasingly vulnerability to fascism. Theorists concur on four theoretical elements of the post-political condition: a) the neoliberal economy has subjected governments to the demands of corporations, and this renders governments powerless to deal with macro issues such as climate change; b) in consensual postpolitics, government is a social administrator, replacing an institution where activists debate ideologies of equality and justice; c) neoliberal governments create an environment of unspecific anxiety to soften public resistance to increased surveillance; and d) in the ideological vacuum created since the Left ceded victory to capitalism, there are few Western political organizations advocating an alternative economic ideology based on an agenda of equality and eco-social justice (Dean; Swyngedouw; Žižek).

Politics has become a public relations game with large corporations owning and controlling the media—both the message and medium. Mainstream media witness the perversions of necrophilia but keep citizens in the dark about the real state of the world by producing propaganda that keeps citizens in their place. The voyeurism of the media can be compared to the observers of a date rape who videotape the necrophilic act and then post it on social media to extort the victim who never gave consent, was not a participant and is powerless to reclaim her privacy. Human rights are eroding through distortion. Pipeline protesters are criminalized as domestic terrorists. Religious organizations demand the "right" to exclude based on religion. Activists are regarded as domestic terrorists if they obstruct the economy with blockades or boycotts. Governments perpetuate the oppression of vulnerable citizens, silencing them and instilling fear by invading their privacy. Governments promote consumption, which has the strategic effect of numbing the shrinking middle class into giving up privacy and freedom if it means greater protection for their lifestyle. This then is the perverted necrophilia of democracy in the dominant Western paradigm.

In contrast, Arendt's political philosophy of natality celebrates collective imagination and initiative in generating new ideas and enacting them. Repoliticization disrupts the post-political consensus that has silenced and numbed citizens. The political task is to enlarge equality and freedom by acknowledging Earth as a political being that is meting out justice through climate change and mocking those who co-opt the notion of sustainability in order to prolong capitalism's tyranny. For models on acknowledging nature's rights, we look to Indigenous philosophies of vitality and regeneration including *Buen Vivir* or *Sumak*

Kawsay in Andean cultures, *Ubuntu* in southern Africa and *mino bimaadiziwin* in Anishinaabeg (Simpson qtd. in Klein np).

CONCLUSION

The dominant Western paradigm is experiencing massive systems failure as it faces self-inflicted death by anthropogenic climate change. The urgent need to deconstruct the dominant necrophilic paradigm is not driven by reformist ideology. This deconstruction of necrophilia clears a space to imagine a different paradigm of natality that offers us a radically different way to think and be in eco-social communities committed to the vitality and regeneration of the ecosphere—our planetary home. In researching a paradigm of natality, I discovered that it not a new concept; it is, in fact, a living philosophy in many Indigenous cultures whose resilience in adapting to climate change spans millennia.

REFERENCES

Arendt, Hannah. *The Human Condition.* 2nd ed. Chicago: University of Chicago Press, 1958.

Barber, Katherine, ed. *The Canadian Oxford Dictionary.* Oxford, UK: Oxford University Press, 1998.

Cavarero, Adriana. *In Spite of Plato: A Feminist Rewriting of Ancient Philosophy.* Cambridge: Polity Press, 1990.

Daly, Mary. *Gyn/Ecology.* Boston: Beacon Press, 1978.

Daly, M., J. Caputi, and S. Rakusin. *Websters' First New Intergalactic Wickedary of the English Language, Conjured in Cahoots with Jane Caputi.* Boston: Beacon Press, 1987.

Dean, Jodi. "Post-politics? No, thanks!" Future Non Stop, 2012. Web.

Haarmann, Harald. *Foundations of Culture: Knowledge-Construction, Belief Systems and Worldview in their Dynamic Interplay.* Bern: Peter Lang, 2007.

Haraway, Donna. *Staying with the Trouble: Making Kin in the Chthulucene.* Durham, NC: Duke University Press, 2016.

Jantzen, Grace M. *Becoming Divine: Towards a Feminist Philosophy of Religion.* Bloomington: University of Indiana Press, 1999.

Klein, Naomi. "Dancing the World into Being: A Conversation with Idle No More's Leanne Simpson." YES! *Magazine*, 5 March 2013. Web.

Maffi, Luisa. "Language, Knowledge and Indigenous Heritage Rights." *On Biocultural Diversity: Linking Language, Knowledge and the Environment.* Ed. Luisa Maffi. Washington, DC: Smithsonian Institution Press, 2001. 412-432.

Mathews, Freya. *The Ecological Self.* London: Routledge, 1991.

Merchant, Carolyn. *The Death of Nature: Women, Ecology and the Scientific Revolution.* New York: Harper & Row, 1980.

Merchant, Carolyn. *Radical Ecology: The Search for a Livable World.* London: Routledge. 1992.

Passman, Tina. "Out of the Closet and into the Field: Matriculture, the Lesbian Perspective and Feminist Classics." *Feminist Theory and the Classics.* Eds. N.S. Rabinowitz and A. Richlin. London: Routledge, 1993. 181-208.

Plumwood, Val. *Feminism and the Mastery of Nature* 2nd ed. London: Routledge, 1994.

Plumwood, Val. "Nature in the Active Voice." *Climate Change and Philosophy: Transformational Possibilities.* Ed. Ruth Irwin. London: Continuum International Publishers Group, 2010. 32-47.

Sardar, Ziauddin. *The Revenge of Athena: Science, Exploitation and the Third World.* London: Mansell Publishing, 1988.

Shiva, Vandana. *Staying Alive: Women, Ecology and Development.* London: Zed Books, 1988.

Shiva, Vandana. "Everything I need to know I learned in the forest." *YES! Magazine,* Winter 2012. Web.

Simpson, Leanne. "Birthing an Indigenous Resurgence: Decolonizing Our Pregnancy and Birthing Ceremonies." *"Until Our Hearts Are on the Ground": Aboriginal Mothering, Oppression, Resistance and Rebirth.* Eds. Dawn M. Lavell-Harvard and Jeannette Corbiere Lavell. Toronto: Demeter Press, 2006. 25-33.

Spivak, Gayatri C. "Can the subaltern speak?" *Marxism and the Interpretation of Culture.* Eds. C. Nelson and L. Grossberg. University of Illinois Press, 1988. 271-313.

Suzuki, David. *The Sacred Balance: Rediscovering Our Place in Nature.* 2nd ed. Vancouver: Greystone Books, 1997.

Swyngedouw, E. "Whose Environment? The End of Nature, Climate Change and the Process of Post-Politicization." *Ambiente & Sociedade* 14.2 (2011): 69-87.

Wolfstone, Irene Friesen. "Transgressive Learning: Journey to Becoming Ecocentric." *Global Citizenship, Common Wealth and Uncommon Citizenships.* Eds. Lynette Schultz and Thashika Pillay. Brill Sense, Rotterdam: Sense Publishers, 2018. 191-205.

Žižek, Slavoj. "Censorship Today: Violence, or Ecology as a New Opium for the Masses." 2008. Web. 29 June 2013.

4.
The Guardians of the Conga Lagoon
Defending Land, Water, and Freedom in Peru

ANA ISLA

THE EXPLOITATION OF MINERALS in Peru since colonial times has
reduced the concentration of these resources in both quantity and
quality. What remains are low concentrations of dispersed particles in
areas that are rocky, icy, forested, or mountainous, and that make it
impossible to extract the minerals using traditional deep-pit mining
methods. Therefore, open-pit mining is the current technological
method available. Open-pit mining removes entire mountains, forests,
and glaciers, with the aim of finding rocks with gold, silver, and other
metallic and non-metallic minerals. Open-pit mining uses dynamite to
kill the surface matter (e.g. forests, mountains, glacier covers, lakes,
jalcas or spring water sources three thousand metres above sea level).
Moreover, its heavy machinery eliminates biological diversity (e.g. flora,
fauna, and micro-organisms) and scars the landscape with giant craters.
The shattered rock, combined with a cyanide and water mixture to
remove gold, destroys ecological cycles and contaminates ecosystems,
poisons the hydro resources, and pollutes the atmosphere through the
release of poisonous substances, thereby affecting all life. This process,
known as lixiviation, has a strong impact on communities that live close
to mining operations, as it also competes for water and energy. Cyanide
lixiviation contaminates permanently, as it continues leaking into the
land, water, air, etc. Governments or mining corporations seldom
consider the changes brought on by this chemical cocktail. Abandoned
mining projects all over the world leave a legacy of permanent water
contamination from cyanide, metals, and non-metals.

In 1987, *The Brundtland Report* (World Commission on Environment
and Development), also called *Our Common Future*, entangled the
international debt crisis with the ecological crisis and suggested
"sustainable development" as a means to eliminate poverty and to
contain environmental disaster. At the Earth Summit in Johannesburg

94

in 2002, mining was defined as sustainable development. The advocates of the Global Mining Initiative, who wanted it defined as sustainable development, were the International Chamber of Commerce, the World Business Council for Sustainable Development, and Business Action for Sustainable Development. Thirty mining corporations and several NGOs—among them the International Union for Nature Conservancy, Conservation International, and others—sponsored this initiative. A key tactic of mining supporters is to portray mining as a way to bring investment, create jobs, and reduce poverty. In the sustainable development (SD) paradigm, despite on-going debates and the search for alternatives, economic growth remains a dominant objective. Development theory rests on a binary world in which subsistence economies are constructed as "undesirable" and "undignified" (Esteva). In fact, the ongoing destruction of subsistence economies is the central elementiof what is today understood as "development."

Moreover, the neoliberal agenda also established globalization as an open field for corporations in which there are no legal, social, ecological, cultural, or national barriers. Bilateral free-trade treaties and agreements, signed between 1995 and 2010, such as NAFTA, institutionalized neo-liberal reforms by reducing tariffs and export taxes on investments. Private-sector-friendly legislation and codes regarding the rights of foreign investors were incorporated into free trade agreements, providing additional legal protection to corporations, including mechanisms for suing governments that rescind permits for operations.

However, pressured by Indigenous activists, the International Labour Organization Convention 169 has recognized the ancestral rights of Indigenous populations since 1989. Despite the fact that the Convention was signed by many Latin American governments, these same countries are ignoring it or are nullifying its effects (Rodriguez-Pinero Royo). In addition, "in 1999, the International Finance Corporation (IFC), the financial arm of the World Bank, created the position of Compliance Advisor Ombudsman to monitor obedience with the social and ecological conditions attached to World Bank finance" (De Echave 20). In a related move, the United Nations General Assembly in July 2011 recognized water and sanitation as human rights but at the same time introduced an official forum on water as a vital "commodity." The Council of Canadians has ncountered the definition of water as commodity and built a movement of global scale under the slogan *Water is Life* and created an Alternative World Water Forum (Forum Alternatif Mondial de l'Eau [FAME]).[2] It fosters The Blue Planet Project, which promotes "water justice based on the principles that water is a human right, a public trust, and part of the global commons."[3] While critically important for protecting water commons, such initiatives continue to be only partially successful because mining can rely on unregulated conditions and voluntary compliance as the only means of ensuring corporate respect for environmental and human rights.

During the 2002 Earth Summit in Johannesburg, in the absence of a global regulatory or legal system to hold multinational corporations to account for their operations in the periphery, multinational organizations were forced to address the social and environmental repercussions of the expansion of mining and to propose partial solutions. As corporations were confronted with rural and Indigenous community upheavals, they began to formulate their own regulation strategies, referred to as corporate social responsibility (CSR). At the Earth Summit, in Johannesburg, CSR was added to the multi-stakeholder negotiations (MSN). The multi-stakeholders (which include governments, NGOs, and businesses) have come to realize that close relations with the communities (e.g. women, youth and children, Indigenous peoples, local authorities, workers and trade unions, scientific and technological communities, and farmers) are. crucial for their operations. The concepts of CSR and MSN foster the notion that corporation and community interests are compatible—that it is a question of dollars and cents—and that each member of a community supposedly has a common interest, represented by the venture (mining), in fair negotiations. Catherine Coumans argues that in a deregulated framework there is no community right to reject mining investments, and therefore the most tragic outcomes of mining projects in the periphery occur when rural communities refuse to become stakeholders in what they perceive as the plunder of their Indigenous lands and resources (31).

In South America, the most appealing countries in terms of attracting investments in extractive resources are Peru, Colombia, Ecuador, Bolivia, and Chile because of the large land tracts that Indigenous people occupy, meaning that it is not (yet) private property. Consequently, Indigenous peoples and peasants' local economies are under attack. For millennia, these communities have exercised control over the land, water, and livelihoods that corporations now want to expropriate. Enforced by the Johannesburg Earth Summit and deregulation, communities have had little success in their efforts to get mining corporations to uphold local peoples' right to know the impact of the projects, and their right to reject them. Therefore, they increasingly find themselves face to face with the violence of mining operations. As a result, communities are forced to mount a political struggle for their territories as mining operations are transforming their physical, social, economic, and cultural environments. Even though the costs for local communities are high, they have no choice but to take on their own governments, mining agents, and their international backers (i.e., the World Bank, First World governments, new laws, tribunals, middle-class investors and their pension plans, lobbyists, and political parties).

The Guardians of Conga Lagoons, an environmental justice movement in the sense used here, is about recognition that Peru's economy is based on agriculture and livestock production. They understand that the future of their families will be determined by the water quantity and quality produced by the *jalcas* (springs) at the tops of the mountains. This chapter brings to the forefront a nucleus of resistance to "mining as sustainable developmen" by women and men, peasants and Indigenous people of Cajamarca, Peru, caught in the middle of a confrontation to defend the land and water as commons. This chapter also highlights the environmental racism exhibited by a multinational corporation as well as in the functioning of the Peruvian state. Ecofeminists argue that racism denies the essential humanity of people. Indigenous people, similar to women, are naturalized; their land is labelled "unoccupied" or "unused," and is thereby easily appropriated by those who claim they can make it "productive." In this way, suffering and death are rationalized in the name of progress (Mies; Salleh).

On the one side, Yanacocha Gold Mining Corporation shifted the social and environmental costs of mining extraction onto local communities on the grounds of their existing on the periphery, being peasants, and being Indigenous. On the other side, the Peruvian state, under Fujimori's administration (1990-2001), passed a series of measures in favour of mining investment such as ending restrictions on remittances of profits, providing tax stability packages to foreign investors, guaranteeing mining companies exclusive control of land use, and facilitating entry to mineral deposits, among other advantages. Alan Garcia, President of Peru (2006-2011), unabashedly summed up the racism of the State when he called Indigenous people "Perros del Hortelano" ("the gardener's dogs," colloquially meaning that they do not "use" the resources of their territory, but nor do they want others to "use" them), and stated that Indigenous people are "not first class citizens" (*Democracy Now*). Garcia's administration embarked on territorial reordering, invented new decrees, and derogated laws that supported ancestral land claims and rural communities (Isla). His administration ignored socio-environmental conflicts and established relationships with mining corporations based on bribery. Corporations benefit from corruption expressed on the *obolo minero* (one cent payment) (Campodonico). *Obolo minero* stipulated that mining corporations should pay 3.75 percent of their profits in royalties. Since this payment was voluntary, several corporations made no contribution whatsoever. If taxes had been paid, instead of *obolo minero*, the government would have received billions of dollars more than what was collected by passing this hat. During the last presidential elections in 2011, the new president, Ollanta Humala, promised to change the racist state policies that deny social justice and destroy nature. In Cajamarca, referring to the Conga Mining Project, he promised to respect the community decision of "no mining." Instead, after the election, he gave this mining a green light and went after anti-mining communities with bullets and bombs, leaving Cajamarca's population feeling betrayed, but not necessarily disempowered. These "invaded" communities understand that civil unrest is the only option left to those who do not want mining in their locality.

DEFENDING LAND: THE CASE OF YANACOCHA GOLD MINING

Yanacocha began operating in 1992 in the region of Cajamarca, in Northern Peru. It is owned by the Peruvian mining group Buenaventura, Newmont Mining Corporation of Nevada (U.S.), and the International Finance Corporation of the World Bank.

Yanacocha mining operations has already produced environmental liabilities due to the disproportionate use of space and the unpaid socio-environmental damages. Professor Jose Perez Mundaca (*Conflicto Minero*), documented the struggle between Yanacocha mining and the people in Cajamarca. He shows that Yanacocha has brought several negative changes. Socially, Yanacocha turned the city of Cajamarca into a camp for the mine, transforming the city into an entertainment area for miners, and creating social conflict. In the rural areas, it impaired the environment of the peasants by buying their land and/or rendering it infertile due to contamination. Economically, the mining company enjoys exorbitant profits due to the high quality of the ore deposit, low costs of production, and tax exemptions. For instance, Yanacocha production costs are extremely low due to the use of state infrastructure (roads) among other things. For 2001, Chip Cummins estimated this cost at 85 dollars per ounce of gold, compared to the 212 dollars in expenditure that same year, for Newmont in Nevada, United

States. Politically, Yanacocha creates an enclave economy, where decisions are made in the U.S., as its product destination is the foreign market. Its influence has penetrated most of Peru's institutions and organizations, such as ministries municipalities, NGOs, the Chamber of Commerce, universities, and the Catholic Church. Geographically, the open-pit system means the removal of huge amounts of land mixed with huge amounts of cyanide at river and lake sources. Not only was the ground removed and the water contaminated, several hills and lagoons were literally demolished. Environmentally, Yanacocha, due to its magnitude and its location at the source of several streams of regional importance, has generated considerable negative impacts on drinking water, watershed resources, and supply of water for irrigation. The hydrology is rain-fed, nourishing high-altitude grasslands *jalcas* (lakes at an altitude between 3,500 and 4,000 metres above sea level), as well as alpine lakes and wetlands, or *bofedales*. These lakes and wetlands are the sources of all the streams, rivers, and drinking water for the surrounding areas, including most of Cajamarca's 250,000 residents.

These impacts of the Yanacocha mine have generated general resistance that began in Cerro Quillish.

THE GUARDIANS OF CERRO QUILLISH: FIRST STAGE OF ENVIRONMENTAL JUSTICE MOVEMENT 1999-2010

The first stage of the conflict occurred in the immediate rural surroundings of the mineral deposit. Porcón, for example, is a town located at km. 14 of the highway, while the Yanacocha mine is at km. 24. But if you draw a straight line between the two, the mine is four or five kilometres away from Porcón. People from this town can hear the miners working twenty-four hours a day. The main actors during this first stage of resistance against the mining corporation were the farmers that the company bought land from at undervalued prices, who initiated a struggle for fair payment for their land and their incorporation as workers to the mine.

Nelida from Porcón says,

My father was one of the landowners in Quilish. He did highland agriculture. In 1997, he sold twenty-three hectares of land for S/.3,900 (CAD$1500). This represents S/.169 (CAD$65) per ha. He was forced to sell his property to the mine because he was informed that if he did not sell the land the state was *going to confiscate it without paying him a cent. When he sold the land, the company told him that they would give him work. The mine employed him for three months at minimum wage, and then they fired him.*

She adds,

> *Mining authorities have identified the families who are against mining. They call us backwards, because they say we oppose investment and progress. To coopt the opposition, the mine has two strategies. First, some people, including women, are employed for a few months, then they are fired. When they are dismissed, they distance themselves from the community and do not want to participate in the struggle, because they are ashamed of having sold out to the mine. Second, the miners believe that the mountains are not alive and are there to be crushed. However, for us, a mountain means a lot. It is our protector, our guardian, it is what gives us water, medicine, and it is our company.... Miners do not realize that by destroying our mountains, they are devastating us as well as themselves.*
> (Nelida)

This struggle intensified in June 2000 when, in a small town called Choropampa, 151 kilograms of liquid mercury spilled over a forty-kilometre-wide area, contaminating three mountain villages, including Choropampa. Some young people began picking the mercury up with their hands—to disastrous effect. Campesinos learned that contamination kills: "More than 900 people were poisoned from the spill" (Cabellos and Boyd). In an effort to push the company to take responsibility for the health damage, headed by the town's mayor, the villagers blocked the road that connects Cajamarca to Yanacocha. After negotiations between the mine and the community, some Choropampa residents signed individual compensations. Others sued Newmont in the United States Federal District Court in Denver, in 2001 (Johnson and Caceres). Years after the mercury spill in the villages of San Juan, Magdalena, and Choropampa, the health of the population has not been restored, and adequate medical care is still lacking (GRUFIDES).

Professor Perez *(Conflicto Minero)* recounts that during the peasant struggle in Porcón, Marco Arana, a parish priest of Porcón, Nilton Deza, a biologist of the UNC, and Reinhardt Seifert, a German engineer resident in Cajamarca organized the environmentalist association ECOVIDA in 1999. It led the anti-mining opposition. Cerro Quilish became the focus of conflict as it feeds the Quilish, Porcón and Grande Rivers. The struggle was intensified after the mercury spill. Since 2001, the opposition against Yanacocha has become a movement in which varied forms of struggle defend health and life from contamination. The movement organized local, provincial, and regional standstills in defence of the environment; it resorted to a legal and judicial lawsuit against Yanacocha mine after the mercury spill; and municipal ordinances were passed to protect the hills. For instance, the Provincial Municipality of Cajamarca declared Cerro Quilish a Protected Area; committees advocating the protection of the area at different levels were organized, including FARC (Frente Amplio Regional de Cajamarca), FDI (Frente de Defensa de los Intereses, Ecologia y Medio Ambiente de Cajamarca); congresses, such as the First Departmental Congress in Bambamarca (Primer Congreso Departamental), were organized to discuss the environmental impacts and sustainability; new roles for old organizations, such as *rondas campesinas*, appeared; pilgrimages to the aquifers hills threatened by mining were organized; professional bodies such as the College of Physicians, Biologists, and Engineers declared themselves against mining; and Sunday speeches were organized in the atrium of the Cathedral of the city of Cajamarca.

Nongovernmental organizations (NGOs), such as the Asociación de Defensa y Educación Ambiental (ADEA) (Association for the Defence and Environmental Education), hired specialists to assess the Environmental Impact Study (EIS) of Yanacocha, specifically as it related to the water sources in Quillish and Porcón Rivers. This study found "iron and aluminium" beyond the values allowed by the General Law of water for human consumption, agriculture, and livestock (Perez, *Conflicto Minero* 153). Regarding water contamination, Nelida says,

Women are very aware of the water problem, because we are in the kitchen, we have to do laundry, and if there is no water we have to find it somewhere for the animals and the family. We know that polluted water is linked to infertility in women and animals. Women can no longer have children, and if they do, some of them are ill. For instance, there is a six-year-old child who was born stunted. The cuys *[guinea pigs] and the cattle are sometimes stillborn or born deformed. Two years ago my uncle had a cow whose offspring came out deformed; it had the head of a duck and was stillborn.*

With the publication of the results of the EIS, ADEA and other environmental groups pressured the central government authorities to take corrective actions and threatened the government with a departmental strike. This confrontation has escalated since 2011 when the owners of Yanacocha hoped to extend their mining project to what has been denominated Conga Mining.

CONGA MINING PROJECTS

The Conga Mining project is several times larger than the initial Yanacocha mine. The Conga project is presented as one mine project though it proposes the installation of nine more mining projects in this region.[4] Plans would consume 3,069 hectares of land to extract the gold and copper that lies beneath, and would affect between 3,000-16,000 hectares of fragile mountaintop wetlands including numerous lakes, rivers, and marshes that supply the region's drinking water (Bernard and Cupolo). According to Newmont, "The Conga Project in Peru involves surface mining of a large copper porphyry deposit also containing gold that is located 24 kilometres northeast our Yanacocha Gold Mine."[5]

The taking of land and water are inseparable. According to Fidel Torres and Marlene Castillo, in the immediate area surrounding the operations of the Project Conga, close to 700 *jalcas* (springs), 96 percent of which have flow rates important for agricultural and human use, have been documented. Of these, 398 (59 percent) are between 3,500 and 4,000 metres above sea level; 133 (19.7 percent) between 3500 and 3000; and 145 (21.4 percent) between 3000 and 2500 metres above sea level. There is a complex system of underground water flows connected with the above ground water that makes up the aquifer in the area. *Jalcas* or spring water sources are common spaces controlled democratically by peasants living in the area who use the water and are organized in *Junta de Regantes* (Collective Association of Water Users). They maintain the ecological health of the aquifers and the equitable distribution of its benefits around the knowledge of the dynamic of the water. This social pact is respected by its membership, and it is around this community knowledge that the environmental justice movement is organized.

According to the International Institute on Law and Society, this mega-project would directly affect peasant communities and *ronderos campesinos*, and they are entitled to the same rights and benefits that Indigenous peoples are awarded under constitutional and legal mandates.[6]

Professor Wilder Sanchez maintains that:

the Conga project is unworkable because it is located at an altitude ranging from the 3,700 to 4,262 meters above sea level, at the headwaters of five river basins: 1) Jadibamba River; 2) Chugurmayo River and 3) Chirimayo River (both tributaries of the Sendamal, which is attached below with the Jadibamba, originating La Llanga River); (4) Chaillhuagon River, which feeds Rio Grande and Chonta River; (5) Toromacho Creek, which feeds the Pachachaca River and Llaucano River. Three large river basins will suffer severe impacts: the Rio La Llanga, of Celendín, which irrigates the Valley of Llanguat and flows into the Marañón; of the Llaucano River, which irrigates the valleys of Bambamarca and Chota, and that of the Cajamarquino River, which irrigates the valleys of Cajamarca, Llacanora, Namora, Matara and San Marcos.

Since, the majority of people in Cajamarca fiercely oppose this six-billion-dollar project, the Peruvian national police and the Peruvian army—on orders from the central government—have heavily suppressed their resistance. As a result, they have suffered: declaration of emergency, police and military presence, repression, harassment, tear gas, burning of camps close to the lagoons, police monitoring, fiscal persecution, attacks, illegal detentions, and blows. They have also endured violent encounters that caused the deaths of five people, including a minor, and left a *rondero* paraplegic and another blind.

DEFENDING WATER AS COMMONS — ¡CONGA NO VA!

Professor Jose Perez (*Agua-Procesos*), argues that until 1980, the highlands and *jalcas* were commons, that is, open to common use for grazing livestock and subsistence agriculture. The impact of peasant agriculture was minimal because they use organic fertilizer. However, between 1987 and 2007 there was a loss of 75,454 ha. of *jalca* mainly due to two factors: 1) the expansion of agricultural and livestock derived by the demand of multinational factories of milk (PERULAC now INCALAC, and GLORIA S.A.); and 2) the mining area of Yanacocha, which removed several thousand hectares of land in the area of the *jalca*. It mixed water with cyanide, degrading the soil and contaminating the water. All these processes have meant greater pressure on the soil with damaging effects on the environment in general and water in particular.

103

In addition, reforestation projects by the international cooperation through the planting of eucalyptus and pine trees created new water problems. Instead of sowing water, eucalyptus trees consumed water from the *puquios* (smaller areas of fresh water) and *jalcas*. Finally, since 2000, they now plant trees that are native species. FONCODES has also built reservoirs to store water, and CARITAS has built family wells.

So how will Conga mining affect water quality and quantity? In an interview with Professor Wilder Sanchez, he stated that:

Conga Projects will draw at least six million tons of wetlands (similar to marsh or swamp) that today occupy 103 ha., thus destroying the [earth's] water sponge, which stores water from rain and mist [and] filters it slowly to the lakes, streams and groundwater that give rise to the springs. Since there are several mining projects, the cumulative impact of craters will destroy the groundwater and seriously alter the normal flow of the hydro-geological system; due to its huge depth, it will cause the disappearance of the springs and other lakes in the surrounding area, and contaminate groundwater with sediments, heavy metals, and acid water infiltration between the rocks removed.

In addition, the water can be contaminated with chlorine by-products. Chlorine is used to treat gold mine wastewater and remove cyanide, which helps extract gold from mining ore in a process called heap leaching. The city's water treatment systems are currently inadequate to remove these by-products, which can include the carcinogen THM (trihalomethane).[7]

Professional and community knowledge added to the protests. Unrest continued to mount: on September 29, 2011, a strike took place in eight rural areas (Namococha, Quengorio Alto, El Alumbre, Corralpampa, San Antonio, and others); on October 14, 2011, another strike took place, this time in Encañada; the Cajamarca Regional Government called for a Regional Stoppage on November 9, 2011; and on November 24, 2011, several other strikes were organized in the Region of Cajamarca.

As the protests increased, conflicts emerged at every level of government. At the national government level, Conga has removed two Prime Ministers, Lerner and Valdez, and several Ministers and Vice-Ministers because they confronted the protest with bullets, creating national unrest.

In the midst of this national disagreement, the region of Cajamarca elected as Regional President a Communist Party Patria Roja leader who had not been corrupted or intimidated by Yanacocha. His policy defends the headwaters of watersheds, prohibits mining operations, guarantees the right to defend the water resources, and supports the defence of ecosystems as commons.

In an interview registered by Bernard and Cupolo ("Cajamarca"), the vice-president of Cajamarca region said:

In June of 2011, we (the Regional Government) visited the lagoons on the Conga site to conduct a general overview of the land and found its ecosystems to be too fragile for mining activities. Shortly after, we reviewed the environmental impact study (EIS) that approved the project in 2010 (during Alan Garcia's administration) and found serious deficiencies. As the protests became more and more frequent, we felt obligated to respond to our residents' concerns and represent their voice, so we put together the 036 regional ordinance.

However, on April 17, 2012, Peru's Constitutional Tribunal ruled against Ordinance 036. The court said Cajamarca government officials overstepped their powers by establishing regulations against the mining project (Bernard and Cupolo "Cajamarca"). Following this decision, the central government and the corporations demanded prison for the regional president of Cajamarca.

The water struggle in this region is at every level. Women in Cajamarca city told me that their home in the city only has running water for as little as a few hours per day due to ongoing mining in the headwaters of the Rio Grande. They have to wake up before three or four a.m. to collect water if they want to cook or take showers. Therefore, water pollution has made women play an important role in the process of ¡Conga No Va!, despite the fact that women in the mountains did not have a tradition of participating in the protests. The defence of water, in the year 2012, made visible the participation of the peasant women, the professional women, and the Sisters of St. Francis. For instance, on June 19, 2012, pregnant women from the region marched on the streets against Conga mining.[8] These women have been playing central roles in deflecting confrontations. During moments of tension between the police and the men, they physically put themselves in between the two to stop the violence. In other situations, women rescued their men from the hands of the police.

THE GUARDIANS OF THE LAGOONS: SECOND STAGE OF THE ENVIRONMENTAL JUSTICE MOVEMENT 2011-2014

Patria Roja facilitated the work of ecologists organized in ECOVIDA. In this rural area, this political party provided the critical mass through their teachers' organizations—SUTEC (Sindicato Unico de Trabajadores de la Educación de Cajamarca)—and its organization of *ronderos campesinos*. SUTEC and ECOVIDA created the "Front for the Defence of the Interests, Ecology and Environment in Cajamarca" (FDI stands for Frente de Defensa de los Intereses, Ecologia y Medio Ambiente de Cajamarca) and organized the First Departmental Congress in Bambamarca. Nelida from Porcón spoke at this congress, saying,

> *Campesinos in the countryside, who survive mainly through agriculture and cattle rearing, have reported high levels of animal deformities, huge amounts of fish washing up dead, a severe water shortage leaving them unable to irrigate their crops, skin deformities on themselves and their children, and unusually high rates of cancer and birth defects.*

As new groups joined the struggle, FDI changed into FUD (Frente Unico de Defensa de la Vida y el Medio Ambiente de Cajamarca). Then several other struggle fronts were organized, among them Frente de Defensa de Cajamarca (FDC) and Frente de Defensa Ambiental de Cajamarca (FDAC). These organizations educate, organize, and mobilize the people to defend their rights.

Since October 8, 2012, Bambamarca's *ronderos* (a peasant institution that provides security in rural areas) is taking care of the lakes of Mamacocha, Mishacocha, Laguna Negra, and Laguna Seca, while in Celendín, peasants, ronderos, and teachers are guarding the lakes of Perol and Azul.

The teachers in Celendín and the *ronderos* in Bambamarca are the backbone of the environmental justice movement in the region. I conducted several interviews with them in Bambamarca and Celendin.

In Bambamarca, Eddy Benavides, president of the Frente de Defensa de la Provincia de Bambamarca (FDPB) (Defence Front of the Province of Bambamarca) stated:

> *I represent FDPB in the Frente de Defensa de Cajamarca (FDC), which is a conglomerate of organizations that defend the natural resources and the water resources. We are well*

organized, well prepared, and cannot be stopped by anyone. You know why? Because apart from the consciousness that we have acquired, we have deep wounds that were caused by the Socabón (deep-pit) mining in Hualgayoc fifty years ago. These corporations have environmental liabilities as two rivers (Tingo Maigasbamba and Arazcorge Hualgayoc) are dead, and there is a lack of water. Now we only have two living rivers (Yaucano and Pomagon) that come from the lakes of Conga. But in spite of that, the Government has accepted two new open-pit mining projects in our area. The amount of water used for leaching is enormous. Water for mining is like blood to the body. In essence, our fight is for the defence of water, life, and the future of our peoples. So, we are not going to surrender ever. This struggle is emblematic in Bambamarca.

My first question was about what are the *rondas campesinas or ronderos*. Benavides commented that:

The birth of the rondas campesinas *was in Quillamarca–Chota, and one year later in San Antonio and Bambamarca. RCs in San Antonio is the second iteration throughout Peru. Rondas campesinas are legally recognized in the Constitution. Rondas started more or less thirty-six years ago in order to stop the abigeato (land and livestock theft). The role of the* rondas campesinas *was to bring justice to the rural community. Justice was a result of a discussion of the case directly by the claimant and the respondent in front of ronderos. The application of justice is fast and free, because it is community justice. In an hour, we solved land disputes that had sometimes taken several years in courts. But there are times that the case should go to the Prosecutor. In the countryside,* rondas campesinas *expanded to resolve issues of injustice, be it land, deaths, violence against women etc. With the destruction of land and water, a new role for the* rondas campesinas *emerged.*

"The peasant is very intelligent," Benavides says, whereas, if before "the role of the RCs was to exterminate the small thieves (*abigeos*), now our role is to capture and punish the big thieves (miners)."

Benavides argued that FDC knew of the possible destruction of the lakes, but that the rest of the peasants did not know. Benavides then stated that:

*Conga was going to complete the looting performed by
Yanacocha if the peasants and ronderos did not go to the lakes.
It was only when peasants and ronderos got to the lakes and saw
the great wonders we have, that we finally understood. We now
have a better reading of how mining destroys the headwaters.
And this awareness makes peasants and ronderos travel to all
the villages to inform and build this solid popular mass.*

In Bambamarca, *rondas campesinas* have provincial, district, zonal,
and base committees. I interviewed three Bambamarca women *ronderas*,
who actively participated in all sorts of things, from counting how
many people there are in each town (demographic work), supervising
government programs to determine whether they are working properly
(i.e., food programs), to solving violence against women and organizing
the patrol schedule. The agendas of men and women *ronderas* are the
same, but women have emphasized legal issues to defend land, pensions
for children, to escape domestic violence, and to seek allies. The role
of women in the Assemblies is to bring order and punish the guilty.
The women have argued that the men in Bamabamarca have changed.
Before *rondas campesinas*, the men in their homes and the police on the
streets were sexist. Women were afraid of both. Since women entered
the ranks of the *ronderas*, men in their households say, "Go to your
organization and I will stay here with the children. When you return let
me know what agreements you made." In this sense, the women's lives
have become more equitable in their houses as well as on the streets as
they now work in cooperation with the police and the judicial system.
Barbarita says:

*Las ronderas are organized from Bambamarca city to small
villages in order to defend their communities. Our authority
comes from the community assembly that elected us. Ronderas
are women from sixteen to sixty years of age. Women ronderas
patrol nightly with men. We do not have weapons. We walk
with sticks and a penca to punish the bad of society. We never
attack but we do defend ourselves.*

Barbarita articulates that,

*[S]ince November 24, 2012, the Ronderos/as are taking care of
the lakes to prevent the mining industry from starting any of
its work. The miners are building two reservoirs—Chailhuagon*

and Perol—to stop us from using our lakes. On the day miners come and bring their machinery all the farmers stop working and go up to surround the lakes. The lakes are almost 4,000 m above sea level, and it is cold there. We take turns to look after them. In an organized manner, we have installed ourselves around the lakes with plastic tarps that cover us from the rain and night. Since a number of members have to work every day for their livelihood, each Wednesday about forty community members from each community go to the lakes and come back on Sunday. Others leave on Sunday and return on Wednesday. So, all the members of the different communities are involved in the care of our lakes. The women in general are responsible for cooking the food, either fiambre-style or communal pot. But the purchasing of food is taken care of by all community members. By taking care of the lakes, the peasant wins, gains authority and earns respect. For us water is life. We will win, because this fight is for justice, so that we can leave a future for our children. The rondas are united and it will be very difficult for the miners to defeat us. They know that we are in the thousands.

Maria adds:

For women, the rondas *have been our university where we learn. Women were the most devalued beings until we joined the* rondas. *We did not know how to read or write, we did not have identification (DNI), and we were not registered as citizens anywhere. We would die and it would not be known who had formally died. With the support of men, in our organization of* rondas campesinas, *women have learned to speak, defend themselves, and gain authority.*

She argues further that,

Since ronderos *declared mining as stealing, and as an activity against nature and people, we are more active. For instance, when the mining industry entered Huambamarca, which is a livestock area, we organized and seized their cars and personnel. In one car, we found some mining personnel. We made them do rounds, without shoes, for five hours, from hamlet to hamlet. They were carrying signs that read we are miners, delinquents;*

we killed the waters, children, and pregnant moms. At the end of the walk all those workers vowed to never return to this town.

As the mining dispute grows, the Peruvian government wants to take over what the *ronderos* have been doing. For instance, the Ministry of the Interior has taken the power of some *rondas campesinas* away through the Ley de Rondas. The government has been trying to neutralize the autonomy of the *rondas campesinas* by creating false *rondas*. But, in Bambamarca, the government and the corporation have not succeeded in fracturing the *rondas campesinas*, and they continue to be autonomous. Therefore, Edy Benavides says not to confuse *rondas campesinas* in Cajamarca with false *rondas*:

1) The real *rondas campesinas* are led by people who have *rondado* or were born in peasant communities; and
2) The false *rondas*, such as the *Federación de Rondas Campesinas* organized by the mining companies, have no history. They are working with the government, particularly the *Ministerio del Interior*. (Benavides)

I also interviewed around ten members from the Plataforma Inter-institucional Celendina (PIC), who are teachers, musicians, and peasants. Milton Sanchez states that:

The organization of the guardian of lagoons was not easy. The mining company argued that Bambamarca only wanted more money from mining. To learn more, we went to Bambamarca, where we were told that they were not going anywhere with us if we do not first resolve the boundary problems. Since Bambamarca is the major force, we went to FDAC for help. This organization proposed to have encounters to discuss mining in our backyards. A first encounter took place in Celendín. At least fifty leaders from all provinces of Cajamarca met in the Parish House, where we wrote our first manifesto, The Charter of Celendín. It prohibited mining activities in the area of Pozo Seco, Lagunas de Alto Peru in San Pablo; Minas Conga in Celendín; Mogol, El Vaquero, El Clarinero, Colesmayo in San Marcos; Tantahuatay in Hualgayoc; La Zanja in Santa Cruz; La Shacsha in Baños del Inca; Cerro Negro and Quillish in Cajamarca (San Pablo Declaration,

November 20-21, 2011). At this meeting, we also decided to have meetings in every town so that the people would become familiar with the problems.

Sanchez adds that:

The second meeting was in San Marcos, in March 2011. People in San Marcos said, "our rivers also are born up above, thus it is also our struggle." At this meeting, we decided to contact the Regional President, who already had been in office for three months. We sent two letters that the Regional President did not respond to. Consequently, we decided to convene ourselves and go to the Regional Government. Upon the arrival of eighty leaders, we sat in the auditorium, the president refused to meet us, but the Regional Vice President, who is from Celendín, came out to meet us. He told us that the President was in Chota Province, but when he heard we were going to occupy the place and that we would not move until the President met with us, he arrived in five minutes. The Regional President, Gregorio Santos, had been a rondero *in San Marcos. A* rondero *peasant said to him "Goyo, remember when we took care of the lakes in San Marcos. Remember we slept there and watched the stars together." He remembered the episode and agreed that we had to do an inspection of the lakes. We went to do an inspection of the watersheds with the Regional Vice President. Then we requested a review of the EIA, which reveals details on what they will do to the headwaters of the watershed. After this episode, we had the third interprovincial encounter in Bambamarca, where eleven out of thirteen provinces attended. There Gregorio Santos, the Regional President of Cajamarca, realized the magnitude of the movement and formed a Frente de Defensa Regional (FDR) with members of his party.*

Milton Sanchez stated:

In Celendín, we were only three members aware of the Conga Mining Project. Celendín is where more than 90 percent of the Conga project could take place. To expand the knowledge, we decided to have a forum. After the forum, in February 2011, PIC [Inter-institutional Celendín Platform (ICP)] was born. As the forum was a success, everyone in Celendín, including newspapers, began to talk about Conga. The Ronda Campesinas in Celendín joined after learning about the experiences in Hualgayoc.

Sanchez recalled their first encounter with mining authorities in Encañada, a province of Cajamarca, during its first EIA presentation.

The mining company released their first EIA in an assembly. For it, corporations took the workers and their families in dump trucks and trucks. On one platform of a tent was located the staff of the Ministry of Energy and Mines, and the mining executives. In another tent there were their workers. Over a thousand police surrounded the two spaces of the auditoriums where there were about 5,000 people attending. Without an invitation, seventeen of us arrived in a van to the assembly. Government and corporation personnel tried not to let us in the auditorium where the authorities were located. But we prevailed. I was allowed to ask one question to the mining company. And the question was what do you mean when you say that the lakes will be moved? The response was that the lakes were going to be moved from one place to another. When we left the tent, we saw in the hills hundreds of people from different villages that had come walking to reject the project. But they were intimidated by the numbers of police that they had never seen. We returned home indignant.

Sanchez continued with his account:

With the help of GRUFIDES and the regional government, we learned that the lakes would disappear and become open pits; other lakes would be converted into their landfills; that the dimensions that they referred to were enormous and they would destroy our water. We started to ask why would mining commit such cruelty against nature and people. After this information, we went to the lakes, we photographed them, and we put them on the Internet. When we became familiar with the lakes we realized that they were part of the ecosystem in which we live, that we were interconnected, and we began to love our lakes. That's where magically the lakes were born, because we do not have snowcaps, we have jalcas. We learned that what one does not know one does not defend.

He further commented,

The work we did had negative effects on the mining project. When the company released its EIA in Celendín it realized that its acceptance level had dropped dramatically, and it started giving away backpacks, jackets, hats, and caps in the schools.

The movement initiated another regional strike. Meanwhile the offensive from the central government against the protests increased. On December 3, 2011, the central government declared a State of Emergency in four provinces (Cajamarca, Celendín, Hualgayoc, and Contumaza). The following day, on December 4, the government sent their ministers to talk with representatives of the provinces. Sanchez informed us that:

Prime Minister Valdez came first to say who could come in to talk and who could not. In Celendín, we decided that if any of us were unable to talk in there nobody would go in. In the end, we all were accepted and went in. As a representative of the Platform, I presented three demands:

1) To review the EIA,
2) To do an audit of Yanacocha,
3) That Conga should be presented as nine projects in the Headwaters of the Watershed, instead of one as it is presented. Thus, we wanted to know what would be the impact of the nine projects, not just one.

The government rejected any discussion on these three points. Here government representatives say the first thing you need to do is to finish the strike. We responded that we do not make decisions; that we only represent the will of the people. Thus, we requested time to go to the lakes to ask our constituents' opinions. The government rejected our proposition, instead it said that if we did not sign an Act of Surrender it would invoke a State of Emergency and we would not be allowed to leave. At the end of the negotiation, we were allowed to leave, and we hid. Since we could not do anything from hiding, we made contact with the Congressman of Cajamarca and requested a session in the Congress. Wilfredo Saavedra from FDAC and I from PIC went to the Congress. On December 6, 2011, we got

113

five minutes to expose the issue, and to explain what Conga meant for us. Upon leaving the Congress, the press surrounded us to ask questions to Saavedra to whom the government gave leadership and had directed its attacks, because he was in jail accused of terrorism, and the government wanted to equate the water movement with terrorism. When we advanced half a block, State Security came and took us to the DIRCOTE *(National Counter Terrorism Security). We spent ten hours in* DIRCOTE *until ten Congressmen came and stayed with us until they let us go.*

THE NATIONAL MARCH FOR THE WATER, AND THE SECOND REGIONAL STRIKE.

The Cajamarca region rose up in defence of water, its economy, its dignity, and aspirations. The *Marcha por el Agua* (March for Water) was convened to take place on February 1ˢᵗ, 2012, by the Fronts of Defence against mining, by men and women of the city and the countryside, from Celendín, Bambamarca, San Marcos, San Pablo, and Cajamarca, and from other inland areas of Cajamarca. Marco Arana, director of GRUFIDES, led the protest. A national announcement was made to join the Conga movement because the conflict had taken on national dimensions. People were mobilized in other regions such as Cuzco, Arequipa, and Piura.

The March demanded the following from the central government of Peru:

- •respect for the rights of peoples to prior and informed consultation in strict observance of the 196 International Convention of the ILO;
- •compliance with international treaties and national laws, with respect to the conservation of fragile ecosystems;
- •protection and conservation of the headwaters of the basin, as water springs and sources;
- •return of powers of municipal and regional institutions to regulate mining activities on a large scale;
- •recognition of water as a human right;
- •requirement to not have any more open-pit mining, or worse, mining using cyanide and mercury.

Eddy Benavides stated:

In Bambamarca, we began the March for Water from the lakes. People were joining as the march passed by the districts affected by mining (Choropampa, Chilete, Yonan [Tembladera] and the Village Ciudad de Dios). By the time the march arrived at the border between Cajamarca and La Libertad, thousands of women and men from other regions, such as Amazonas, San Martin, Piura, Lambayeque, and Ancash gathered together and marched toward Lima, the Capital city. We arrived in Lima on February 10 in what we called ten wonderful days of solidarity.

Milton Sanchez stated,

In Celendín about 3,000 [people] were brought together for the March. People in Cajamarca were waiting for us to join the march to Lima. We arranged strategies; we thought the police at some point were going to bomb us, so people had to reorganize themselves by looking for the location of the flags of their villages. In one town, Celendín's flag broke. I stood in one place while a lady sewed the flag. While I was waiting the people from all walks of life started asking about our struggle. By listening to what we were doing, people brought us food and water. One lady told me that she had nothing to give, but that we should take her daughter with us, and the young lady accompanied us for a stretch of the walk. In our walk we received gifts, pharmacies gave us throat medicine, shoemakers gave us shoes for those whose shoes had been worn out on the road. By the time we arrived in Lima, we were 40,000 people. On the way, thousands joined in. That march made millions of Peruvians aware of our struggle.

Maria from Bambamarca shared some insights about the march:

We, women with our families walked to Lima. We were stimulated to see our leaders walking without a break. Women with their young children went by bus, but in the villages and cities we walked the streets asking people to accompany us to Lima. I walked until my shoes fell apart. On our walk, the villages would get up to greet us, they would give us food, water, clothes, shoes, etc. The solidarity of the villages gave us encouragement to continue. While I walked, I thought, it is our life that is at stake, because our villages are agricultural and livestock producers. If Conga destroys our spring waters, just like Yanacocha had already done to another part of our land, we are condemned to death. For example, we use the water directly from the lakes to irrigate our lands, for our kitchens and for our animals. These lakes do not require any work, because they are there only to give life. For us it is cruel that corporations want to destroy our waters that require no financial investment from the government. So, for us, ¡Conga No Va! means that our own lives are at stake. We, women, think not only of ourselves, but also in mothers and their children, the children of our daughters and so on. It would be cruel to leave them with nothing.

After the march, Celendín and Bambamarca organized the Comando Unitario de Lucha (CUL). On March 31, 2012, the National Assembly of the Peoples in Celendín, hosted by the local *rondas campesinas,* convened a Permanent Regional Strike. The Second Regional indefinite strike was organized in the provinces of Cajamarca, Hualgayoc, Celendín, San Pablo, San Marcos, and other areas. It began on May 31, 2012, with mass demonstrations, rallies, candlelight vigils, and soup kitchens. In the city of Cajamarca, at Plaza Bolognesi, police lashed out against the women who were preparing the common pot. They emptied the contents of the pots, which contained the food for peasants who came from other areas, and then beat those peasants. Groups in favour of mining also organized a "March for the peace and development of Cajamarca," and declared that it was looking for a dialogue about the development of one of the poorest areas of the country and in rejection of the guilds, which were against the Conga project.[9] According to Milton Sanchez,

In Celendín, the mayor was in support of mining; to find an excuse to end our strike, the mayor sent people to burn part of a municipal office with the purpose of accusing us as terrorists. On July 3, 2012, after thirty-four days of strike, helicopters shooting from the air at protesters confronted a march on the public plaza. In this mass killing, four members of Celendín died (Antonio Joselito Sanchez Huaman, 29; Faustino Silva Sanchez, 35; Cesar Medina Aguilar, 16; and Paulino Leonterio Garcia Rojas, 48), and two hundred of us took refuge in the Virgen del Carmen Church and others went to the lakes.

Nineteen were arrested on the street and taken to Chiclayo city
for fifteen days where they were beaten.

The next day, with an already declared sixty days state of emergency, the population of Bambamarca decided to light candles in the church in Plaza de Armas, in memory of their companions fallen in Celendín. Once again, the police and army lashed out at citizens, killing Joselito Vásquez Campos, who was twenty-eigh years old. Milton Sanchez argued that,

> *During the state of emergency, soldiers stationed in Celendín committed numerous abuses and raped girls sixteen and seventeen years old. Many of them became pregnant and do not even know who the fathers of their children are. For instance, one night the* rondas campesinas *caught four soldiers abusing a girl. They were captured and taken to the house of the rondas who called the commanding officer of that group. As a result, the* rondas *were accused of kidnapping the girls and the military asked for members of the* rondas *to be convicted to thirty-two years in prison. This is another example of the criminalization of protest. We all have been charged; for instance, I have about forty charges against me. Several other abuses have been committed against us. We do have a camp on Laguna Azul. One time when we went down to a meeting, the police burned our food and our camp. But in front of Laguna Azul there is a family—the Chaupes. They have refused to sell their land to the corporation. Since 2011, police have been trying to remove them by killing their sheep, and burning their house, but they are still there, living in terrible conditions, withstanding the cold of winter. The* ronderos *supported them by rebuilding their home. This family is the power we have up in the lakes.*

DEFENDING FREEDOM: CRIMINALIZATION OF THE PROTEST

When Yanacocha arrived in Cajamarca, many hoped that mining could have been done without severe impairment of the waters and agriculture, with respect to local populations, and with the possibility of the creation of job opportunities. Instead, during eighteen years of mining, Cajamarca has become the Peruvian region of greatest socio-environmental conflicts because mining has destroyed the area's rivers and lagoons, and contaminated water for animal and human consumption. Furthermore, those who have protested and defended water have ended up dead or in prison.

As the struggle increased, NGOs in favour of Yanacocha, such as PRO NATURALEZA, arrived in Cajamarca, and others against Yanacocha, such as GRUFIDES, solidified the anti-Conga movement. Marco Arana, the head of GRUFIDES, says,

> At GRUFIDES we speak of a socially-just development. This means that there cannot be development against people, and much less against the most vulnerable groups. Let the people have the right to health under the conditions they want. As for the definition of basic needs, GRUFIDES has three mandates: a) defence of human rights, b) technical assistance to communities with environmental impacts, c) political advocacy to change the legal framework in relation to mining activities. ... GRUFIDES brought technical support to review the Conga Mining Project EIA study, including Robert Moran who, in March 8, 2012, presented a study entitled "Peru, Conga Mining Project: Comments on the Environmental Impact Study and Related Issues." Since the social protest has been criminalized, GRUFIDES oversees the technical and legal issues, and acts as an advisor to the community. In sum, GRUFIDES is a technical-legal organization for the defence of human rights.

Mirtha Vasquez, GRUFIDES' lawyer, describes how the central government established new legal features in Cajamarca, by which environmental advocates and regional authorities who support the popular struggles against mining are prosecuted. These can be summarized as:

> 1) Selective persecution: people who are prosecuted are those who assume some public leadership.
> 2) The evidence for prosecution is based on pictures of the leaders addressing the public.
> 3) Freedom of expression has become criminalized. If I disagree I have no right to say anything.
> 4) Criminal laws have been modified in an outrageous manner. Since the neoliberal globalization agenda that deregulated the nation/state legal system, a new legality has been established. For instance, since Toledo's administration (2001-2005), the

118

penalty against protest was increased to six years in jail. During Garcia's administration (2006-2011), the law defined protest as "Organized Crime" (July 2007), aimed to categorize it as a crime of extortion, and the penalty reached twenty-five years.
5) Humala's administration (2011-2016) promulgated Law No. 30151 in January 2014. This law grants the armed forces, as well as the national police, a license to kill. They are exempted from criminal responsibility.
6) Further, Humala has created a resolution against the participants of Conga. They are not processed in Cajamarca, but rather in Chiclayo or Lima, taking away GRUFIDES' ability to defend the accused. In this case, GRUFIDES is presenting allegations that the people can go to jail not because of the veracity of the allegations, but because the people are poor and cannot afford the cost of transportation, food, hotel, etc.
7) Today, prosecution is used to pursue social protest leaders. The mining and state prosecutors legally denounce those who announce a protest march on the radio or in the newspaper. Anyone can face detention and prosecution on charges of "terrorism."

Mirtha Vasquez also argued that,

Mining officials could legally denounce a person to prevent the possible disturbances that might occur. This might be something that the company is imagining could happen. Until December 2013 there were 303 leaders denounced. Each leader has over twenty complaints. Most cases must be filed away because there is no evidence. And the judges call the accused sometimes a year after the possible disturbance happened that are the subject of the complaint. These complaints serve to deflate the movement. Due to sexism, women are less vulnerable to be persecuted, and to be legally denounced. The police and prosecutors say "they are just women." Nevertheless, there are ten percent women defendants.

GRUFIDES is helping mainly rural communities to bring their cases to the Inter-American Court of Human Rights (CIDDHH). On March 18, 2014, the *Coordinadora Nacional de Derechos Humanos* (National Coordinator of Human Rights)—a civil society body aimed at protecting human rights in Peru, gave evidence to the *CIDDHH* in Washington. The complaint raises the violations of rights of Indigenous and tribal people as enshrined in ILO Convention 169 and the UN Declaration on the Rights of Indigenous Peoples, and the violation of human rights of the Guardians of Conga Lagoons by the state and the police. This complaint was resolved in part. On May 7, 2014, CIDDHH issued precautionary measures in favour of a guarantee of the lives and safety of forty-six leaders in the fight against the Conga mining project. Among the forty-six leaders are the Chaupe family in its entirety, a journalist, and the leaders of defence fronts, among others. The leaders of the movement are going to ask the government to shelve more than two hundred complaints against all those who have been participating in the struggle. The second part of the complaint, still pending resolution, submitted in April 2012, is a decision on the illegal presence of the Yanacocha in territory of the lagoons of Conga.

Further, on June 26, 2014, the Justice System forced the Regional President of Cajamarca, Gregorio Santos, to respond in Lima to undefined complaints. On the same day of his appearance in court, the judge ordered preventive prison for fourteen months in what has been called a "political ambush," organized by President Humala, to stop Santos' candidacy and re-election as Regional President. Raul Wiener, a journalist from *La Primera*, wrote,

> Everyone knew that Gregorio Santos was going to be stopped. This was not a new issue, but it became clear in the year 2012, when Cajamarca challenged Humala for not keeping his promises to the water of the lagoons from the eagerness for gold from Yanacocha, which was one of his election banners and helped him achieve the majority vote of the region. Already at the beginning of this year, the controller's office sent an ever-increasing number of auditors to find offences to accuse the regional president, who had faced the investment door that apparently was the beginning of a new cycle of expansion of mining investments in the country.

While in prison, despite having no charges laid against him, Gregorio Santos was re-elected as Regional President of Cajamarca in October 2014. As a result, Humala's government extended Santos' incarceration for another twenty-five months. In the elections of 2016 for President of the Republic, Santos obtained 4.1 percent of the national vote, making him the only leader, in Peru today, with an anti-mining political profile.

CONCLUSION

Mining has been portrayed as a way to bring investment, create jobs, and reduce poverty. Instead, as this chapter has shown, mining corporations, using deregulation and free trade agreements to enhance massive profits, are destroying the water systems of the Cajamarca community. Conga, as an environmental justice movement, has demonstrated that teachers and *rondas campesinas* are forces that do not negotiate water, because water is life for their communities. Despite government brutality, their courage has not been compromised.

At this time, *Conga No Va!* is a triumphant movement. On the one hand, the Guardians of the Lagoons—in defending land, water, and freedom—have defeated President Ollanta Humala, and banned him from Cajamarca for falsely promising to terminate Conga Project in order to win the national election. On the other hand, the Guardians of the Lagoons—in organizing a successful movement—have successfully put the Conga Mining Project on hold for the last six years. Moreover, this movement is demanding a clear accounting from the Earth Summit in Johannesburg that declared mining as an acceptable and viable form of "sustainable development," when this is so obviously not at all the case.

This research was funded by Brock SSHRC Institutional Grants (BSIG) and CRISS Research, Brock University.

[1]The North American Free Trade Agreement, Chapter 11, was added to the bilateral free trade agreements in Latin America. The dispute settlement provisions of NAFTA Chapter 11 allow investors to pursue legal remedies against government measures under the laws of that country, but also provide the option for investors to bring their claims against a State directly to international arbitration through the procedures and services of several international arbitration entities. It means that in the case of conflict between a country and a corporation, the World Trade Organization (WTO) is in charge of applying "justice" ("Bill Moyers Reports").

[2]See Water Justice: Resource Centre on Alternatives to Privatisation.

[3]See "Blue Planet Project."

[4]Yanacocha South, Carachugo, San José, La Quinua Sur, Capping West, Cerro Negro, Cerro Quilish, Chaquicocha and Yanacocha Green.

[5]See Operations: Yanacocha, Peru ("South America: Regional Key Facts").

[6]Constitution: arts. 89, 149; 4th DFT, Law of Ronderos Campesinos, art.1; Convention 169 of the ILO on Indigenous and tribal peoples in the countries independent; United Nations Declaration on the rights of indigenous peoples; and doctrine and jurisprudence of the IACHR (IILS 1).
[7]See Geredien, "An update from the field on the Conga Mine."
[8]See on YouTube, "Marcha de las mujeres."
[9]See more at: "Demostrado: Cajamarca quiere Paz y Desarrollo."

REFERENCES

Arana, M. Personal interview, 2013. Cajamarca.
Barbarita. Personal interview, 2013. Bambamarca.
Benavides, E. Personal interview, 2013. Bambamarca.
Bernard, Alice and Diego Cupolo. "Cajamarca Anti-Mining Movements Celebrate and Question Study Results for Peru Conga Gold Mine. " Upside Down World. 20 April 2012. Web. Retrieved June 2014.
Bernard, Alice and Diego Cupolo. "Peru: Cajamarca Protests Continue as Conga Gold Mine Awaits Green Light." Upside Down World. March 29, 2012. Web. Retrieved June 2014.
"Bill Moyers Reports: Trading Democracy." Transcript PBS.org (n.d.). Web. Accessed July 2010.
"Blue Planet Project," The Council of Canadians (n.d.). Web.
Cabellos, E., and S. Boyd. The Price of Gold. DVD video. Brooklyn, NY: Icarus Films, 2003..
Campodonico, Humberto. "Inaceptable prórroga del óbolo minero." La República. 9 February 2011. Web. Retrieved on July 2014.
Coumans, C. "Alternative Accountability Mechanisms and Mining: The Problems of Effective Impunity, Human Rights, and Agency." Canadian Journal of Development Studies 30.1-2 (2010): 27-48.
De Echave, J. "Mining and Communities in Peru." Community Rights and Corporate Responsibility: Canadian Mining and Oil Companies in Latin America. Ed. L. North, T. D. Clark and V. Patroni. Toronto: Between the Lines, 2006: 17-36.
Democracy Now. "As Tensions Flare in Peruvian Amazon, Award-Winning Actor Q'orianka Kilcher Heads to Peru to Support Indigenous Rights." 2009. Web. Retrieved July 2010.
"Demostrado: Cajamarca quiere Paz y Desarrollo." Panorama Cajamarquino 30 May 2012. Web.
Esteva, G. "Development." The Development Dictionary: A Guide to Knowledge as Power. Ed. W. Sachs. London: Zed Books, 2010.

Geredien, Ross. "An update from the field on the Conga Mine." Earthblog. Earthworks.org, 30 January 2014. Web.

GRUFIDES (Producer). "Caso: Yanacocha:Choropampa—Diez años después del derrame de Mercurio." June 2014.

International Institute on Law and Society (IILS). "Obligaciones Internacionales del Estado Peruano ante el Megaproyecto 'Conga.'" December 6, 2011 . Web. Retrieved July 2014

Isla, A. "The Eco-class-race Struggles in the Peruvian Amazon Basin: An Ecofeminist Perspective." *Capitalism Nature Socialism* 20.3 (2009): 21-48.

Cáceres Neyra, Jorge Luis. "The Rule of Law and Environmental Justice in Perú: Lessons Learned from the Choropampa Mercury Accident." Monografias.com. Web. 2008.

Líderes y Lideresas de Comunidades Campesinas y Rondas Campesinas de Cajamarca MC-452-71 República de! Perú, 2014.

"Marcha de las mujeres embarazadas en Cajamarca en rechazo al proyecto Conga." Prensa Vision. 19 June 2012. YouTube.

Maria. Personal interview, 2013. Bambamarca.

Mies, M. *Patriarchy and Capital Accumulation. Women in the International Division of Labour*. London: Zed Books Ltd., 1998.

Nelida. Personal interview, 2013. Cajamarca.

Perez, J. *Conflicto Minero en el Peru. Caso Yanacocha—Cambios y Permanencias*. Cajamarca: Universidad Nacional de Cajamarca, 2012.

Perez, J. *Agua-Procesos Sociales, Desarrollo y "Guerra" Mediatica en Cajamarca*. Cajamarca: Universidad de Cajamarca, 2012.

Rodriguez-Pinero Royo, L. "La Internacionalización de los derechos indígenas en América Latina: ¿El fin de un ciclo?" *Pueblos indígenas y Politica en América Latina*. Ed. S. Marti i Piug Barcelona: Bellatierra-Cidob, 2007. 181-200.

Salleh, A. *Ecofeminism as Politics. Nature, Marx and the Postmodern*. 1997. London: Zed Books, 2017.

Sanchez, M. Personal interview, 2013. Celendin.

Sanchez, W. Personal interview, 2013. Cajamarca.

"South America: Regional Key Facts. Operations: Yanacocha, Peru." Newmont (n.d.). Web.

Torres, F., and M. Castillo. *El Projecto Minero Conga, Peru: Riesgo de Desastre en una Sociedad Agraria Competitiva*. Cajamarca, 2012.

Vasquez, M. Personal interview, 2013. Cajamarca.

Water Justice. Resource Centre on Alternatives to Privatisation. WaterJustice.org (n.d.) Web.

Wiener, R. "Detengan a Santos." *La Primera*. 26 June 2014. Web.

World Commission on Environment and Development. *Our Common Future*. New York: Oxford University Press, 1987.

5
Ecofeminism, Commons, and Climate Justice

PATRICIA E. (ELLIE) PERKINS

IN MUCH RECENT WORK ON ECOLOGICAL ECONOMICS, degrowth, and the transition to more sustainable socio-economic systems, "commons" is emerging as a paradigm for future economic institutions. This goes beyond the idea of a commons as a common-property regime with the socio-political structures required to prevent open access. The vision more broadly is one of people working together, cooperatively, to build methods of production, service provision, and exchange which create value and well-being while integrating ecological care, justice, and long-term planning to the best of diverse communities' abilities. This includes institutions such as co-ops, land trusts, and non-market or beyond-market collective ways of organizing production, distribution, consumption, and waste or materials management.

Commons often function better and more sustainably than either private property and markets, or state governance, for a whole range of reasons: markets create strong incentives to over-exploit resources, exclude some users whose needs must then be met in other ways, generate pollution, ignore ecosystem services and long-term impacts, and otherwise "externalize" crucially-important costs of resource use while undercutting society's ability to address those costs and manage human development sustainably. States may be seen as inherently corrupt and inefficient, inflexible and unskilled.

Preventing the so-called "tragedy of the commons" by controlling open access through strong social institutions requires a high level of general civic consciousness, co-operation, the ability to listen and mediate differing goals, conflict resolution, flexibility and good will throughout society, especially in the context of social dynamism and diversity.

As Elinor Ostrom said in her 2009 Nobel Economics acceptance speech in 2009:

a core goal of public policy should be to facilitate the development of institutions that bring out the best in humans. We need to ask how diverse polycentric institutions help or hinder the innovativeness, learning, adapting, trustworthiness, levels of cooperation of participants, and the achievement of more effective, equitable, and sustainable outcomes at multiple scales.

Aboriginal traditions of hospitality, sharing, potlatch (or giving away material wealth as a sign of moral and community standing), humility, and reverence for the earth, its creatures, and life systems are central to the locally appropriate commons governance processes. First Nations also had nested governance hierarchies that seem to me to correspond with what Elinor Ostrom has cited as successful ways to govern large-scale commons.

The interdisciplinary International Association for the Study of the Commons was formed in 1989, building on the Common Property Network, which was formed in 1984. IASC now has over one thousand institutional members and has sponsored twelve international conferences, with the most recent in Japan in June 2013 and in Alberta in May 2015. There are regional meetings, an online digest, a digital library and bibliographies, and discussion groups.

New books on commons appear every week, and the idea that commons governance represents something fundamentally different from "the Market" or "the State" is becoming well-known and widely accepted.

DEFINITIONS AND TYPOLOGIES OF COMMONS

So what exactly is a commons? The word is a somewhat odd collective noun, pluralized but singular—how do we understand and use this idea? There is a risk, noted already in the literature, that "commons" will become the latest glom-on term, co-opted and vague, obscuring more than it conveys. However, "commons" starts out more overtly oppositional to capitalism than other terms like "sustainability" or "development," focusing as it does on ownership and property, land, resources, and assets that are explicitly NOT privately owned (Linebaugh).

Commons take a big step towards internalizing externalities to use neoclassical terminology—and towards discourse-based valuation of ecological and social goods and services, bringing politics together with economics, in the best alternative or heterodox traditions of political ecology and feminist ecological economics.

Ideas on common goods and their governance have a long history. The Justinian Code of CE534 divided things into *"res privatae, res publicae, res communes, res nullius, and res sacra. Res comunes* included earth, water, air, sky, flora and fauna and navigable waterways" (Ricoveri 37). In Europe and elsewhere, common land was long maintained for agricultural use, including hunting, foraging, and pasturing animals (Thompson). Worker and housing cooperatives, guilds, community barn-raisings, "mutual aid," and repeated examples worldwide of crises bringing out altruism, solidarity, generosity, and courage in stricken communities are indications that people's desire to act communally is ever-present (Cato 9-12; Ricoveri 63).

A recent book on commons and ecological governance says,

> The commons is a term that applies to the resources utilized, owned, or shared by multiple individuals on a group basis.... The traditional commons had to do with the management of resources on a local, not global, level. Those resources were not comprehensible if removed from the micro-societal context in which they existed.... Current-day, widespread use has diluted the formerly rigorous definition of the term "commons," ... and fostered a vast expansion in the scope of those resources now considered worthy of research within a commons-related context. (Suga 4-6)

The book's editors state,

> This volume rests on the perspective that modern society is composed of three elements: a public sector, common sector and private sector.... If humanity were a society driven by the profit motive alone, it would be a society of disparities highlighted by unbearable levels of inequality. That is why society demands the existence of a public sector committed to the redistribution or balancing of income and assets through the power of taxation.... Modern societies also incorporate a common sector that is neither public nor private ... that operates independently of the profit motive or the interest in upholding public authority. Structures or communities of this nature are typically composed of households, various cooperatives or non-profit organizations ... (and) international

volunteer associations…. Cooperation and/or coordination are the driving principles on which these organizations operate.. (Murota and Takeshita xxii)

International legal scholar Shawkat Alam claims:

Collective rights are often affiliated with Indigenous people, as they are defined as rights held by groups—"a collection of persons that one would identify as the same group even under some conditions in which some or all of the individual persons in the group changed" (Xanthaki 13). It follows that collective rights are connected to a community or group, which is often of minority status. However, it has been argued that the 'recognition of collectivities and collective rights is one of the most contested in international law and politics." Indeed … this concept of collective rights can be seen to conflict with Western ideas of individual freedom and liberty…. Collective rights have been seen to foster tolerance, and diversity of culture and knowledge. To this end, many Indigenous peoples view the recognition of their cultural rights as "of paramount importance" or "as a token of respect towards their identity and communities as well as the only way for their survival and development" (Xanthaki 13). (Alam 588)

Elinor Ostrom and Charlotte Hess, long-time commons researchers, define the term as follows:

Commons is a general term that refers to a resource shared by a group of people. In a commons, the resource can be small and serve a tiny group (the family refrigerator), it can be community-level (sidewalks, playgrounds, libraries, and so on), or it can extend to international and global levels (deep seas, the atmosphere, the Internet, and scientific knowledge). The commons can be well founded (a community park or library); transboundary (the Danube River, migrating wildlife, the Internet); or without clear boundaries (knowledge, the ozone layer). (4-5)

In a recent book on commons, David Bollier and Burns H. Weston use the following definition: "A commons is a regime for managing common-pool resources that eschews individual property rights and

State control. It relies instead on common property arrangements that tend to be self-organized and enforced in complex, idiosyncratic ways" (347).

Italian commons activist Giovanna Ricoveri's definition is:

> The commons are goods or means of subsistence which are not commodities, and therefore they constitute a social arrangement that is the complete opposite of the one created by the market economy.... The commons are local systems that can be managed effectively only by those who have a precise and detailed knowledge of the area and who know its history, language, culture, vegetation, mountains and other physical attributes.... Thus there does not exist, nor can there exist, a general law that is valid for all systems of the commons for the very reason—contrary to what is generally believed—that they are open local systems, receptive and adaptable to the local "whims" such as climate, the different attributes of the localities in terms of natural resources, the knowledge of the inhabitants, their professionalism—all elements that cannot be defined in law. (31-36)

Elinor Ostrom too has emphasized the importance of locally-constructed governance processes, local monitoring, and enforcement of environmental quality and access to the resource. This makes monitoring more efficient, cost-effective, and accurate (Ostrom "The Future" 83).

"New commons" go beyond common-property regimes, with their socio-political structures required to prevent open access. The vision more broadly involves people working together, cooperatively, to build methods of production, service provision, and exchange which create value and well-being while integrating ecological care, justice, and long-term planning to the best of diverse communities' abilities (Hess 37). Examples include institutions such as co-ops, land trusts, and non-market or beyond-market collective ways of organizing production, distribution, consumption, and waste or materials management.

In the face of climate change, as Leigh Brownhill and Terisa Turner explain, movements in the Global South and North, largely led by women, are resisting ongoing enclosures for extraction and fossil fuel industries and, in the process, reclaiming commons.

To the extent that the capitalist energy system is seized and redirected towards commoning, actors within it have reduced dangerous emissions and elaborated an alternative system premised on sustainable energy.... This "actually existing" movement of commoners is the result of the exploited taking over some of the organizations of capital and using them to (a) undermine profit and at the same time (b) negotiate and construct means for satisfying universal needs. (16)

For example, La Via Campesina's Declaration at the International Forum for Agroecology stated, "Collective rights and access to the commons are a fundamental pillar of agroecology. We share access to territories that are the home to many different peer groups, and we have sophisticated customary systems for regulating access and avoiding conflicts that we want to preserve and to strengthen" (Giacomini 98). La Via Campesina also notes, "As savers of seed and living libraries of knowledge about local biodiversity and food systems, women are often more closely connected to the commons than men" (98). Turner and Brownhill's definition of "civil commons" is "the organized provision of the essentials of life to all" (806).

For Mies and Bennholdt-Thomsen, necessary steps in this process include:

defending and reclaiming of public space, and opposition to further privatization of common resources and spaces; ... (localized) production, exchange, and consumption; ... decentralization; reciprocity (instead of) mechanical mass solidarity; ... policy from below, as a living process, instead of policy from above; ... (and) manifold ways of realizing a community and a multiplicity of communities. ("Defending" 1021-1022).

To add some detail and groundedness to these definitions, here are some Canadian and international examples of commons. Following a bit of history to set the context, I will discuss these examples at increasing scales from local to global.

EXAMPLES OF NEW COMMONS

Co-operatives and Credit Unions

There is a long history in Canada of communities developing creative ways of securing social livelihood and building community resilience through cooperation. Canada still has the highest per capita credit union membership in the world: 35 percent of Canadians are credit union members. According to the Canadian Co-operative Association (CCA), there are approximately nine thousand co-operatives and credit unions in Canada which provide products and services to eighteen million members in all economic sectors—agriculture, retail, financial services, housing, child care, renewable energy, etc. Co-ops have more than $370 billion in member-owned assets, employ 150,000 people, and have strong links with their local communities via volunteerism, community donations, and sponsorships. Their survival rate is higher than that of traditional businesses (62 percent are still operating after five years, compared with 35 percent for traditional businesses; after ten years the figures are 44 percent and 20 percent respectively) (see also, "Co-ops & Cooperation").

In Canada, mutual insurance companies were founded in the 1840s; dairy producer co-operatives in central and Atlantic Canada in the mid-1800s; the first known consumer co-operative in Stellarton, Nova Scotia, in 1864; a co-operative bank at Rustico, Prince Edward Island, also in 1864; and worker co-operatives connected with the Knights of Labour in the 1880s.

University of Victoria emeritus history professor Ian MacPherson, who has recently written a history of the Canadian co-operative movement, says,

It should be noted that all these beginnings took place before there was specific, enabling co-operative legislation; before there was any general acceptance of international co-operative principles; and before regulators had any significant understanding about the nature of co-operative enterprise. In short, the early experiments were just that—experiments undertaken by groups working within flexible and developing company law to create institutions to meet their needs and likings; in some instances at least, though, they were attempting to imitate European precedents.... A significant issue in thinking particularly about beginnings, but also about the sustained ongoing strength of co-operatives, is the association with traditional co-operation (e.g., the ritual co-operation typically found in most rural areas at the time of planting and harvesting) and spontaneous co-operation (when groups, perceiving opportunities, collaborate for joint purchase of supplies or the sale of produce). Much of this kind of co-operation is informal, but it is important

131

for the beginnings and the subsequent development of formal co-operative institutions. It provides context, networks, and bonds of association without which many co-operatives would not have succeeded, particularly in their formative and stabilizing phases. In that sense, it is misleading to think that an institutional approach to understanding co-operative movements is fully satisfactory. The "movement" has a life beyond institutions, often stretching deeply into cultural, community, kinship, and class relationships. The movement is not easily measured. (2-3)

More recently, Macpherson states,

During the last two decades, there has been a steadily widening and deepening interest in the development of different kinds of co-ops. Perhaps the most common area of interest has been in co-ops that provide "slow food," food produced locally as much as possible, preferably organic, so as to lessen dependence on food produced elsewhere and brought to Canada in ways that seriously impact the environment. Across the country, too, there is a significant rise in transportation co-ops (e.g., car share co-ops, bike co-ops) and energy co-ops based on wind power or the production of biodiesel fuels. Many young people have found it useful to develop worker co-ops in the high-tech industries or to seek alternative forms of housing. Communities facing health issues because of declining support of governments and aging populations have organized different kinds of health or service co-ops. These co-ops are similar to the new co-ops found around the world, a modern rebirth. (18-19)

Local Commons in Toronto

Here are a few examples of organizations and projects in Toronto which are building local commons. I am sure that similar examples exist in most communities around the world.

Not Far From The Tree (which was started by York University Faculty of Environmental Studies graduate Laura Reinsborough in 2008) puts Toronto-grown fruit to good use by picking and sharing it locally. Fruit trees planted long ago in the city are still producing lots of apples, pears, cherries, berries, and other fruit. According to the organization's website:

When a homeowner can't keep up with the abundant harvest produced by their tree, they let us know and we mobilize our volunteers to pick the bounty. The harvest is split three ways: a third is offered to the tree owner, a third is shared among the volunteers, and a third is delivered by bicycle to be donated to food banks, shelters, and community kitchens in the neighbourhood so that we're putting this existing source of fresh fruit to good use. It's a win-win-win situation! This simple act has profound impact. With an incredible crew of volunteers, we're making good use of healthy food, addressing climate change with hands-on community action, and building community by sharing the urban abundance.

The Yes in My Backyard program similarly links volunteers and landowners to grow vegetables in the city. On their website, they state:

Many people would like to garden but live in apartment buildings or do not have access to yard space suitable for growing food. And yet others have access to a yard but do not have the time, interest, or the physical ability to maintain a vegetable garden. Some just like the idea of co-operating with others to create a garden together. Whatever the motivation for participating, YIMBY is working to build community and strengthen relationships between people who might not have otherwise met.

Located on eight acres of city-owned conservation floodplain land in North Toronto, the Black Creek Community Farm helps build community food security and food justice by producing healthy vegetables which are sold locally through harvest shares, farmer's markets, and volunteer programs. Its mission is "to engage, educate, and empower diverse communities through the growing and sharing of food" (see also, Slaughter; "Urban farm").

Community supported agriculture farms exist across Canada and in many other countries around the world. Food consumers purchase a share of each year's mixed vegetable crop at the beginning of the growing season, providing cash up-front for farmers and spreading the risks and rewards of agriculture. In some CSAs, consumers also help out in the fields. An Ontario website provides a directory of CSA farms across the province so that potential customers can find one in their area (Community Supported Agriculture website).

Anarres Worker Co-operative, formed in 2003, provides affordable technology services and online communications tools for the non-profit social sector, including website development, hosting, and IT support. Their website says,

We ... believe computer technology and the web should primarily be tools for community building. We are passionate about using opensource software for reasons of both utility and ethics. We believe in its affordability, flexibility, and effectiveness.... We are activists and social advocates in our own right, and we strive to bring this aspect of ourselves to our work as much as we do our technical competence and experience. (Anarres)

The Co-operative Housing Federation of Toronto represents more than 45,000 people living in more than 160 non-profit housing co-operatives. Since 1975 it has provided development assistance for new housing co-ops, as well as education and services, a bulk-buying program for its members, information for the public in eight languages, diversity education, and policy support (Coop Housing Federation).

Regional and International Commons
The 885-kilometre Bruce Trail extends from Queenston to Tobermory, Ontario. It was built and is maintained by nine regional clubs of the Bruce Trail Conservancy, which maintain a conservation corridor and public footpath along the Niagara Escarpment—a UNESCO World Biosphere Reserve—through the "kind permission" of private landowners, coordination with public lands and roadways, and the gradual purchase of land through a charitable preservation fund (Shimada). The regional clubs also organize volunteer-led nature walks, hikes, and excursions, including a series of hikes where participants meet at Toronto subway stations and go by bus to the hike site.

The Great Lakes Commons Initiative, begun in 2010, is "a cross-border grassroots effort to establish the Great Lakes as a commons and legally protected bioregion" (Great Lakes Commons). One of its projects is the participatory development of an online map of the Great Lakes that links stories and crowdsourced information, creating a shared space for dialogue and exploration (Great Lakes Commons Map). The Great Lakes Commons Initiative is a collaborative, incubated project of On the Commons, a commons movement strategy centre founded in 2001 that publishes a magazine and online newsletter, and hosts a resource centre and network of commons animateurs.

The nonprofit Marine Conservation Institute brings together scientists, local conservation groups and activists, and governments to advocate for transboundary protection of oceans, and is working with government officials, activists, and conservation organizations to publicize and begin organizing a "Baja to Bering" ocean conservation corridor, including important offshore biological diversity conservation sites in the Pacific (Marine Conservation Institute).

GLOBAL COMMONS

The Sky Trust is a proposal to establish a governance structure to control and charge polluters for their atmospheric emissions. Proceeds would accrue to the Trust, which would use them for clean energy investments or dividends.

> Sky Trust … would encourage less pollution because it would reward the commons owners—all of us—for tough emission limits…. For decades we have been told that there are only two choices for the management of scarce resources: corporate self-seeking or the bureaucracy of the state. But there is another way. Commons management has worked for centuries and is still working today. It can be adapted to the most pressing global problems, such as climate change. A new phrase is about to enter the policy realm. To "market-based" and "command-and-control" we can now add "commons-based." (Rowe n.p.)

Creative Commons is a nonprofit organization based in Massachusetts that helps to distribute and manage shared creativity and knowledge. Their website says,

> The idea of universal access to research, education, and culture is made possible by the Internet, but our legal and social systems don't always allow that idea to be realized. Copyright was created long before the emergence of the Internet, and can make it hard to legally perform actions we take for granted on the network: copy, paste, edit source, and post to the Web. The default setting of copyright law requires all of these actions to have explicit permission, granted in advance, whether you're an artist, teacher, scientist, librarian, policymaker, or just a regular user. To achieve the vision of universal access, someone needed

to provide a free, public, and standardized infrastructure that creates a balance between the reality of the Internet and the reality of copyright laws. That someone is Creative Commons." (Creative Commons)

These very brief examples indicate, at different scales, how commons can be assembled, managed, enjoyed, and governed by groups of people using a combination of NGO, government, and private structures, rules, and incentives. Each is different, each has its own constituency and provides distinct services or generates value for its members or "commoners." When considered broadly, these benefits extend beyond the commoners to others in society, which is partly what motivates the commons' development and existence, and also shows why commons fill important gaps in state or private/market forms of governance.

In the next section, I explore some ecofeminist insights regarding the skills and social education that are needed to help commons grow and flourish.

COMMONS FOR CLIMATE JUSTICE

An ecofeminist methodology begins close to home, for both theory and activism; looks closely at the boundary between the paid and the unpaid, and at the relation between social and material value and political power; and finds strength, resilience, and sustainability in diversity.

Ecofeminists have a great deal of experience with the challenges of finding common cause, building movements, and overcoming barriers to inclusion. At the best of times, we do this by recognizing the importance of identity, welcoming diversity, listening to everyone's viewpoints, respecting diverse knowledges, recognizing commonalities that often appear and manifest themselves in unexpected ways, and building on strengths to create a strong political force.

It is exactly the social and economic assets that are most important for subsistence that remain commons (unprivatized) in most of the world (Mies and Bennholdt-Thomsen *The Subsistence Perspective*). These include water air, forests, and pastures in many places (sources of forage and biofuels), language, and many aspects of popular culture. For land and intellectual property, commons may be contested but by no means surrendered. The many advantages of collective interdependence, especially in times of heightened risk and uncertainty due to climate change, lead people to fall back on tested and familiar methods of mutual aid: culturally-reinforced commons governance. Global studies confirm that women are almost always the leaders, participants, and muscle behind environmental justice movements (Kirk; Kuester; Kurtz; Merino; Perkins; Verchick; Weiss). Feminist climate groups include the Women's Environmental Network, Idle No More, MADRE, Women's Earth and Climate Action Network, Via Campesina, Women's Environment and Development Network, Our Land Our Business, System Change Not Climate Change, and Gender CC—Women for Climate Justice (Awadalia).

Women's gendered social roles, economic positions, and expertise derived from paid and unpaid work responsibilities are logical reasons for their leadership. As a result, environmental and climate justice movements often employ organizing and activist techniques developed within the feminist movement, such as consciousness raising, unmasking patriarchy, and contextual reasoning—the grounding of the movement's theorizing in women's lived experiences rather than abstractions (Weiss). Moreover, environmental and climate justice activism changes the lives of the women involved and, by extension, other women, forcing them to confront the constraints they face—time, work and other opportunities, political agency, etc.—and thereby creating the conditions and potential for more radical change (Weiss 6). This seems to describe the process which has been playing out in Canada since the late 1990s, with a huge push from Indigenous women's grounded, culturally embodied activism (Nixon).

The "green transition" includes many examples: urban food provision (community and rooftop gardens, urban fruit harvesting, local and slow food movements, community shared agriculture, collective food box programs, etc.), bike and car sharing, co-operative housing, senior and child care, tool banks, skill share and repair workshops, freecycle goods exchanges, etc.

Without a centralized strategy or plan, people worldwide are creating collaborative ways of meeting their basic needs that are far closer to commons than to impersonal, marketed private property.

GENDER CONSIDERATIONS

Echoing Silvia Federici, the major ecofeminist commons theorist, Herbert Reid and Betsy Taylor explain how patriarchy and dualisms have been central to the enclosure of commons, both historically and currently (26-27, 84-85). They find hope, however, in the global justice movements:

Beneath the political and ideological turmoil, what must not be missed is that people from many diverse places and regions are seeking new ways to integrate nature, human sociability, and the creative arts. Out of a remarkably clear determination to reclaim to commons, they affirm the possibility of building new worlds. Body-place-commons is a radical theory of subjectivity as intersubjectivity. As such, one of its vital messages is that social hope and democratic change inhere in collective agency. (217-218).

Gender considerations permeate all the proposals and discussions related to building commons. If microcredit schemes are seen as a way to allow local communities to (re)gain control of communal assets, then women's access to microcredit becomes a key issue. When political capital is seen as the constraint on communizing resources, women's differential political and social capital, and the relationship between financial and other forms of capital, assume importance. Land ownership, where women often face extreme discrimination, is obviously a factor in establishing control of commons. The scale at which resource control is considered also has gender implications; women may have more or less political influence at different scales (Dolšak and Ostrom 337-357).

Even the themes that Dolšak and Ostrom generated through empirical research on commons governance challenges are a ripe terrain for gender analysis:

1.The increased interconnectedness of the biophysical world across scales and institutions across levels requires that adaptation to challenges occur at multiple levels.
2. The interests of resource users at these multiple levels are often in conflict.
3. Allocation of rights to resources (individual rights for privatization of a resource or community rights in the process of devolution) is a political process.
4. Access to this political process is limited by the structure of the macro institutions and also by the human, political, and social capital available to each group of actors.
5. More open political systems and more interconnected economies provide a larger set of adaptation strategies.
6. Adopted policy solutions are incremental and not linear. (338)

As ecofeminists well know, discriminatory institutions do not just disappear, and those interested in (re)building commons must critically engage with these institutions as part of the process of politically driven socio-economic change.

BUILDING COMMONS: EDUCATION, SKILLS, POLICIES

Elinor Ostrom's research has demonstrated that successful commons governance institutions share several characteristics:

- •they face uncertain and complex environments;
- •the local population is stable over long periods of time; people care about their reputations and expect their descendants to inherit the land;
- •norms have evolved which allow individuals to live in close interdependence with each other and the community is not severely divided;
- •the resource systems and institutions have persisted over time;
- •they are robust and sustainable.

Ostrom developed a set of "design principles" that help to account for the success of those commons governance institutions that have proven to work well:

- •clearly defined boundaries for the commons
- •congruence between appropriation and provision rules and local conditions (local appropriateness)
- •collective-choice arrangements (individuals can participate in modifying the rules)
- •monitoring of the rules by members takes place
- •there are graduated sanctions for violations of rules
- •rapid, low-cost conflict-resolution mechanisms exist
  rights to organize are recognized, at least minimally (outside
- •authorities do not challenge the rights of members to devise their own institutions);
  and, for larger systems:
- •there are multiple layers of nested enterprises which perform governance functions. (Ostrom *Governing* 89-90; see also Łapniewska 90).

Tiered and nested organizational layers exist in many co-operative federations and credit unions, as Jack Quarter et. al. note in their study of the social economy in Canada:

> The tiering arrangement represents a type of functional integration in which co-operatives with common needs co-operate with each other through an apex organization that helps them with their service provision. Often apex organizations serve as the voice of the sector (its members) to government, seeking to represent their needs. Sometimes they provide practical services to member organizations such as assistance with loans, loan guarantees, and information ... (or as) brokers for national and international markets ... (and) business associations. (67)

This shows how commons management is qualitatively different from both state/government organization and market rationality.

What are the attributes and skills required in the general populace for commons to be managed well, and for this paradigm and framework to spread? It should be obvious by now that I am not talking about a wholesale, sudden substitution of commons-type goods and service provision for everything done by the market; rather I see this as an inexorable progression where commons of various kinds gradually expand in the interstices and to meet the many gaps in the global and local economy, whenever (and exactly because) commons meet some needs better than any other provisioning system. It's possible to envision a nearly-infinite overlapping set of communications and governance structures covering all kinds of commons and groups of people, from watersheds, airsheds, agricultural areas, and political jurisdictions to epistemic commons, information commons, housing and shelter networks, community-shared agriculture and food box groups, arts and culture groups of all kinds, and other networks that create social, political, ecological, and economic communities. This addresses social and psychological needs for belonging which may be as important as material needs in keeping a socio-economy running well, reducing material throughput while maintaining health and well-being.

Ken Conca, in writing on how to nurture improved institutions for global water governance, states,

> Scholarship on the effective sustained management of common-property resources has shown the importance of institutions as second-order public goods that help to provide the underprovided

good of social co-operation. One obvious area in which such second-order public goods would facilitate the nurturing of institutions is resolution of environmental disputes.... The dispute-resolution approach could also be linked to growing interest in the idea of environmental peacemaking ... processes such as cooperative knowledge ventures and the emergence of regional-scale identities might help to transform situations of conflict and insecurity using environmental relationships as catalysts, with non-state channels as important venues. (384-5)

Bollier and Weston speak of innovations in law and policy being needed in three areas to foster commons governance: general internal governance principles and policies for commons, building on the work of Elinor Ostrom and the Indiana University Workshop in Political Theory and Policy Analysis where she carried out much of her research; macro-principles and policies that the state/market can embrace to develop commons and "peer governance"; and catalytic legal strategies to validate, protect, and support commons (Bollier and Weston 349). As examples, they cite conceptualizing commons as equal and legitimate partners with the state and the private sector—a triarchy of state/market/commons for governance options; adapting private contract and property law to protect commons, as in the GPL or General Property License which copyright owners can attach to software to assure that the code and any future modifications to it will be forever accessible to anyone to use, and the Global Innovation Commons, a huge international database of lapsed patents; "stakeholder trusts" to manage and lease ecological resources on behalf of common groups and distribute revenues to them, such as the Alaska Permanent Fund or a Sky Trust; re-localization and "transition towns" movements; Community Supported Agriculture and Slow Food movements assisted by government policies; expansion of the public trust doctrine of environmental law to include atmosphere and water; wikis and crowd-sourced platforms to include citizen experts in policymaking and enforcement, participatory environmental monitoring of water quality and biodiversity, etc. (Bollier and Weston 351).

Computer technologies, online organizing, and communications now allow people to create participatory communities and commons of many new kinds. According to legal scholar Beth S. Noveck, these forms of collective action are potentially vibrant and efficient, and should be recognized and encouraged in law by allowing legitimate, decentralized self-governance.

Carol Rose, in a classic 1986 paper, showed that the legal status of commons is well-represented and understood in modern Western legal traditions. Thus, recognition of the importance of commons has long existed in Western legal traditions as well as those that have resisted the colonial imposition of Western governance institutions, where commons, often protected through women's work and leadership, have safeguarded many communities' resilience in the face of capitalism and colonialism.

CONCLUSION

Scholarship and activism on commons of all kinds is growing exponentially. There are many ways that all of us can contribute, participate, and share our own skills and knowledge.

Ecofeminist and feminist political ecology theory and practice are consistent with building commons in many ways. Much of what communities are already doing in the face of climate change can be seen as advancing the development of the participatory, locally appropriate governance institutions that are working to protect commons. Here are a few examples:

•We can spread the knowledge of commons, the work of Ostrom and others, the importance of this "third way," not market, not state, but drawing from each—in our classes, consulting, government, and activist work.
•We can build the skills needed for sustainable commons governance at the local level—respectful communication, dispute resolution, shared provisioning, transmission of ecological knowledge and care—seeking inspiration in ecofeminist and Indigenous traditions
•We can conduct research analyzing perverse subsidies and barriers to commons governance models to provide policy advice, and work for the broader acceptance of commons as legitimate and valuable
•We can foster and demonstrate the use of discourse-based collective valuation processes to build local democracy and co-responsibility
•We can assist new co-ops and commons initiatives to support co-operative growth in all sectors.

An earlier version of this paper was presented at a special session on

"Gender and 'Global Ecological Crisis'" at the International Feminist Journal of Politics *conference, University of Southern California, Los Angeles, May 9-11, 2014.*

REFERENCES

Alam, Shawkat. "Collective Indigenous Rights and the Environment." *Routledge Handbook of International Environmental Law.* Eds. S. Alam, J. H. Bhuiyan, T. M. Chowdhury, and E. J. Techera. London: Routledge, 2012. 585-602.

Awadalia, Cristina, Piper Coutinho-Sledge, Alison Criscitiello, Julie Gorecki, and Sonalini Sapra. "Climate Change and Feminist Environmentalisms: Closing Remarks." *FeministWire* 1 May 2015. Web. Accessed 8 March 2019.

Anarres. Web. Accessed 6 Jan. 2014.

Bakker, Karen (2007). "The 'Commons' Versus the 'Commodity': Alter-Globalization, Anti-Privatization and the Human Right to Water in the Global South." *Antipode* 39.3 (2007): 430-455.

Black Creek Community Farm. 2013. Web.

Bollier, David and Silke Helfich, eds. *The Wealth of the Commons: A World Beyond Market and State.* Amherst, MA: The Commons Strategies Group / Levellers Press, 2012.

Bollier, David and Burns H. Weston. "Green Governance: Ecological Survival, Human Rights and the Law of the Commons." Eds. David Bollier and Silke Helfrich. *The Wealth of the Commons: A World Beyond Market and State.* Amherst, MA: The Commons Strategies Group / Levellers Press, 2012. 343-352.

Bond, Patrick (2009). "Water Rights, Commons and Advocacy Narratives." *South African Journal on Human Rights* 29.1 (2013): 125-143.

Brownhill, Leigh and Terisa Turner. "Commoners Against Climate Change." *Europe Solidaire sans Frontières* 25 February 2008 (n.p.). Web. Accessed 8 March 2019.

Canadian Co-operative Association (CCA). "Co-ops in Canada." Web. Accessed 8 March 2019.

Cato, Molly Scott. "The Earth is our Mother—What are the Economic Implications?" *Gaia Economics,* 2004. Web. Accessed 28 Oct. 2013.

Community Supported Agriculture. (n.d.). Web. Accessed 8 March 2019.

Conca, Ken. *Governing Water: Contentious Transnational Politics and Global Institution Building.* Cambridge, MA: MIT Press, 2006.

"Co-ops & Cooperation." PlanetFriendly.net (n.d.). Web.

Co-operative Housing Federation of Toronto. 2013. Web. Accessed 6 Jan. 2014.

Creative Commons. 2013. Web. Accessed 6 Jan.2014.

De Peuter, Freig and Nick Dyer-Witheford. "Commons and Cooperatives." *Affinities: A Journal of Radical Theory, Culture, and Action* 4.1 (Summer 2010): 30-56. Web. Accessed 8 March 2019.

Dolšak, Nives and Elinor Ostrom. The Commons in the New Millennium: Challenges and Adaptation. Cambridge, ma: mit Press, 2003.

The Ecologist (1992). Special issue: "Whose Common Future?" 22.4 (1992).

Federici, Silvia. *Caliban and the Witch: Women, the Body, and Primitive Accumulation.* Brooklyn, NY: Autonomedia, 2014.

Federici, Silvia. "Feminism and the Politics of the Commons." *The Commoner* 15 (2011): n.p. Web. Accessed 8 March 2019.

Giacomini, Terran. "Ecofeminism and System Change: Women on the Frontlines of the Struggle Against Fossil Capitalism and for the Solar Commons." *Canadian Woman Studies/les cahiers de la femme* 31.1/2, (2016): 95-101.

Great Lakes Commons. 2013. Web. Accessed 6 Jan. 2014.

Great Lakes Commons Map. 2013. Web. Accessed 6 Jan. 2014.

Hess, Charlotte. "Mapping the New Commons." Presented at the twelfth biennial conference of the International Association for the Study of the Commons, Cheltenham, UK, 14-18 July 2008. Web. Accessed 8 March 2019.

Hess, Charlotte and Elinor Ostrom. *Understanding Knowledge as Commons: From Theory to Practice.* Cambridge, MA: MIT Press, 2007.

Hyde, Lewis. *Common as Air: Revolution, Art and Ownership.* New York: Farrar, Straus & Giroux, 2010.

International Association for the Study of the Commons (IASC). Web.

Kirk, Gwyn. "Ecofeminism and Environmental Justice: Bridges Across Gender, Race, and Class." *Frontiers* 18.2 (1997): 2-20.

Kurtz, Hilda E. "Gender and Environmental Justice in Louisiana: Blurring the Boundaries of Public and Private Spheres." *Gender, Place & Culture: A Journal of Feminist Geography* 14.4 (2007): 409-426.

Łapniewska, Zofia. "Reading Elinor Ostrom through a Gender Perspective." *Feminist Economics* 22.4 (2016): 129-151.

Linebaugh, Peter. *The Magna Carta Manifesto: Liberties and commons for all.* Oakland: University of California Press, 2009.

Marine Conservation Institute. 2013. Web. Accessed 30 Oct. 2013.

Macpherson, Ian. "The History of the Canadian Co-operative Movement: A Summary, a Little Historiography, and Some Issues." Social Economy Hub. Web. Accessed 29 Oct. 2013.

Mellor, Mary. "Cooperative Principles for a Green Economy." *Capitalism Nature Socialism* 23.2 (2012): 108-110.

Merino, Jessic. "Women Speak: Bringing Gender to the Forefront in Environmental Justice." *Ms. Magazine* 1 November 2017. Web. Accessed 8 March 2019.

Mies, Maria and Veronika Bennholdt-Thomsen. *The Subsistence Perspective*. London: Zed Books, 1999.

Mies, Maria and Veronika Bennholdt-Thomsen. "Defending, Reclaiming and Reinventing the Commons." *Canadian Journal of Development Studies* 22.4 (2001): 998-1023.

Murota, Takeshi and Ken Takeshita. *Local Commons and Democratic Environmental Governance*. Tokyo/New York/Paris: United Nations University Press, 2013.

Nixon, Lindsay. "Ecofeminist Appropriations of Indigenous Feminisms and Environmental Violence." *Feminist Wire* 30 April 2015. Web. Accessed 8 March 2019.

Not Far from the Tree. 2013. Web. Accessed Oct. 31 2013.

Noveck, Beth Simone. "A democracy of groups." *First Monday* 10.11 (2005) (n.p.). Web. Accessed 8 March 2019.

On the Commons. Web. Accessed 28 Oct. 2013.

Ostrom, Elinor. *Governing the Commons: The Evolution of Institutions for Collective Action*. New York: Cambridge University Press, 1990.

Ostrom, Elinor. "Beyond Markets and States: Polycentric Governance of Complex Economic Systems." Nobel Economics Prize lecture, December 8, 2009. Web. Accessed Oct. 28 2013.

Ostrom, Elinor. *The Future of the Commons: Beyond Market Failure and Government Regulation*. London: Institute of Economic Affairs, 2012.

Perkins, Patricia E. "Environmental Activism and Gender." *Gender and Economic Life*. Eds. Deborah Figart and Tonia Warnecke. Northampton, MA: Edward Elgar, 2013. 504-521.

Perkins, Patricia E. "Canadian Indigenous Female Leadership and Political Agency on Climate Change." *Climate Change and Gender in Rich Countries: Work, Public Policy and Action*. Ed. Marjorie Griffin Cohen. London: Routledge, 2017. 282-296.

Quarter, Jack, Ann Armstrong, and Laurie Mook, Eds. *Understanding the Social Economy: A Canadian Perspective*. Toronto: University of

Toronto Press, 2009.

Reid, Herbert and Betsy Taylor. *Recovering the Commons: Democracy, Place, and Global Justice.* Urbana/Chicago: University of Illinois Press, 2010.

Ricoveri, Giovanna. *Nature for Sale: The Commons Versus Commodities.* London: Pluto Press, 2013.

Rose, Carol. "The Comedy of the Commons: Commerce, custom, and inherently public property." *University of Chicago Law Review* 53.3 (1986): 711-781. Web.

Rowe, Jonathan. "The Parallel Economy of the Commons." *State of the World Report,* 2008. Web. Accessed Oct. 29 2013.

Shimada, Daisaku. "How Can Societies Create Common Access to Nature? The Roots and Development Process of the Bruce Trail, a Canadian Case Study." Post-doctoral research paper. Toronto: York University Faculty of Environmental Studies, 2010.

Slaughter, Graham. "Jane-Finch farm rakes in first harvest." *The Star* 16 October 2013. Web. Accessed 6 January 2014.

Suga, Yutaka "The Tragedy of the Conceptual Expansion of the Commons." *Local Commons and Democratic Environmental Governance.* Eds. Takeshi Murota and Ken Takeshita. Tokyo/New York/Paris: United Nations University Press, 2013. 3-18.

Thompson, Edward P. *Customs in Common.* New York: The New Press, 1993.

Turner, Terisa E. and Leigh S. Brownhill. "Gender, Feminism and the Civil Commons: Women and the Anti-Corporate, Anti-War Movement for Globalization from Below." *Canadian Journal of Development Studies* 22.4 (2001): 805-818.

Xanthaki, Alexandra. *Indigenous Rights and United Nations Standards: Self-determination, Culture and Land.* Cambridge: Cambridge University Press, 2007.

"Urban farm near Jane and Finch to receive $400,000 grant." CBC *News* 16 October 2016. Web. Accessed 6 January 2014.

Verchick, Robert R. M. "In a Greener Voice: Feminist Theory and Environmental Justice." *Harvard Women's Law Journal* (Spring 1996): 23-88. Web. Accessed 8 March 2019.

Weiss, Cathy (2012). "Women and Environmental Justice: A Literature Review." Women's Health in the North (WHIN), Australia. Web. Accessed 20 September 2015).

Yes in My Back Yard. 2013. Web. Accessed 6 Jan. 2014.

6.
Finite Disappointments or Infinite Hope?

Working Through Tensions Within Transnational Feminist Movements

DOROTHY ATTAKORA-GYAN

IN MANY COUNTRIES, WOMEN ARE FIRST to experience an increased workload, health problems, and other damaging effects associated with some of the negative impacts of globalization, deterioration of agriculture, economic instability, and migration (Van Esterick; Horvorka; DeZeeuw and Njenga; Patil, Balakrishnan and Narayan; Perry). From food production and acquisition to food processing and preparation, as well as serving food both within the home and for the public, we know that women play a major role in feeding their families and communities worldwide (Van Esterick; Pandey; Horvorka, DeZeeuw and Njenga; Desmarais; Sachs and Alston; Shiva). In this way, food tends to shape, reflect, and mirror much of human nature and values (Van Esterick). Yet the study of food generally gets relegated to disciplines such as nutrition, economics, and agronomy, which according to Penny Van Esterick are disciplines guided by rules of hard science. Activists, artists, and scholars such as Wilbur and Keene, DeZeeuw and Njenga, Abarca, Sachs, Brandt, and Horvorka continue to redefine the terms under which globalized food systems work. I join these individuals in calling for a redefinition of food systems that take complex power circuits into account. I ask for us to reassess hegemonic food systems to question how power is reinforced, as well as disrupted across dynamic axes of difference.

This paper addresses some identified challenges in working across differences within transnational feminist organizing, with a particular focus on the marginalization and invisibilization of rural, peasant, and Indigenous women. I look specifically to examples of transnational feminists involved in organizing and mobilizing around issues of food sovereignty in North and Latin America, as well as parts of Africa and South Asia.

Despite evidence that shows rural, peasant, and Indigenous women have long organized and advocated for change—transforming lives and policies in their communities—their voices are still largely marginalized. Mainstream constructions of these women often romanticize them and their work (Sachs; Agarwal). These assumptions negatively influence how these women are perceived. It effects how they continue to be left out of decisions (and decision-making positions) that impact their day-to-day lives. For this reason, they continue to be relegated to the margins of society and are still erased from important spaces and dialogues that shape their future. Such tensions expose power differentials that exist both outside *and* inside food sovereignty movements, essentially posing a challenge to grassroots resistance movements. In this sense, transnational feminist networks are an indication of the dynamic nature of feminist solidarity. To conclude, I use Inuit symbolism vis-à-vis Peggy Antrobus to explore the concept of feminist solidarity as a spiral, a model taken from Antrobus. I do so to make the argument that feminists must always be attentive to assessing their own internal incongruencies that give way to power imbalances that mirror the very systems being dismantled. All while prioritizing important work to challenge ideologies, policies, and institutional barriers that infringe on human rights, specifically in this chapter, as it pertains to food.

CHALLENGES OF WORKING ACROSS DIFFERENCE IN TRANSNATIONAL FEMINIST ORGANIZING

The body of literature on transnational feminism illustrates a field that is contested, in flux, and constantly evolving and shifting (Mohanty). Transnational feminist practices, "while they connect collectives located in more than one national territory, also embody specific social relations established between specific people, situated in unequivocal localities, at historically determined times" (Mahler 444). Contemporary understandings of transnational feminism developed as a response to exclusionary practices within feminist and development discourses, globalization, and tensions within NGOs (Mohanty; Mahler and Pessar; Grewal and Kaplan; Chowdhury; Dufour, Masson and Caouette; Patil; Bachetta; Alvaraz; Blackwell; Conway; Dempsey, Parker and Krone; Razack; Hawkesworth). Transnational practices vary and include organizations, networks, individuals and collectives, local and national movements, and international NGOs that work towards addressing gender and feminist issues that intersect with other issues (Dufour, Masson, and Caouette; Patil; Vargas; Alvarez; Moghadam; Basu; Hewitt). Transnational processes are also anchored in *and* transcend more than one nation-state (Mahler and Pessar). This chapter takes interest in transnational feminist circuits that operate under or are mandated by food sovereignty principles.

Neoliberal policies in the past such as Structural Adjustment Programs (SAP) in the 1980s and 1990s restructured many economies around the globe (Deere). SAP-impacted countries in parts of Latin America, Asia, and Africa (Deere; Sachs; Perry). In one Latin American case, rural poverty rose to 65 percent in the 1980s, gradually decreasing to 59 percent by 2005 (Deere 6). In other cases, it threatened the autonomy and livelihood of rural, peasant, and Indigenous women, their families, and communities (Desmarais; Sachs; Patel; Deere). In this way, food sovereignty was a direct response to free market economies, organizations, and governments that supported SAPs (Desmarais; Sachs; Dufour, Masson and Caouette). Such discrepancies led to an uprising of rural, urban, and peri-urban women on the ground (Deere; Desmarais; Brownhill and Turner; Brownhill "The Struggle for Land"; Patel). In Latin America, transnational social movement organizations such as Latin American Coordinator of Rural Organizations (CLOC), the Continental Coordination of Indigenous Nationalities of People of Abya Yala, and La Via Campesina all paved the way for rural and peasant organizing, expanding current definitions of food sovereignty (Deere). La Via Campesina—the "international association that brings together organizations of small and medium farmers, agricultural workers, rural women, and indigenous peoples from four continent" (Deere 3)—has played a monumental role in the emergence of food sovereignty and some of the different ways it has come to be understood since the World Food Summit in 1996 (McMichael; Desmarais; Deere and Royce; Patel; Martínez-Torres and Rosset; Holt-Gimenez).

Phillip McMichael asserts that food sovereignty as a term emerged in the 1980s, and the popularization of it as a practice gained traction by the 1990s. These movements were largely driven by Indigenous women in North America, rural and peasant women in Latin America, and Indigenous, black, and other women of colour in the Global South (Desmarais; McMichael; Sachs; Deere). Food sovereignty is a sociopolitical movement that according to La Via Campesina believes that "every country and people must have the right and ability to define their own food, farming and agricultural policies ... the right to protect domestic markets" (Martínez- Torres and Rosset 160). This is not to be mistaken with food security. Food security stresses the ability to "acquire safe, nutritiously adequate, and culturally acceptable foods in a manner that maintains human dignity" (Van Esterik 227). Food sovereignty on the other hand, stresses the importance of individual and community access to self-determination over their own food, including how, what, where, and who grows or produces it (Wilbur and Keene; Brownhill; Brownhill and Turner; Mcmichael; Desmarais; Sachs). It emphasizes individual and community agency, while placing the voices of peasant and indigenous rural farmers and urban agriculturalists at the forefront of grassroots movements (Wilbur and Keene; Nabulo, Kiguli, and Kiguli; Brownhill; Desmarais; Perry).

DECENTERING POWER(S) IN (AND THROUGH) TRANSNATIONAL FEMINIST ORGANIZING AND BEYOND

Due to stereotyping and assumptions, a feminists' social location can shape how they are perceived as a feminist. Take stereotypical misconceptions of rural and peasant women that typically frame them as illiterate helpless recipients of aid, assistance, and knowledge production imported from the West (Mohanty; Sachs; Deere; Agarwal). Like Indigenous women, rural and peasant women are dislodged into some fixed mythical past that keeps them and their work static (Agarwal; Sachs). Not only must rural women always be made to disappear, Robyn Dallow states that "rural life and some degree of geographical isolation go hand- in-hand" (4). What this means is that rural women are constantly relegated to the margins and outskirts geographically, metaphorically, and literally (Dallow). They are isolated and rendered invisible. Romanticization of rural women legitimizes how power unfolds and flows in unequal ways, often outside of their favour (Sachs). We can think of rural and peasant women as facing multiple threats from across various spaces. They face challenges within their family homes and community; from the larger structural and systemic institutions, policies, and processes; as well as from within feminist movements themselves.

Economists, politicians, decision-makers, as well as men and women in positions of power and privilege tend to provide partial accounts on peasant women's situated perspectives (Sachs). Their activism and community work often goes unnoticed and unappreciated. Instead, they are taken for granted and their labour viewed as an "expected part" of their "natural roles" (Dallow). In addition, Peter Rosset contends that rural women's voices are traditionally excluded from social, economic and political power. They continue to face sexist and misogynist attacks, violence, and assault (Sachs; Sachs and Alston; Hovorka, De Zeeuw and Njenga). These local farmers are often unable to compete with larger multimillion-dollar corporations who lead the way in a capitalist economy (Desmarais; Abarca; Horvorka "Urban Agriculture"). In many places they still play a larger role in subsistence farming than in comparison to their male counterparts who dominate commercial market farming (Horvorka "Urban Agriculture"). Furthermore, food sovereignty movements can also tell us something about the tensions within transnational feminist movements (Heng). Growing debates over questions of power, privilege, and representation has shown that at times, feminists—both academics and activists—may reinforce the very power imbalances feminism seeks to dismantle. What transnational feminism has meant for some women is what Mohanty identified as the reality of international, global processes and organizing, heavily rooted in Western discourses privileging the West (see also, Razack). The idea that only certain issues are taken up in the Global North and eventually travel to the Global South has been a myth that has caused tension within transnational feminist networks (Mohanty). This trend also emerges in food sovereignty movements. For instance, while rural and peasant women have long been resisting and organizing (despite limited dialogue around their agency), some Western scholars in positions of power still render rural women's narratives as alternative or mimicry (Mohanty). And yet, this could not be further from the truth.

In *Gendered Fields: Rural Women, Agriculture and Environment*, Carolyn Sachs urges scholars to understand the daily lives of rural and peasant women, and to highlight their situated knowledges. Sachs argues that this affirms the role of indigenous, rural, and peasant women as knowledgeable plant gatherers and early inventors of horticulture who have long studied plants and crops (See also, Shiva). Understanding these women as historically *already* knowledgeable inventors and scientists (See also, Shiva), Sachs helps shift our understanding of women in the Global South who are depicted as needing to be conquered and saved by the West (See also, Mohanty; Shiva). Instead, Sachs provides an account of rural women and their agency that disrupts mainstream conceptualizations of them as solely recipients of knowledge imported from the West. Drawing from Sachs, I want to highlight some counternarratives that illustrate some of the different ways indigenous, rural, and peasant women have been transforming their lives, as well as those in their communities.

I begin with the role Indigenous/First Nations/Aboriginal, Métis, and Inuit people of North America have played in resisting dominant (read: settler) food systems. On *All My Relations,* a podcast hosted by Swinomish and Tulalip visual storyteller Matika Wilbur and Cherokee activist scholar Adrienne Keene, the pair interview Valerie Segrest, a Native nutrition educator and member of the Muckleshoot Tribal Nation who works as a coordinator for the Muckleshoot Food Sovereignty Project (Wilbur and Keene). During the show, Segrest expands the definition of food sovereignty using Indigenous frameworks she was raised with. To do this, the educator identifies what sovereignty has meant for her and her community, historically and in a contemporary setting. Segrest locates sovereignty as "the right to self- determination," something Elders in her community introduced her to. It is important to note that in the late 1980s, organizations such as the First Latin American Meeting of Peasants and Indigenous Organizations took up demands for indigenous autonomy and self- determination as it related to their people and food systems (Deere). For Segrest, food sovereignty is, "a way of healing, [it is] the remedy ... something to be kept alive in everyday actions such as fishing, drinking tea, or standing up for water rights." The activist later connects the concept of food sovereignty with her ancestral past, the present, and the future, tying ancient Indigenous teachings with present-day struggles. For example, Segrest shares how her people's relationship with camas—a bulb that grows into a blue flower of the asparagus family—has changed due to settler colonialism, which among many things has brought about a decrease in knowledge about this Native starch. According to Segrest, "less than four percent of camas prairies are intact" today. Part of enacting food sovereignty for Segrest is in trying to revitalize camas, ensuring that her community and future generations to come have access to what has always been a staple in her community.

Similarly, Sachs provides an account of North American native women in a unique way that connects their stories with other rural and peasant women globally. Sachs historicizes the genealogy of food movements weaving in three crops that play an important role for indigenous, rural, and peasant women: corn, rice, and coffee. For example, Sachs documents the sacred role corn plays in North American native culture and connects it with women in parts of Africa who also depend on it for sustenance and have a similar relationship with maize. From North America to Africa, Sachs illustrates how the brutal atrocities of colonialism and globalization have disrupted these different communities' relationship with crops they are dependent on (See also, Shiva; Wilbur and Keene). I find Sachs and Segrest relevant because both contribute accounts of food systems and food sovereignty in ways that adds a nuanced layer, one that speaks to Indigenous knowledges and teachings that celebrate the land and the harvest that it provides us.

Furthermore, in some Latin American countries such as Peru, Bolivia, Mexico, and Venezuela, Indigenous and peasant women there have taken their cooking skills and transformed them into "commercialized housework" in order to support their families (Abarca 94). Community kitchens, also known as public kitchens, are complex spaces with the potential to both empower and subjugate women. Meredith Abarca states that public kitchens are grounded in three basic principles: "a space to listen to the voices of traditionally muted people; to recognize the validity of different fields of knowledge; [and to] build on trust which means keeping ourselves honest" (106). Kathleen Shroeder highlights how public kitchens, beyond being safe spaces for women to gather, are prominent places in the community and a source of civic pride. Meredith E. Abarca writes of growing up in Mexico with a mother and grandmother who both owned a public kitchen. Abarca speaks of how these spaces provided more than economic stability for women in her community. For Abarca, they played a social role and were an opportunity for women and local activists to convene and organize. In some cases, they also functioned as a meeting hall for local politics, as well as provided a daycare, playground, and school for children like herself (Abarca). Abarca and Schroeder suggest that public kitchens work as a social device and have allowed women the opportunity to create connections with others, take up leadership positions, find love in some cases, and even explore their creative sides. Other counternarratives draw our attention to the role of peasant women in Africa and South Asia.

While much is known about the Kenyan Green Belt Movement in Kenya, an initiative championed by environmental activist Wangari Mathaai (see Brownhill and Turner; Mawdsley), little is disseminated about rural women pushed to urban or peri-urban settings called to find creative ways to produce sustainable food for their families and the challenges that come with doing so. Some rural and peasant women find themselves in urban centres due to difficulty finding work in rural agriculture (Nabulo, Kiguli, and Kiguli; Hovorka, DeZeeuw and Njenga). Many arrive at urban locations cash poor and houseless (Nabulo, Kiguli, and Kiguli; Hovorka, DeZeeuw and Njenga). Experiences with sexism when seeking land drives these women to borrow and search for free unused plots in different, often dangerous neighbourhoods, garbage dumps, or undeveloped lands in valleys such as deserted or vacant lands to grow their crops (Nabulo, Kiguli, and Kiguli; Sachs; Sachs and Alston; Hovorka, De Zeeuw and Njenga). In these cases, crops are cultivated in people's compounds, along roads, railways, and under power lines; livestock are kept in compounds and slums, or on vacant lands, while others wander freely (Nabulo, Kiguli, and Kiguli; Sachs; Sachs and Alston; Hovorka, DeZeeuw and Njenga). Other locations where women grow their crops include road reserves; drainage channels; wetlands; contaminated scrap yards; dumping sites for solid and liquid waste; vacated industrial areas; family gardens; gardens belonging to community kitchens; community gardens; school gardens; gardens located on private lands and communal areas; public lands; and institutional lands (Nabulo, Kiguli, and Kiguli; Sachs; Sachs and Alston; Hovorka, DeZeeuw and Njenga). Many of these locations come with limited access to clean water and lack electricity (Nabulo, Kiguli, and Kiguli; Sachs; Sachs and Alston; Hovorka, DeZeeuw and Njenga). Simply put, lack of ownership and access to land drives women to grow crops in unsafe areas (Nabulo, Kiguli, and Kiguli). And yet, regardless the risks and harms, urban agriculture has meant that more women in Kenya and Uganda for example, have been able to feed their families, grow medicinal herbs, and save money by avoiding larger market value products (Nabulo, Kiguli, and Kiguli; Sachs; Sachs and Alston; Hovorka, DeZeeuw and Njenga). Shifting out of Africa, other scholars direct our attention to India and the Chipko Movement.

According to Mina Roces, women in Asia have been organizing since as far back as the 1920s and 1930s (7), if not further. Of relevance here are the works of Anupam Pandey, Vandana Shiva, and Emma Mawdsley who add to the literature on women in the forests of the Uttaranchal Himalaya and their involvement in the Chipko movement. During the Chipko Movement in India, Garhwali women revitalized the ancient practice of farming (Pandey 351). It is important to note that the forest in this area provides a means for sustainable food production systems in the form of nutrients and water (Shiva 59). While the Chipko Movement is primarily known as a grassroots resistance movement, I am interested in women in this region as knowledge keepers, protecting not only the forest and its food, but ancestral teachings about the plants in the area that are threatened by cultural genocide. According to Shiva, rural women in the region have long been knowledge producers and continue to hold on to ancestral knowledge that scientists, pharmaceutical companies, and food corporations are unaware of and seek out. Bina Agarwal goes on to state that while rural women are by no means the sole repositories of this knowledge, they are significant bearers of information regarding the plants they collect or use. Additionally, the information they hold about local trees, grasses, and food-related forest produce is severely threatened under food shortage conditions (Agarwal 58; see also, Shiva).

Other challenges these women face include pharmaceutical companies that have threatened their knowledge by introducing patents, as well as an inability to retain the knowledge by passing it down to younger generations, an act of resistance that provides an outlet for survival and sustainability (Shiva).

Inserting narratives of rural women turned urban dwellers in Kampala, Uganda, or the knowledge keepers involved in the Chipko movement in India, as well as Sachs tracing of crops to map out a genealogy of food sovereignty is important in redefining food systems. Such inclusions of feminists not typically centered- those from North American indigenous communities to Indigenous and peasant women in Latin America, Asia, and Africa- these stories are important. Especially so if we are to build solidarity across all of our differences. Returning to Sachs, we are reminded of the valuable role these women have/ and continue to play in shaping how food sovereignty is both defined and practiced. Such accounts shift how we view rural women within agriculture, for instance, as knowledge producers and horticulturalists. It grants indigenous, peasant, and rural women the necessary agency they wield as experts *already* working in careful and complex ways to heal and feed their families and communities. In addition to organizing and mobilizing, they are leaders in their communities rising up against governments and corporations. Presenting their work disrupts the idea that feminism flows from North to South. It asks those outside these communities to confront assumptions that inform how they perceive those inside them.

CONCLUSION

We must always assess where and how we can shift how we conceptualize feminism to include models that create and offer space to those not typically centered. Peggy Antrobus highlights that within Inuit tradition, storytelling takes the form of a spiral. She challenges feminists to perceive the transnational feminist movements as a spiral:

A spiral is open ended, continuous, ever enlarging our understanding of events, our perspectives. The global women's movement can be thought of as a spiral, a process that starts at the centre (rather than at the beginning of the line) and works its way outwards, turning, arriving at what might appear to be the same point, but in reality, at an expanded understanding of the same event. A spiral is dialective, allowing for the organic

growth of a movement of women organizing—a movement in
a state of on-going evolution as consciousness expands in the
process of exchanges between women, taking us backwards
(to rethink and reevaluate old positions) and forwards (to new
areas of awareness). (21)

While tensions will always emerge where people gather to organize,
viewing transnational feminist solidarity as a spiral helps us to visualize
the ebbs and flows that come with the work. It stands as a metaphor
for how we negotiate power that circulates in our movement, as well as
in and between us. Understanding complex power dynamics sometimes
means acknowledging that critical self- reflection is just as important
as mainstreaming gender in urban agriculture efforts (Hovorka) or
ensuring that good policy interventions are put in place (Nabulo, Kiguli,
and Kiguli). It suggests that we have to look both within *and* without
if we are to make transformative change. The future of transnational
solidarities, according to Paola Bacchetta depends largely on "a
continued ability to self-critique and a mutual will to avoid bulldozing,
effacing, distorting and excluding" (970). Similarly, Paola Moya argues
that any attempts to work with others across difference requires that
we examine ourselves, including the parts of ourselves that we are not
always proud of. Moya quotes Chicana feminist writer Cherríe Moraga,
who eloquently and beautifully articulates,

Because the source of oppression forms not only our radicalism,
but also our pain, to do the kind of work self-examination
requires us to do [like] admit how deeply "the man's" words
have been ingrained in us. The project of examining our own
locations within the relations of domination becomes even
riskier when we realize that doing so might mean giving up
whatever privileges we have managed to squeeze out of this
society by virtue of our own social locations. We are afraid to
admit that we have benefitted from the oppression of others. We
fear the immobilizations threatened by our own incipient guilt.
We fear we might have to change our lives once we have seen
ourselves in the bodies of the people we have called different.
We fear the hatred, anger, and vengeance of those we have hurt.
(Moraga qtd. in Moya 150)

As Moya asserts, we have to also be willing to change ourselves. This
means using whatever privileges we have to leverage and highlight

indigenous, rural, and peasant women's voices, specifically as it pertains to food systems (Mohanty; Antrobus; Wilbur and Keene). It requires not only valuing these narratives but implementing them in our work and lives in order to shape feminist futures that are anti-and decolonial. Feminist solidarity as a spiral (Antrobus) means pushing beyond equality towards equity, and away from sameness towards fairness (Hovorka). It requires an acceptance that not all voices are equal or even the same, nor do they speak in unison (Razack; Grewal and Kaplan; Mohanty; Lock Swarr and Nagar). Solidarity as a spiral necessitates mutual stretching that requires an allocation of funds for those who travel long distance or require childcare (Desmarais). It necessarily means we advocate against travel bans that discriminate against certain religions, countries, or populations. It means being aware of how power differentials, unequal access to financial and cultural capital, as well as an imbalance in privilege can make a difference in how rural and peasant women engage in their work (Desmarais). A spiral model of feminist solidarity entails embracing various types of feminist organizing. That some women "organize vis-à-vis non-violent means that are grounded in concrete demands, seeking recognition of basic humans and social rights" (Rosset 7). While others are concerned with "defending rural livelihood and the development of sustainable livelihoods that respect nature and traditional knowledge" (Rosset 7). By shifting the narrative and giving space for alternative ways of knowing and being, unequal hierarchies of power slowly dissolve. Understanding the daily lives of rural women demystifies our understandings of who they are and how they work (Desmarais). Their situated knowledge and experiences open us all up to new ways of building solidarity with one another.

REFERENCES

Abarca, Meredith E. "Chalas Culina-ries: Mexican Women Speak From their Public Kitchens." *Taking Food Public: Redefining Foodways in a Changing World*. Eds. Psyche Williams and Carole Counihan. New York: Routledge, 2012. 94-109.

Agarwal, Bina. "Environmental Management, Equity and Eco-feminism: Debating India's Experience." *Journal of Peasant Studies* 25.4 (1998): 55-95.

Alexander, M. Jacqui and Chandra Talpade Mohanty. "Cartographies of Knowledge and Power: Transnational Feminism in Radical Praxis." *Feminist Genealogies, Colonial Legacies, Democratic Futures*. Eds. M. Jacqui Alexander and Chandra Talpade Mohanty. New York:

Routledge, 1997. 43-84.

Alexander, M. Jacqui and Chandra Talpade Mohanty. "Introduction: Genealogies, Legacies, Movements." *Feminist Genealogies, Colonial Legacies, Democratic Futures*. Eds. M. Jacqui Alexander and Chandra Talpade Mohanty. New York: Routledge, 1997. xiii-xlii.

Alvarez, Sonia E. "Feminist Politics of Translation in the Latin/A Américas." *Translocalities/Trans-localidades: Feminist Politics of Translation in the Latin/A Américas*. Eds. Sonia E. Alvarez and Claudia de Lima Costa. Durham, NC: Duke University Press, 2014.

Alvarez, Sonia E. "Introduction to the Project and the Volume: Enacting A Translocal Feminist Politics of Translation." *Translocalities/Trans-localidades: Feminist Politics of Translation in the Latin/A Américas*. Eds. Sonia E. Alvarez and Claudia de Lima Costa. Durham, NC: Duke University Press, 2014. 1-17.

Alvarez, Sonia E. "Translating the Global: Effects of Transnational Organizing on Local Feminist Discourses and Practices in Latin America." *Meridians: Feminism, Race, Transnationalism* 1.1 (2000): 29-67.

Antrobus, Peggy. *The Global Women's Movement. Origins, Issues, and Strategies*. London: Zed Books, 2004.

Bachetta, Paola. "Re-Scaling Trans/national 'Queerdom': 1980s Lesbian and 'Lesbian' Identitary Positionalities in Delhi." *Antipode* 34 (2002): 947-973.

Basu, Amrita. "Women's Movements and the Challenge of Transnationalism." *Encyclopedia of Women's Studies*, vol. 1, *Women's Movements*. Ed. Subhadra Channa. New Delhi: Cosmo Publications. 259-66.

Blackwell, Maylei. "Weaving in the Spaces: Transnational Indigenous Women's Organizing and the Politics of Scale." *Dissident Women. Gender and Cultural Politics in Chiapas*. Ed. R. Aida Hernandez et al. Austin: University of Texas Press, 2006. 115-154.

Brownhill, Leigh. "Gendered Struggles for the Commons: Food Sovereignty, Tree-Planting and Climate Change." *Women and Environments International: Special Issue on Women and Global Climate Change* 74/75 (2007): 34-37.

Brownhill, Leigh and Terisa E. Turner. "The Struggle for Land and Food Sovereignty: Feminism in the Mau Mau Resurgence." *Globalization and Third World Women: Exploitation, Coping and Resistance*. Eds. Ligaya Lindio-McGovern and Isidor Wallimann. New York: Ashgate Publishing Ltd., 2011. 81-105.

Brownhill, Leigh. S., Wahu M. Kaara and Terisa E. Turner. "Gender

Relations and Sustainable Agriculture: Rural Women's Resistance to Structural Adjustment in Kenya." *Canadian Woman Studies/les cahiers de la femme* 17.2 (1997): 40- 44.

Chowdhury, Elora Halim. "Locating Global Feminisms Elsewhere: Braiding U.S. Women of Color and Transnational Feminism." *Cultural Dynamics* 21.1 (2009): 51-78.

Conway, Janet M. "Geographies of Transnational Feminisms: The Politics of Place and Scale in the World March of Women." *Social Politics* 15.2 (2008): 207-231.

Conway, Janet M. "Transnational Feminisms Building Anti-Globalization Solidarities." *Globalizations* 9.3 (2012): 367-381.

Dallow, Robyn. "The Role of Women in Agriculture and Rural Settings." *Australian Journal of Rural Health* 1.1 (1992): 3-10.

Deere, Carmen Diana. "Women's Land Rights and Rural Social Movements in the Brazilian Agrarian Reform." *Journal of Agrarian Change* 3.1,2 (2003): 257- 288.

Deere, Carmen and Frederick Royce, Eds. *Rural Social Movements in Latin America: Organizing for Sustainable Livelihoods.* Gainesville: University Press of Florida, 2009.

Dempsey, Sarah E., Patricia S. Parker and Kathleen J. Krone. "Navigating Socio-Spatial Difference, Constructing Counter-Space: Insights from Transnational Feminist Praxis." *Journal of International and Intercultural Communication* 4.3 (2011): 201-220.

Desmarais, A. "Chapter 6: Co-operation, Collaboration and Community." *La Via Campesina: Globalization and the Power of Peasants.* Black Point, NS: Fernwood Publishing, 2007. 161-189.

Dufour, Pascale, Dominique Masson and Dominique Caouette, Eds. *Solidarities Beyond Borders: Transnationalizing Women's Movements.* Vancouver: University of British Columbia Press, 2010.

Grewal, Inderpal and Caren Kaplan. "Transnational Feminist Cultural Studies: Beyond the Marxism/ Poststructuralism/ Feminism Divides." *Scattered Hegemonies: Postmodernity and Transnational Feminist Practices.* Eds. Indepal Grewal and Caren Kaplan. Minneapolis: University of Minnesota Press, 1994.

Hawkesworth, Mary E. *Globalization and Feminist Activism.* Lanham: Rowman and Littlefield, 2006.

Heng, Geraldine. "A Great Way to Fly? Nationalism, the State, and the Varieties of Third World Feminism." *Feminist Genealogies, Colonial Legacies, Democratic Futures.* Eds. M. Jacqui Alexander and Chandra Talpade Mohanty. New York: Routledge, 1997. 30-45.

Hewitt, Lyndi. "Framing Across Differences, Building Solidarities:

Lessons from Women's Rights Activism in Transnational Spaces." *Interface: a Journal For and About Social Movements* 3.2 (2011): 65-99.

Horvorka, Alicia. "Gender in Urban Agriculture: An Introduction." *Women Feeding Cities: Mainstreaming Gender in Urban Agriculture and Food Security*. Eds. Alicia Hovorka, Henk De Zeeuw, and Mary Njenga. Warwickshire: Practical Action Publications, 2009. 1- 32

Horvorka, A. "The No. 1 Ladies' Poultry Farm: A Feminist Political Ecology of Urban Agriculture in Botswana." *Gender, Place & Culture: A Journal of Feminist Geography* 13.3 (2006): 207-225.

Hovorka, A. "Urban Agriculture: Addressing Practical and Strategic Gender Needs Development." *Practice* 16.1 (2006): 51-61.

Hovorka, Alica, Henk De Zeeuw and Mary Njenga, Eds. *Women Feeding Cities: Mainstreaming Gender in Urban Agriculture and Food Security*. Warwickshire: Practical Action Publications, 2009.

Ishani, Zarina. "Key Gender Issues in Urban Livestock Keeping and Food Security in Kisumu, Kenya." *Women Feeding Cities: Mainstreaming Gender in Urban Agriculture and Food Security*. Eds. Alicia Hovorka, Henk De Zeeuw, and Mary Njenga. Warwickshire: Practical Action Publications, 2009. 105- 21.

Lock Swarr, Amanda and Richa Nagar. "Continuing Conversations: Critical Transnational Feminist Praxis Contributors." *Critical Transnational Feminist Praxis*. Eds. Amanda Lock Swarr and Richa Nagar. Albany: State University of New York Press, 2010.

Mahler, Sarah and Patricia Pessar. "Gendered Geographies of Power: Analyzing Gender Across Transnational Spaces." *Identities* 7.4 (2001): 441-459.

Martínez- Torres, María Elena and Peter M. Rosset. "La Vía Campesina: The Birth and Evolution of a Transnational Social Movement." *The Journal of Peasant Studies* 37.1 (2010): 149-175.

Mawdsley, Emma. "After Chipko: From Environment to Region in Uttaranchal." *Journal of Peasant Studies* 25.4 (1998): 36-54.

McMichael, Phillip. "Historicizing Food Sovereignty." *The Journal of Peasant Studies* 41.1 (2014): 17-41.

Mendoza, Breny. "Transnational Feminisms in Question." *Feminist Theory* 3.3 (2002): 295-314.

Moghadam, Valentine M. *Globalizing Women: Transnational Feminist Networks*. Baltimore: John Hopkins University Press, 2005.

Mohanty, Chandra Talpade. "Under Western Eyes Revisited: Feminist Solidarity Through Anticapita-list Struggles." *Feminism Without Borders: Decolonizing Theory, Practicing Solidarity*. Durham: Duke

University Press, 2003.

Moya, Paula M. L. "Postmodernism, 'Realism' and the Politics of Identity: Cherrie Moraga and Chicana Feminism." *Feminist Genealogies, Colonial Legacies, Democratic Futures*. Eds. M. Jacqui Alexander and Chandra Talpade Mohanty. New York: Routledge, 1997.

Nabulo, Grace, Juliet Kiguli, and Lilian Kigul. "Gender in Urban Food Production in Hazardous Areas in Kampala, Uganda." *Women Feeding Cities: Mainstreaming Gender in Urban Agriculture and Food Security*. Eds. Alicia Hovorka, Henk De Zeeuw, and Mary Njenga. Warwickshire: Practical Action Publications, 2009. 79- 91.

Njenga, Mary, Nancy Karanja, Kuria Gathuru, Samuel Mbugua, Naomi Fedha, and Bernard Ngoda. "The Role of Women-Led Micro-Farming Activities in Combating HIV/AIDS in Nakuru, Kenya." *Women Feeding Cities: Mainstreaming Gender in Urban Agriculture and Food Security*. Eds. Alicia Hovorka, Henk De Zeeuw, and Mary Njenga. Warwickshire: Practical Action Publications, 2009. 167-179

Pandey, Anupam. "Globalization and Ecofeminism in the South: Keeping the 'Third World' Alive." *Journal of Global Ethics* 9.3 (2013): 345-358.

Patil, Rajeev, Radhika Balakrishnan and Uma Narayan. "Transgressing Rights: La Via Campesina's Call for Food Sovereignty." *Feminist Economics* 13.1 (2007): 87-93.

Patil, Vrushali. "Transnational Feminism in Sociology: Articulations, Agendas, Debates." *Sociology Compass* 5.7 (2011): 540-550.

Perry, Keisha-Khan. *Black Women Against the Land Grab: The Fight for Racial Justice in Brazil*. Minneapolis: University of Minnesota Press, 2013.

Razack, Sherene. "'Your Place or Mine'? Transnational Feminist Collaboration." *Anti-Racist Feminism: Critical Race and Gen-der Studies*. Eds. A. Calliste and G. Sefa Dei. Halifax: Fernwood, 2000. 39-54.

Roces, Mina. "Asian Feminism: Women's Movements from the Asian Perspective." *Women's Movements in Asia: Feminism and Transnational Activism*. Eds. Mina Roces and Louise Edwards. New York: Routledge, 2010.

Rosset, Peter. "Agrarian Reform and Food Sovereignty: An Alternative Model for the Rural World." *Rural Social Movements in Latin America: Organizing for Sustainable Livelihoods*. Eds. Carmen Deere and Frederick Royce. Gainesville: University Press of Florida, 2009. 55-78.

Sachs, Carolyn. *Gendered Fields: Rural Women, Agriculture and Environment.* Boulder, Colorado: Westview Press, 1996.

Sachs, Carolyn and Margaret Alston. "Global Shifts, Sedimentations, and Imaginaries: An Introduction to the Special Issues on Women and Agriculture." *Signs* 35.2 (2010): 277-287.

Shiva, Vandana. *Staying Alive: Women, Ecology and Development.* London: Zed Books, 1989.

Schroeder, Kathleen. "Feminist Examination of Community Kitchen in Peru and Bolivia." *Taking Food Public: Redefining Foodways in a Changing World.* Eds. Psyche Williams and Carole Counihan. New York: Routledge, 2010. 510-601.

Vargas, Virginia. "Feminism, Globalization and the Global Justice and Solidarity Movement." *Cultural Studies* 17.6 (2003): 905-20.

Wilbur, Matika and Adrienne Keene. "Ep #2 Food Sovereignty: A Growing Movement." *All My Relations Podcast,* 2009. Web. Accessed March 8, 2019.

On the Land Question

7.
Sasipihkeyihtamowin[1]

Niso Nehiyaw Iskwewak

MARGARET KRESS

*Through my language I understand I am being spoken to, I
am not the speaker.
The words are coming from many tongues and mouths...
and the land around them.
I am a listener to the language's stories, and when my words
form
I am merely retelling the same stories in different patterns.*
                              —Jeanette Armstrong, "Land Speaking"

NÊHIYAWÎWIN AS JUSTICE

This storywork of resistance lives in the experiences, spiritual narratives, and the blood memories of two Cree *onîkâniwak* (leaders), and within the bodies of the two *Nêhiyawak* knowledge keepers I have come to know[2] (Archibald *Indigenous Storywork*; Anderson *Recognition of Being, Life Stages*; Burnouf; LaDuke "Forward"; McCall; McLeod). I am a listener, and I pay attention to the *Nêhiyaw* sounds, words, and phrases which Stella Blackbird, Tammy Cook-Searson, and other Cree linguists have shared with me on this journey. On this day, I give gratitude for their land lessons, their life lessons, and, most importantly, their heart lessons.

As I accentuate *Nêhiyawîwin* within this storywork, I do so to signify an ontological redemption of Indigenous women's knowledge as environmental justice. Words found in Cree narrative memory or oral histories (McLeod) are tied to ancestral knowings and traditional lands, and as such, *Nêhiyawîwin* (the language of the people) with all its intricacies is entwined by subsistence and survival, ceremony, and spirit. In this revitalization of tribal identity and place, words and phrases reclaim a pedagogical and spiritual space for what comes alive in Cree knowledge systems. In this context, place is land, spirit, and body, and by using, defining, and re-establishing the nuances within the language and culture, a *Nêhiyaw* worldview is reclaimed, preserved, and elevated within academic discourses. Within the dominant narrative of environmental justice, an adoption of *Nêhiyaw* decolonizes the euro-centricity of such a framework as it creates space for Indigenous voices and Indigenous bodies. The principles of natural laws (Michell) and Indigenous ethics (Brant) within a *Nêhiyaw* worldview help dismantle this dominant narrative, and bring truth to the imparting of colonial histories and the Indigenous resistances that followed.

For many Canadians, the viewing of "land justice" through an Indigenous lens is both foreign and fresh. To move past a framework of Western justice work, and an analysis of such, to acknowledge the truth of colonialism, and to make an effort to truly understand the state of environmental racism and its antidote, Indigenous environmental justice, one must embrace the counter position of this westernism by reclaiming *pimâtisowin*: the sustenance, the good life, and the spiritual drive of life for *Nêhiyawak* (Ermine "Aboriginal Epistemology").

Past histories have shown us numerous examples of Indigenous peoples recognizing value in contemporary justice—those pieces of reconciliation have held up both settler acknowledgments and apologies—and yet many scholars and activists believe these apologies to be watery and thin. It seems an authentic segue, and the recourse necessary, must include the ceremonial and traditional knowledges and languages of First peoples so all persons' sensibilities, and especially those of Métis, Inuit, and First Nations peoples, are intact (LaDuke *On Redemption*; Sinclair; TRC). In this way, the Cree concept of *Pimâcihiwêwin* ("a giving or a saving of life") (Solomon qtd. in Kress 250) or, in a Western sense, this idea of an Indigenous redemption, prompts both settler and Indigenous peoples to become partners in peace, which can be achieved by embracing the other through sincere acknowledgement, reconciliation, mutual respect, and collective action (Kress; LaDuke *On Redemption*; TRC). Within this Canadian salvage, settlers must be first to extend the "olive branch" as they recognize and accept that Indigenous peoples are the ones to lead in this renewal; Indigenous peoples will be the ones to set the tone in restoring language, ceremony, oral history, culture, and tradition as an ontology of environmental justice through the harmony of political, earthy, and spiritual acts. This reclaim exposes perpetrators' actions and actively shares historical and contemporary realities of land and body conflicts; however, its primary function helps to dismantle the systemic defences of environmental racism through truth telling, as it makes way for a return to Indigenization and *pimâtisowin* (Kress; LaDuke "Forward," *On Redemption*).

Environmental racism discourse has not graced the lexicons of Indigenous peoples until recently (Blackbird and Richardson cited in Kress); further, the phrase "environmental racism" is not held as a commonality among First peoples. However, these *Nêhiyawak* women recognize, feel, and know the explicit and subtle examples of such infractions and crimes. As Stella and Tammy speak of their ancestors' endurances, including those of "whipped bodies" and "stolen lands" (qtd. in Kress 68), I am reminded of how environmental racism is

> an assault on Indigenous Peoples' human rights and public health including their right to their unique special social, cultural, spiritual, and historical life ways and worldviews. Environmental racism results in the devastation, contamination, dispossession, loss, or denial of access to Indigenous peoples' biodiversity, their waters, and traditional lands and territories. Environmental racism is now the primary cause of human health effects of Indigenous Peoples and the forced separation and removal of Indigenous Peoples from their lands and territories, their major means of subsistence, their language culture and spirituality all of which are derived from their cultural, physical, and spiritual relationship to their land. (International Indian Treaty Council, 2008, para 6, cited in Kress 35)

The women in this story see justice as an Indigenization of Canadian history, a spirit-filled prophesy, uplifting sacred dialects, nuances, and rhythms of the *Nêhiyawak* in the reclamation of territory, kinship, and *pimâtisowin*. They resist "the pressure to participate in academic discourse that strips Indigenous intellectual traditions of their spiritual and sacred elements. [They] take the stand that if the spiritual and sacred elements are surrendered, then there is little left of our philosophies that will make any sense" (Hart qtd. in Kress 87).

Embedded in the life works of Stella Blackbird and Tammy Cook-Searson is the understanding of *Kanawayhitowin* (the Cree word calling for the spirit in each of us to come forth to protect each other, and in a sense, all living entities) (Kress). The English translation of *Kanawayhitowin* generates an idea that all peoples are responsible to care for each other's spirit and, in turn, for all life forms and Earth Mother herself. This concept of caring is central to the evolution of understanding how and why Indigenous knowledges and languages are necessary, critical, resistive, and empowering in a spirit of reconciliation and environmental justice. *Kanawayhitowin* is supported by *Wahkohtowin* (the Cree word meaning kinship); as an Indigenous pedagogy, *Wahkohtowin* (O'Rilly-Scanlon, Crowe, and Weenie) supports "kindredness" (Anderson *Recognition of Being*), tradition, language, and ceremony (O'Rilly-Scanlon, Crowe, and Weenie). This *being in relation*, young and old, learning and living with each other, is integral to *Nêhiyaw* knowledge and the lifeworks of these women.

Further to this, the ability to open oneself to the Cree language, and to the knowledges of these *Nêhiyaw women* brings one to the place of *Wahkohtowin*. This state of being related is fundamental to Indigenous culture and traditional beliefs (Ermine) and to the redress or Indigenization of environmental justice. Language, land, and love are at the core of who *Nêhiyaw* are. Through an embracement of the oral, of Cree cosmology and story, and of relationship or *Wahkohtowin*, one moves to a place of understanding, hope, generosity, solidarity, and reciprocity (Senehi; Wilson). I believe this fusing of Cree narrative memory (McLeod) and storywork (Archibald *Indigenous Storywork*), shows a collective understanding of justice, reconciliation, and peace generation.

In understanding *kâkinow ni wâgômâkanak*[3]—the degree to which we are all related—I recognize that regardless of where we sit, on the borders of one knowledge or another, we must support each other through this learning (Kress). The foundation of *Kanawayhitowin*, this place of protecting each other's spirit (Kress) not only provides a context of caring when retelling stories, it frames many Indigenous women's concerns for future generations, and at the same time demands the protection of traditional ecological knowledges, cultures, and languages as pillars of survival and environmental justice. As a non-Indigenous speaker, I follow the protocols of Indigenous Elders and knowledge keepers (Blackbird, personal conversation, 18 July 2015; Burnouf, personal conversation, 25 May 2012; Fitznor, personal conversation, 14 May 2015; Ratt, personal conversation, April 13, 2014; Wilson, personal conversation, 18 March 2012; Wilson, personal conversation, 2014) to assert *tâpowakeythi tamowin* (truth), *kisewâtisowin* (kindness), *asakîwin* (sharing/caring), and *tâpwîwin* (honesty) within this work (Michell). By sharing *Nêhiyaw* in the way I have learned, by simply offering pieces of the language, I keep the embodied knowings of Chief Cook-Searson and Elder Blackbird alive within this understanding of a critical Indigenous justice:

When one Indigenous language slips away, it is as if heavy doors, once open and giving us access to a particular understanding of this place, have slammed shut, shutting us out forever. Part of our shared understanding is gone. That most of us do not speak these languages is irrelevant. Each of them is a passageway into the meaning of this place. Each one lost is a loss of meaning and possible understanding. (Saul 106)

It seems scholars who dilute storywork (Archibald *Indigenous Storywork*; Fitznor) by neglecting to recognize and honour the languages of Indigenous nations fracture the spirit of Indigenization. This action not only signifies the differences in sensibilities between euro-settler identities and discourses, and those of Indigenous knowledge keepers, it accentuates the dominant view of justice or environmentalism, while dulling or perhaps even avoiding the Indigenous and collective history of our nation state. Only with the acceptance of ancestral and contemporary traditional knowledge entwined in territory, language, and spirit, can one come to a complete vision of Indigenous environmental justice and *pimâtisowin*. An Elder has this to say:

We cannot intellectualize Spirit; allow yourself to live from the heart. You cannot have truth without respect, you cannot have truth without love, you cannot have truth without courage, you cannot have truth without honesty, you cannot have truth without wisdom, you cannot have truth without humbleness.... The mind is concerned with power, control, ownership ... but we owe our existence to the land itself.... We [Indigenous peoples] have stayed here.... We still have memory of the land, a memory, duty and sacred responsibility to our people. (qtd. in Courchene 10)

In this justice story of Indigenization, reclamation, and redemption, I give you a glimpse of the leadership held by two women who embody *Nêhiyawêwin*: Elder Stella Blackbird and Chief Tammy Cook-Searson share a blood memory of territory and language infused within the borderlands and waters defined in Treaty Six. As long living, thriving traditionalists and contemporary visionaries, each is *okimâwiw*, an honest, trustworthy person, a trailblazer, and a worthy provider who consistently thinks about the future and the sustainability of her kinship (Fitznor).

SASIPIHKEYIHTAMOWIN: STRONG WOMEN SPEAK

Truth, understood in our language ... is the spirit of grandmother turtle. —Elder Courchene (2015)

When Indigenous women speak, they do so in a multitude of ways. Old stories become new, ancestors guide words, and reverences and wisdoms, gifted by spirits and children, are deep and wide. Jo-ann Archibald uses the composition of a cedar basket and the symbolism of the basket's strips to help us synthesize an understanding of the knowledges embedded in the peoples and the lands: "the pieces of cedar sometimes stand alone, and sometimes they lose their distinctiveness and form a design" ("Indigenous Storywork" 373). And so it is in this knowing, with the stories of each of these *okimâwiw*, sometimes they "are distinguishable as separate entities, and sometimes they are bound together" (373). As I define the positionalities of Stella Blackbird and Tammy Cook-Searson and re-tell their stories, I reflect on the teachings of *respect, responsibility, reciprocity, reverence, holism, interrelatedness,* and *synergy* found within Archibald's *Indigenous Storywork*. With their permissions, I use *Askîwina* to define their personas: "over the years" signifies their history, traditionalism, and resilience (Cuthand). These women are long living, deep of thought, and grounded within *Nêhiyaw* accessible to outsiders only through the fluidity of blood memory and story.

This understanding of memory as story is vital to understanding *Nêhiyaw* history, governance, language, traditions, ceremony, and survival (McLeod). "Cree narrative memory is more than simply storytelling. It involves the collective, intergenerational memories of many skilled storytellers [and] [t]hrough the examinations of family, spirituality, identity, and connections through time and space of the Plains Cree people" (Nickels 154). I share the stories of Elder Blackbird and Chief Tammy Cook-Searson, their lived histories, and their resistances of environmental racism as I examine the necessity of *Nêhiyaw* preservation through the context of the language, cultural revitalization, and Indigenous women's leadership.

When I first heard the word *Sasipihkeyihtamowin*, I understood it as resilience, and at that time, felt it was a fitting word to use for a collective description of the women in my dissertation research. It was also a word of *Nêhiyawîw* (the Cree language) and the ancestral language of Elder Stella Blackbird, who was first among Indigenous women to recognize my relationality, to acknowledge it and extend her acceptance of me as she showed me how I was "one of them." When Denesuline Elder Marie Adam asked me about the depth of this Cree word, I put out a query to several Elders, traditional peoples, and Cree linguists including Elder Stella Blackbird, Chief Tammy Cook-Searson, Cree knowledge keeper Joseph Naytowhow, Métis/Cree educator Laura Burnouf, and Elder Stan Wilson. These are their responses:

Strong willed. Nothing can interfere with your actions. (Blackbird qtd. in Kress 129)

I have been told that the word means "resilience, great patience, stubbornness even" and a fortitude to keep on trying and never give up. This means a person will have a lot of resilience. (Burnouf qtd. in Kress 129)

The person has the ability to see things through and has enough perseverance to make it through times of turmoil and hardship. [It implies that] one has the will to keep on going. (Cook-Searson qtd. in Kress 129)

Definitely resilience, patience. One must be strong in character to have this quality. Willpower certainly fits as well. Definitely perseverance. (Naytowhow qtd. in Kress 129)

It means patience, stick-to-it-ness, persistence, striving, unswerving mind.... *Sasophita* is the not letting go part and the *tahimowin* is the strong-mind part. It seems that this word is referring to relationships that are connected to especially women and could mean "to be long suffering" and to be able to overlook slights. (Wilson, S. qtd. in Kress 129)

When examining the context of *Sasipihkeyihtamowin*, I classify it as a resistance, and place it alongside the words and phrases describing the resilience and conscientization of Woodland Cree Chief, Tammy Cook-Searson of the Lac La Ronge Indian Band (Treaty Six) and Cree Elder, Stella Blackbird of Beardy's and Okemasis Willow Cree First Nation (Treaty Six) and Keeseekoowenin Ojibway First Nation (Treaty Two). In these threads of story, it seems blood memory instigates a continued protection of place. This form of spatial justice becomes aligned with the traditional ecological knowledges these women hold, as well as with their connection to *pimâtisowin* (the good life).

KWAKWU AS REFLECTION

Let me tell you a story about a revelation.
It's not the colour of a nation that holds a nation's pride. It's
imagination.
It's imagination inside. —Andrea Menard

In the heart of the Métis nation, I was introduced to Elder Stella Blackbird while looking for healing. Here I received new life, recognition from a woman of affluence, and a renewed commitment to *pimâtisowin*. Although Stella Blackbird was a woman of reverence, I did not immediately see the power of this Elder healer. Recently, past Grand Chief Ovide Mercredi shared with me that he holds Stella Blackbird in great esteem and that "she is one of the few old-time traditional medicine women practicing today" (Mercredi, personal conversation April 13, 2014). Since our meeting almost a decade ago, I have come to understand both her and Chief Tammy Cook-Searson through the knowledges of *Kwakwu* (Cree word for porcupine [Lincoln]. For you see, "*Kwakwu* is known as the little carrier of the medicines; she is revered in Indigenous culture and her quills are found within their decorative symbolisms of strength, trust and faith" (Kress 208). The imagery of *Kwakwu* symbolizes re-emergence; a powerful place one comes to after entering and then separating from a travesty or hardship. Revered among Eastern and Western peoples, *Kwakwu* carries energies of the sacred. As they practise the sacredness of life, these *okimâwiw* live in the spirit of *pimâtisowin:*

Like *Kwakwu*, they have been begged to listen—*nutoka moo*—
and often, they step back from their situations to look forward,
usa puyew usu wapiw, just as the porcupine does (Lincoln).
The recounts of their own histories, the respect for ancestral
wisdom and traditional knowledge, and their persistent vision
to move forward within a contemporary world is reflected in
[their] *Storywork....* The complexities of these women emerge
as they engage in their work by "reversing and suspending
historical time, [and as they] re-enter that protective burrow
of tradition looking out on the future" (Lincoln 127). Like
Kwakwu, these women defend their territories in quiet and
non-confrontational ways. However, when provoked, they
will do everything in their power to protect themselves, their
kinships and their territories. Joyful calmness, youthful thinking

and open negotiation are among their strongest medicines of defence. Through cooperative and somewhat quiet interactions laced with humour, playfulness, and humility, these women exhibit the traits of *Kwakwu*. They are fearless, confident, and relaxed, and as they trust in their own abilities to protect themselves, they know others recognize their strengths. (Kress 208-9)

Through *Kanawayhitowin*, Cook-Searson and Blackbird protect, love, and understand. Their embodiment of this natural law gifts them power to lead and heal, not only themselves, but also others they encounter. As they look to unearth environmental racism and neo-colonialism, they find solutions to kinship wellness in the knowledges of their ancestors and in the collective. Although it may seem simplistic, Cook-Searson and Blackbird lead through movement, ceremony, listening, and loving. This back and forth from the "burrow" helps them generate a milieu for the sanctity of life; they respect each role, large or small, and their actions elevate intergenerational respect, interdependence, and equanimity among peoples, other living entities, and the land, waters, and cosmos. Their resistances against injustices impacting both Indigenous lands and bodies are offered by their physical and vocal presence in homes, lodges, and community halls, and in the classrooms, assemblies, board rooms, and courtrooms of our nation, through arrangements of the political, cultural, and spiritual. As circumstance presents, they have collectively opposed and often cooperated with governments and corporations on countless occasions, and both have fought personal demons and have endured deep pain. As survivors, they support kinships across the woodlands and plains in addressing issues of mental health conveyed through the abuses of residential schools, intergenerational trauma, environmental racism, and neo-colonialism, as well as the psychological and physical poverty these elements present. Both women recognize what has to change. Tammy elaborates:

I always feel bad when I see a young man about eighteen to twenty-five or thirty walking downtown or just a young guy walking down because the young men had such an important role in our society, when they were hunters, trappers, they still are, because they have so much potential ... but then maybe they get frustrated and keep running into different obstacles. They want to do something. Then you have literacy [levels] where people can't read or write.... The land ... they've lost the will of how to survive off the land. There is so much violence in our community. So much sexual abuse, it still continues to happen, but how do you control it and minimize it? So you know, you just know, you see the court systems that are still there, our people are filling up the jail system, and it has to do with alcohol, either alcohol or drugs. We do our best to protect the children, our youth and we try to provide programming for them, try to give them hope. But there is so much to be done. (Cook-Searson qtd. in Kress 210)

As *Nêhiyaw* speakers and traditional knowledge keepers, these women are community leaders of men, women, and children. Nomadic and of the land, these *onîkânîwak* manoeuvre in two worlds as they persist and thrive. In 2005, Cook-Searson was the first woman to be elected chief of the Lac La Ronge Indian Band, a position she has held since. Stella Blackbird's life, defined by her lineage—the unwavering Chief Little Pine was her grandfather—and her *storywork* and healing actions, shows the significance of her infinite relationship with *pimâtisowin*. Prior to leading others, her personal work came with her own conscientization—critical reflection, critical action, and, it seems, critical prayer. She has told me she has been on a healing journey for almost a half century: "And finally going through a healing, lots of healing, I found my name. I was given a name by the spirits. My name is Mihko kihêw iskwêw—Red Eagle Woman. My name Stella is in Latin, it actually means star. And so my colours are here, blue, yellow, white, red, and I have black for the turtle and Turtle is my clan" (Blackbird qtd. in Kress 219).

Although Tammy and Stella continue to live within the limits of the *Indian Act*, they reflect on the intergenerational and familial trauma of residential schools and other limiting reserve policies and confinements, but focus on their resolve to push through those boundaries and protect the *Nêhiyaw* knowledges embedded in their psyches and upon their lands. Personal acknowledgements of their grandparents show how both women identify with the customs of the Woodland and Willow Cree; and how they preserve even the old parts of their language, *Nêhiyaw*. These *iskwêwak* know the properties of the medicines within their territories distinctly, as well, some might say, as they know the terrains of the boreal and the parklands where they protect, honour, and harvest to sustain their kinships. Both women have fought internalized patriarchy of Indigenous and settler governance, and they acknowledge their roles in resisting this entanglement of environmental racism, colonialism, and patriarchy against Indigenous women and children in particular. One example of this lays in the misogynistic behaviour of men who claimed they spoke for the Elders of Treaty 4 territory. On her first official occasion as Chief for the Lac La Ronge Indian Band, these men challenged Cook-Searson by requiring her to report to the Elders, specifically because her embodiment signified the feminine and she did the unspeakable; she wore her headdress in their territory (Cook-Searson cited in Kress). Cook-Searson removed her headdress in this instance (while male chiefs did not), and learned quickly to redress the continuation of this misogynistic action by what is known as "soft power." By consulting a number of Elders from the Prince Albert Grand Council, and by receiving their sanction for the wearing of the headdress while acting in an official capacity, she felt redeemed: "Chief Wesley Daniels from Sturgeon Lake ... was very supportive. 'Nobody, nobody,' he said, 'only your people can take that off your head, not anyone else but your people, your people put that on you'" (Cook-Searson qtd. in Kress 243). Following this, she encountered another incident during which she was asked to remove her headdress, a demand which she bluntly refused unless all chiefs did the same; this resulted in everyone's headdress being removed and blessed along with the sacred bundle. With her actions, this type of misogynistic behaviour was exposed, and although it prevails in some territories, it has never reared its ugly head again in her presence. Today she freely and proudly wears her headdress and traditional clothing as she partakes in local and global ceremonial governance as a proud *Nêhiyaw* woman. Like Blackbird, who learned long ago that women are the leaders, she did this in a quiet and strategic manner.

Today, Stella Blackbird leads many male counterparts, chiefs, government officials, educators, and other spiritual warriors. She muses about the time years ago when she was taught by a male elder to take her place in a circle of men:

> ... years back women weren't given that voice. But when I received my pipe, not my pipe, the pipe I carry for the people, 'cause I don't own anything. I carry my bundle for the people. But, the Elder told me, "Now, you're ready." He taught me. I did ceremonies with him. He taught me how to do naming and other ceremonies. And he said, "You take your tobacco when you see men sitting in a circle and doing, you take your pipe and go and sit with them, put your tobacco there." (Blackbird qtd. in Kress 247)

Blackbird's lifework reverberates with the hope of this directive and in her teachings for young peoples (Mihko kihêw iskwêw). This same hope is evident in the actions of Chief Tammy Cook-Searson as she campaigns tirelessly for the mental health of her ten thousand plus band members. As I observe her, I see this hope in the respect that she extends to all peoples, western and Indigenous, contemporary and traditional:

I talk to Elders and I learn from them because they have so much experience. I go berry picking with my grandma. My grandma is very patient. That's what I learned from my grandma, 'cause I'm always in a rush, always in a hurry, but my grandma will just be really, really patient. She would have one cigarette a night, that's all she ever had; she would sit on the floor and clear everything, have her ashtray there and have her one. That's all she would have. I learn [also] from the healers. Both my parents do medicine, so they heal people and they pick medicines. They know where to get it and how to mix it and stuff like that. I am always learning and I continue to learn from others. (qtd. in Kress 212)

As she explores how one creates healing spaces for children, young people, women, and men, Cook-Searson expressed interest in meeting Elder Stella Blackbird to learn more. Stella's ground-breaking work around healing and environmental redemption sees followers around the world revere her traditional ecological knowledge of medicines and her love of all things living. Since the mid-1990s she has filled the role of Elder-in-Residence for the Urban Circle Training Centre in Winnipeg; it was her vision that spurred both it and Makoonsag, the intergenerational childcare centre attached to the post-secondary learning centre. As an elder, traditional healer, and medicine teacher, Stella has facilitated healing programs throughout Canada and into the United States, and one of her great joys was founding the Medicine Eagle Healing and Retreat Place with Elder Audrey Bone on the sacred and reclaimed territory bordering Riding Mountain National Park. Here at *Wasagaming,* she harvests and prepares medicines and leads ceremonies to heal the hearts, minds, and bodies of many peoples. It is the children, however, who keep her most focused. In *Kwakwu* form, her stamina, quiet persistence, playfulness, and humour ground her during sunny days and dark encounters. Like Chief Cook-Searson, Elder Blackbird has endured the loss of family: intergenerational trauma from residential school left has had its impact, and her kin, her children, and grandchildren have succumbed to diseases, accidents, suicide, and even murder. Her trials have hurt, but they've also strengthened her resolve— she exudes joy, love, and peace, and each day she continues to love and to give all that has been gifted to her.

PIMÂCIHIWÊWIN: REDEMPTION

Leadership is just about helping people.
—Chief Tammy Cook-Searson (qtd. in Kress 2014)

By their very presence, these women hold *Kanawayhitowin* as responsibility. This natural law clears a collective pathway to both reconciliation and redemption as it adheres to a foundation of Indigenous justice (Kress). For Cook-Searson and Blackbird, Indigenous governance embodies a landscape of sacred teachings; the laws of love, respect, wisdom, courage, humility, honesty, and truth are a part of their environmental justice. Their protection of traditional lands, waters, plants, and animals, denotes more than a Western notion of environmentalism or eco-justice can encompass. And while I understand a Western reconciliation to be atoning for a mistake or clearing a debt to make something better, I also know redemption to be something more. It brings peoples together and allows for testimony and listening, while inching toward settling a difference through the alliance of both apology and forgiveness.

Cree knowledge keeper Solomon Ratt shared with me the *Nêhiyaw* word which supports redemption: *pimâcihiwêwin* (Alberta Elder's Cree Dictionary)—however, it seems much richer, and deeper, and transcendent in its stance. *Pimâcihiwêwin* signifies a giving of life or perhaps a saving of life (Kress). Through *Nêhiyaw* eyes, one begins to see how a conscious action associated within the meaning of this word might move beyond a settler's appreciation of reconciliation. *Pimâcihiwêwin* applies the association and infusion of spirit, place, and culture within *Wahkohtowin,* and it instigates a public, physical, intellectual, or spiritual honouring, as this is what gives it life. In fact, *"pimâcihiwêwin is to give life"* (Ratt qtd. in Kress 251), and as such it holds close the *Nehiyawak* understanding of *Kanawayhitowin,* in protecting and honouring the spirit. Elder Blackbird knows full well how one counters environmental racism in the protecting of bodies and spirits, peoples and lands. The following blatant example compels each of us to understand how vast parcels of traditional lands have been stolen from First Nations to cushion the lifestyles of the white settler:

In 1936, the National Parks Branch evicted the Keeseekoowenin Ojibway First Nation from a small reserve "within" the park boundary in "response to pressure from both local and departmental tourism boosters who hoped to create an attraction for automobile travellers from within the province and from the United States" (Sandlos i). This, however, was not the only motivation. The Department of Indian Affairs supported this move from the Band's rich hunting and fishing territory as "they thought such a move would bolster the department's program of assimilating Native people through immersion in the supposedly more civilized occupation of agriculture" (Sandlos i). (Kress 67)

Stella Blackbird has told me personally of the hurting hearts and bodies of her peoples as they were forcibly moved off their lands. In protest, many were whipped while witnessing the burning of their homes. Although Parks Canada acknowledged this wrongdoing and a small parcel of land was returned to the Keeseekoowenin First Nation in 1986, Elder Blackbird and others took upon themselves the role of "watchdog" as the colonial effects of racist policies continued to rear their ugly heads. Historically, the Keeseekoowenin members were banned from the Park and punished when they sought to pick medicines or to look for sustenance. This and other historical accounts show how Parks Canada and several provincial park and town authorities within our country are guilty of environmental racism by the sheer expropriating of lands and the application of police and military force for the leisure needs of settler peoples (Kress; Sandlos; Westra "Environmental Racism," *Environmental Justice*).

Today, almost thirty years after Parks Canada first issued an apology (1986), this national body has a different kind of relationship with the Keeseekoowenin First Nations and Elders Stella Blackbird and Audrey Bone, based upon the principles of *pimâcihiwêwin*. Through a policy agreement signed in 1998 between Parks Canada and the Keeseekoowenin First Nation, members and those of neighbouring reserves may enter the park and harvest medicines under the guidance of Elders Blackbird and Bone. In this same year, the Canadian government acknowledged their wrongdoing in the 1936 expropriation of these reserve lands, and they returned all 435 hectares of former lands including the lakefront access lands to the First Nation along with a twelve-million-dollar compensation package. This historic action set an example for all First Nations and provided impetus for Parks Canada to continue the work of reconciliation. Cheryl Penny, from Parks Canada, comments on the will of Stella Blackbird and other members: "This would not be happening without the tremendous insight and tremendous good will of Keeseekoowenin Ojibway First Nation, without their commitment, their willingness to let us learn, to help us learn, and then find ways of working together" (Indigenous and Northern Affairs Canada). Recently, Parks Canada contracted a young filmmaker, Christopher Paetkau, to tell the world this story of cultural reclamation and redemption. *Makwa Mee Nuun* tells the story of a dark history and the light of today. In this story, three generations of medicine women, one *Nêhiyaw* woman (Stella), and two *Anishinabe* women (Audrey and America), ensure *Washagamee Sagee* or Clear Lake remains a sacred place for the Keeseekoowenin First Nation as they teach people from around the world about the medicines and ceremonies that come from on these reclaimed lands.

Cook-Searson has many thoughts regarding land use specifically around resource development, leisure, and agriculture. When I met with her she voiced concern over the political process of land distribution, and she spoke of *tipascanikimow*,[4] "those that measure up the land" and the *Saskatchewan Homesteaders Act* as realities which systematically colonize Indigenous peoples:

> The ways the permits are issued by the provincial government regardless if we have opposition to it or not, I think that's environmental racism.... It's almost like there's a total disregard of us being the first people here, and having a treaty and agreeing to share the land, and then just the way the resources were given to the province in the 1930s. That [is a] total disrespect of the First Nations people, saying this is all of Saskatchewan's lands and resources. (Cook-Searson qtd. in Kress 186)

In addressing the social and environmental ills of these colonial impacts, Cook-Searson has her own set of knowledges and practical experiences to bring into the circle. She follows the wisdom of Sharon Venne, Cree legal scholar, who is sure the Elders have not "ceded, surrendered, and forever given up title to the lands" (192). She states:

> We are always asserting our rights. [Sometimes] you also have the push back. You know, you have Premier Brad Wall saying "No way in my time as the Premier will there be a resource revenue sharing for any special groups." He refers to us as a

special group and we are not a special group. We have a treaty, an inherent right to these lands, and we agreed to share these lands. Somehow, I think our lands have been "legally" taken. Because [the governments] developed the rules and regulations on how to take over control of the lands. (Cook-Searson qtd. in Kress 267)

She expressed concern over the infusion of a euro-centric notion of leisure and the leasing of lands for tourism. As Tammy Cook-Searson shared her apprehension of federal and provincial park strategies, I heard about the dismissive nature of officials, and the policy which limits her membership's access to the territory of these "protected lands," specific to the termination of harvesting rights. At that moment, I shared with her the story of Elder Stella Blackbird and Riding Mountain National Park, the historical infusion of assimilation, the dislocation of a people, and land encroachment for white settler privilege and enjoyment. I also reflected on the counter-story of resistance and the redemption found through the efforts of both the peoples of Keeseekoowenin Ojibway First Nation and Parks Canada itself. This retelling functions as a redeeming act. It is this sharing that gives Chief Tammy Cook-Searson hope because it signifies authentic reconciliation encompassed in redemption; for redemption is both an apology and a forgiveness, albeit, one that moves parties to rightful action.

This explicit example of environmental racism within Riding Mountain National Park history encompasses the physical removal and dislocation of Indigenous peoples for settler industry, and although perhaps hidden from settler eyes, it shows the fractured spiritual realm of a people. When Indigenous peoples are removed from sacred lands, waters and sites of ceremony, a disembodied spiritual life is left for families of today. In this case, the settler industry of leisure and recreation, and those engaged in it foisted what is known as a cultural genocide upon this Ojibway nation. This act of environmental racism affected human, plant, and animal life, and moreover, the spirit and biodiversity of the land and living entities within and outside the park boundaries. This story is an important story in and of itself, however, it also gives life to current realities, and to the cases of environmental racism Indigenous peoples face and resist today. It resonates with the women across Turtle Island who protect the spirit of *Pimâcihiwêwin* through measures of truth telling, apology, forgiveness, and mutual action and healing.

SEEKING *MAMÂTOWISOWIN*:
RESURGENCE THROUGH ANIKI KÂ-PIMITISAHAHKIK
PÊYÂHTAKÊYIMOWIN

Power is in the earth, it is in your relationship to the earth.
—Winona LaDuke (*Sacred Ecology*)

Cree philosopher and scholar Willie Ermine describes the depth of these *Nêhiyaw* women in the conceptualization of Earth energy: "*mamatowisowin* is the capacity to connect to the life force that makes anything and everything possible" (110). I argue the embodiment of *mamatowisowin* is the key to Blackbird and Cook-Searson's resurgence and vitality. This energy of the *Nêhiyaw* feminine gives hope to kinships as these women believe all living entities have the ability to "be in tune with the universe" (Naytowhow qtd. in Faith 24). The positionality of feminine Indigeneity found in the solidarity of these *Nêhiyaw* women is grounded in the foundation of *mamâtowisowin*; it reverberates for all, and it makes their work a spiritual and peace-giving leadership. Stella has this to say:

One morning I woke up and realized I was part of this, the creation. I am related to the grass, the trees, the sky, the water. This was my awakening and that's when things began to change. (Status of Women Manitoba qtd. in Kress 218)

"Within the energies of *mamâtowisowin*, are *aniki kâ-pimitisahahkik pêyâhtakêyimowin*[5]—'those that follow peace'" (Ratt qtd. in Kress 271). My fresh awareness of Haudenosaunee customary law shows me how women often "carry the burden of peace," (Gabriel qtd. in Kress 271), however, in *Nêhiyaw* natural law, there is an acknowledgement of women who follow peace (Kress). Upon discovering this distinct difference within Indigenous knowledges, it seemed to me, each experience of coming to *Nêhiyaw* justice is a teaching, a lesson, a coming out, or a resistance, rather than a burden. The value of each person, each mother and daughter, and how Blackbird and Cook-Searson honour and validate roles in a kinship, should perhaps be considered in the ways in which we analyze our experiences when searching for justice, or peace: "Perhaps it is that peace is not a burden at all. Perhaps it is that peace is not carried, but rather, journeyed. Perhaps, in fact, peace is a journey of love, in which a place upon the path is set for each member of a community" (Kress 272).

When I think about the lives of Stella Blackbird and Tammy Cook-Searson, and about how their body politic impacts wellness in their communities, and equally for themselves, I see clearly how *Nêhiyaw* is critical to the dignity of a peoples, and the sustainability of their territories and culture. Alex Wilson references the importance of women's work through the Idle No More movement and its adherence to *sakihiwawin* (Kress). So it is this natural law, the one of *sakihiwawin*, I believe, which validates the positions, the voices, and the actions of these *onîkânîwak*, Elder Stella Blackbird and Chief Tammy Cook Searson. Their peaceful resistances and their attention to the love of what they do for those they serve is a vital piece of redemption and resurgence. The upholding of *sâkihitowin* by these teachers of peace is critical, timely, and necessary. Their actions have paved the way for all to become *Aniki kâ-pimitisahahkik pêyâhtakêyimowin* in this collective journey.

CONCLUSION

As I draw this effort to a close, I reflect on the gift of awakening I have received from Elder Stella Blackbird and Chief Tammy Cook-Searson. I think about my conversations, the questioning and the listening, and the spaces in between. I reflect on the spirit of the land and that of these *Nêhiyaw iskwêwak* as one and the same. I think about their voices, the intonation and softness, and the privilege of listening to their stories, their language, and truly learning to hear. Through *Wahkohtowin*, I believe I have come to understand some pieces of *Nêhiyaw*, what this ancient language means for kinship wellness and for Indigenous women's leadership. I now see the path ahead to *Pimâcihiwêwin* and what I must do as my part in the search for an Indigenous environmental justice. *Wecatoskemitotan mena setoskatotan*. Let's work together and support each other (Cook-Searson cited in Kress 157).

ENDNOTES

[1] *Sasipihkeyihtamowin* (two Cree women).
[2] I have come to this storywork under the guidance of Elder Dr. Stella Blackbird, *Mihko kihiw iskwêw* (Red Eagle Woman) of the Turtle Clan. It is she who propelled my understanding of *Nêhiyawak*, along with the redemption found within this Indigenous language, and a Cree worldview of *Sasipihkeyihtamowin*. As I present this story, I do so in her honour, and in the honour of all Indigenous women leaders who have gifted me on this journey. On this life path, I honour all Cree

language keepers, and the oral histories and dialects preserved in the Swampy, Woodland, and Plains Cree knowledge systems.
[3]Swampy Cree word gifted by Alex Wilson.
[4]Woodland Cree word gifted to me by Cook-Searson, 2012.
[5]Cree word gifted to me by Solomon Ratt.

REFERENCES

Anderson, Kim. *Recognition of Being: Reconstructing Native Womanhood.* Toronto: Sumach Press, 2000.
Anderson, Kim. *Life Stages and Native Women: Memory, Teachings and Story Medicine.* Winnipeg, University of Manitoba Press, 2011.
Archibald, J. "An Indigenous Storywork Methodology." *Handbook of the Arts in Qualitative Research.* Eds. J. G. Knowles and A. L. Cole. London: Sage Publications, Inc., 2008. 371-384.
Archibald, Jo-Ann. *Indigenous Storywork: Educating the Heart, Mind, Body, and Spirit.* Vancouver: University of British Columbia Press, 2008.
Armstrong, Jeannette. C. "Land Speaking." *Speaking for the Generations: Native Writers on Writing.* Sun Tracks: An American Indian Literary Series 35. Ed. Simon J Ortiz. Tucson: Arizona Press, 1998. 175-94.
Blackbird, Stella. Personal conversation, July 20, 2015.
Brant, Clare C. "Native Ethics and Rules of Behaviour." *Canadian Journal of Psychiatry* 35.6 (1990): 534-39.
Burnouf, L. Personal conversation, May 25, 2012.
Courchene, D. "Whose Truth, How Reconciliation?" *University of Manitoba News Today,* 4 July 2014. Web.
Cuthand, Doug. *Askiwina: A Cree World.* Saskatoon: Coteau Books, 2007.
Ermine, Willie. "Aboriginal Epistemology." *First Nations Education in Canada: The Circle Unfolds.* Eds. M. Battiste and J. Barman. Vancouver: University of British Columbia Press, 1995. 101-112.
Ermine, W. "Ethical Space: Transforming Relations." Paper presented at the National Gatherings on Indigenous Knowledge, Rankin Inlet, NU, 2005.
Faith, Erika. "Seeking 'Mamatowisowin' to Create an Engaging Social Policy Class for Aboriginal Students." *First Peoples Child & Family Review* 3.4 (2007): 22-29.
Fitznor, L. Personal conversation May 14, 2015.
Fitznor, Laura. "The Power of Indigenous Knowledge: Naming and

Identity and Colonization in Canada." *Vitality of Indigenous Religions Indigenous Peoples' Wisdom and Power: Affirming Our Knowledge through Narratives.* Eds. J. E. Kunnie and N. I. Goduka. Oxford, England: Ashgate Publishing Limited, 2006. 51-77.

Gabriel, Ellen. "Those who carry the burden of peace." Sovereignvoices1. Ellen Gabriel Blog Pages, 5 January 2014. Web.

Hart, Michael Anthony. *Seeking Mino-Pima-tisiwin: An Aboriginal Approach to Helping.* Halifax: Fernwood Publishing, 2002.

Indigenous and Northern Affairs Canada, with Parks Canada. *Partnership in Action: The Keesee-koowenin Story.* 2010. Archive-ca. com. Web.

Kanawayhitowin: Taking Care of Each Other's Spirit. Archive-ca.com. Web

Kress, Margaret. *Sisters of Sasipih-keyihtamowin – Wise Women of the Cree, Denesuline, Inuit and Métis: Understandings of Storywork, Traditional Knowledges and Eco-justice among Indigenous Women Leaders.* Unpublished Dissertation. Faculty of Education, University of Manitoba, 2014.

La Duke, Winona. "Forward." *New Perspectives on Environmental Justice: Gender, Sexuality, and Activism.* Ed. R. Stein. New Brunswick, NJ: State University of New Jersey, Rutgers, 2004.

LaDuke, Winona. *Winona LaDuke on Redemption.* Sacred Land Film Project, 2011. [Video file].

LaDuke, Winona. *Sacred Ecology: Honour the Earth.* 2014. Web.

LeClaire, Nancy, and George Cardinal. Ed. Earle H. Waugh, and Thelma J. Chalifoux. *Alberta Elders' Cree Dictionary/alperta ohci kehtehayak nehiyaw otwestamakewasinahikan.* Calgary: University of Alberta Press. 2002.

Lincoln, Kenneth. "Trickster's Swampy Cree Bones." Chapter 6. *Native American Renaissance.* Oakland: University of California Press, 1985. 122-147.

Manitoba Status of Women. *Honouring Our Gifts: Celebrating the Contributions of Aboriginal Women in Manitoba.* 2011. Web

McCall, S. "My Story Is a Gift: The Royal Commission on Aboriginal Peoples and the Politics of Reconciliation." Chapter 4. *First Person Plural: Aboriginal Storytelling and the Ethics of Collaborative Authorship.* Vancouver: University of British Columbia Press, 2011. 109-136.

McLeod, Neal. *Cree Narrative Memory: From Treaties to Contemporary Times.* Saskatoon: Purich Publishing, 2007.

Mercredi, Olive. Personal conversation March 23, 2014.

Menard, Andrea. "The Halfbreed Blues." *The Velvet Devil*. Velvet & Hawk Productions. 2000. CD.

Michell, Herman. "Nehithawak of Reindeer Lake, Canada: Worldview, Epistemology and Relationships with the Natural World." *Australian Journal of Indigenous Education* 34 (2005): 33-43.

Mihko kihêw iskwêw—Red Eagle Woman. "Medicine Wheel Teachings." Elder presentation for (Northern Teacher Education Program–NORTEP). La Ronge, Saskatchewan, 15 March 2010.

Nickels, B. "Review of *Cree Narrative Memory: From Treaties to Contemporary Times*." *Great Plains Quarterly* (Spring 2009): 153-155.

O'Rilly-Scanlon, Kathleen, Christine Crowe, and Angelina Weenie. "Pathways to Understanding: *Wahkohtowin* as a Research Methodology." *McGill Journal of Education* 39.2 (2004): 29-44.

Ratt, S. Personal conversation April 13, 2014.

Sandlos, John. "Not Wanted in the Boundary: The Expulsion of the Keeseekoowenin Ojibway Band from Riding Mountain National Park." *The Canadian Historical Review* 89.2 (June 2008): 189-221.

Saul, John Raulston. *A Fair Country: Telling Truths about Canada*. Toronto: Viking Canada, 2008.

Senehi, Jessica. "Constructive Storytelling: A Peace Process." *Peace and Conflict Studies* 9.2 (2002): 41-63.

Sinclair, Murray. "Education: Cause and Solution." *The Manitoba Teacher* 93.3 (December 2014): 6-10. Web.

Truth and Reconciliation Commisison of Canada (TRC). *Honouring the Truth, Reconciling for the Future*. Summary of the Final Report of the Truth and Reconciliation Commission of Canada, 2015. Web.

Venne, Sharon H. "Understanding Treaty 6: An Indigenous Perspective." *Aboriginal and Treaty Rights in Canada*. Ed. M. Asch. Vancouver: University of British Columbia Press, 2002. 173-207.

Westra, Laura. "Environmental Racism and the First Nations of Canada: Terrorism at Oka." *Journal of Social Philosophy* 30.1 (1999): 103-124.

Westra, Laura. *Environmental Justice and the Rights of Indigenous Peoples: International and Domestic Legal Perspectives*. London: Earthscan, 2008.

Wilson, A. Personal conversation March 18, 2012; and 2014.

Wilson, Alex. "Idle No More: A Revolutionary Love Song." *The Winter We Danced: Voices from the Past, Present and Future of the Idle No More Movement*. Eds. Kino-nda-niimi Collective. Winnipeg, Manitoba: ARP Books, 2014. 327.

8.
Climate Change and Environmental Racism

What Payments for Ecosystem Services
Means for Peasants and Indigenous Peoples

ANA ISLA

THE FIFTH ASSESSMENT REPORT (AR5) of the Intergovernmental Panel on Climate Change (IPCC) released to the United Nations Framework Convention on Climate Change (UNFCCC) in November 2014 confirmed that "human influence on the climate is clear, and [that] recent anthropogenic emissions of greenhouse gases are the highest in history. Recent climate changes have had widespread impacts on human and natural systems" (IPCC "Fifth Assessment Report" 2). The Fifth Assessment Report also predicted further increases in average surface temperatures, and demonstrates that anthropogenic emissions such as fossil fuels, cement, and flaring, as well as forestry and other land uses, are central in the warming of the climate worldwide. Many scientists agree that the most effective way to respond to the loss and damage of ecosystems produced by global warming is to reduce greenhouse gas emissions made by coal, oil, gas, and cement in large corporations located in the U.S. (21), Europe (17), Canada (6), Russia (2), Australia (1), Japan (1), Mexico (1), and South Africa (1) (Heede). As the Fifth Assessment Report states, "Limiting climate change would require substantial and sustained reductions in greenhouse gas emissions that, together with adaptation, can limit climate change risks" (8). A net reduction in emissions by 2050 could keep global warming below two degrees Celsius (3.6 degrees Fahrenheit). However, scientists argue that two degrees Celcius is a sure prescription for a long-term disaster (Roach).

While the large-scale drivers of deforestation and climate change remain unaddressed, a grim concern at the United Nations Conference on Climate Change (UNCCC) was the development of numerous "payments for ecosystem services" (PES) for carbon capture. In this chapter, I analyze the UNCCC neoliberal political ecology of monetizing nature: first, through the Kyoto Protocol, an international treaty

186

which extends the 1992 UNFCCC that commits State Parties to reduce greenhouse gases emissions; and second, through REDD+ (Reducing Emissions from Deforestation and Forest Degradation), a mechanism that has been under negotiation by the UNFCCC since 2005, with the objective of mitigating climate change through reducing net emissions of greenhouse gases in the industrial world as long as it can purchase carbon credits from indebted rainforest-dense countries. There are several other schemes for payment for ecosystem services resulting from the Conference of the Parties on Climate Change that I do not cover in this piece due to space constraints.

I focus here on the ways that corporations and governments in industrial countries maintain the fiction that they can achieve emissions reductions by buying forestry certifications that attest to the claim that carbon has been absorbed in the forest in question. In this chapter, I draw on ecofeminist subsistence perspectives and ecosocialism to look at the use of the forest as a source of carbon credits. My purpose is to challenge claims often made by the UNCCC that the "green economy" creates social equality, reduces poverty, confronts ecological destruction, and combats climate change. Instead, I argue that the "green economy" is a new stage of capital accumulation, led by the United Nations, which is a result of monetary economics being applied to nature, and which I call "greening." It has nothing to do with cutting greenhouse gas emissions or climate change. Instead, both the Kyoto Protocol and REDD+ program swing the burden of reducing climate change onto the indebted periphery and its inhabitants, thus delaying the decision to move to cleaner energy and technologies. This approach allows industrial countries to continue building polluting infrastructures, and therefore to rely even more on dirty energy. This "greening's" *raison d'être* is the restructuring of indebted peripheral capacities to expand global capital and create new conditions for capital accumulation, while robbing and undermining the original sources of all wealth—the soil and the worker.

First, I present the political ecology of the environmental management perspective, followed by critiques from ecofeminists and ecosocialists that have underlined how the patriarchal capitalist system maintains relationships of domination and subordination of women, peasants, Indigenous peoples, the so-called developing countries, and nature. I then apply these insights to analyze the Kyoto Protocol in Costa Rica and REDD+ in Brazil. I document a war against the subsistence economy of peasants and Indigenous people, women and men, in the peripheral and emerging economies. I conclude by introducing "green" capitalism as another phase of capital accumulation, arguing that its real war is not against poverty and emissions of greenhouse gases but against economies and cultures that centre on subsistence rather than profit.

THE ENVIRONMENTAL MANAGEMENT PERSPECTIVE
OF THE UNITED NATIONS

Since the recognition of the environmental crisis (WCED), nature has become monopolized as a property of globalized capital. Articulated by the World Bank, at the Earth Summits, capitalist development and the environment have become linked. According to World Bank economists, the root of the environmental crisis the absence of prices for biodiversity, air, water, scenery, and all other aspects of the natural world (see Pearce and Warford; Hamilton). They advocate that the "green" economy can resolve this problem through the monetization of nature, meaning the transformation of ecosystem components or processes into products or services that can be privately appropriated, assigned exchange values, and traded in markets. In short, all of nature is transformed into a means of production through this commodification of "ecosystem services."

Key concepts in the "green economy" are: natural capital and payment for ecosystem services. Natural capital refers to the goods and services provided by the planet's stock of water, land, air, and renewable and non-renewable resources (such as plant and animal species, forests, and minerals). "Payment for Ecosystem Services" (PES) is a voluntary transaction in which a buyer from the industrial world pays a supplier for a well-defined environmental service, such as carbon sequestration from a patch of forest or a form of land use, and that supplier effectively controls the service that ensures his supply (Fatheuer 46). PES can be quantified through the calculation of compensation, i.e., the costs and benefits of a decision. In sum, payments for ecosystem services represent a declaration of the reduction of CO_2 emissions and deforestation based on a calculation of the tons of CO_2 absorbed, and/or the number of trees not cut down, in an arboreal project.

The monetization of nature and its services has unified financial institutions, corporations, the industrial world, indebted periphery, emerging markets, environmental non-government organizations (ENGOs), and others. For instance, in Central America, the 1992 Earth Summit in Rio de Janeiro, Brazil, developed Plan Puebla Panama (PPP), also called the Mesoamerican Biological Corridor. The 2002 Earth Summit in Johannesburg, South Africa, outlined the *Iniciativa para la Integracion de la Infraestructura Regional Sudamericana* (IIRSA) (the South American Regional Infrastructure Integration Initiative). The argument in both initiatives was that investment in infrastructure would need to be scaled up to support the broader economic growth and development agenda following from the expansion of the commodification of nature. In 2012, the Earth Summit Rio+20 in Brazil promised to deliver a "green economy" by stimulating reduction of carbon emissions, efficient usage of natural resources, and social inclusion through the monetization of nature (Isla). However, the People's Summit, held as the same time as Rio+20, called the "green economy" a false solution. They maintained that ecological calamity is due to production and consumption patterns developed by the market economy and therefore could not be solved by market mechanisms.

CRITIQUES OF THE ENVIRONMENTAL MANAGEMENT PERSPECTIVE: ECOFEMINISM AND ECOSOCIALISM

Feminists worldwide (Shiva; Bennholdt-Thomsen and Mies; von Werlhof; Salleh "Climate, Water"; "Global Alternatives") challenge the environmental management school by arguing that in the current economic system, in which there are three economies working—the market economy, the subsistence economy, and the natural economy—the clash between these economies has produced the current social and ecological crises.

The *Market Economy* is the visible and regulated part of the growth economy. It counts in the GDP. The growth economy is a cultural, financial, and political system in which the means of production and distribution are privately owned to create surplus value and increase labour productivity. This domain (of men) has created a public world that ignores the physical reality of human existence. It is where pollution is produced by industrialization, urbanization, and consumerism, resulting in the *metabolic rift* expressed in the social and ecological crises.

The *Subsistence Economy or Social Reproduction* is the invisible, unpaid, or poorly paid parts of the economy that sustains metabolic relations with nature. Those in this economy are feminized, primitivized, diminished, and used as a natural resource. But *they are suited for the continuance of life* and do not produce waste. This economy is reproduced by:

• women, through the maintenance of a home, bearing children, socializing children to reproduce labour power, though their

work is enforced by rape, harassment, and sexual assault;
•peasants working on subsistence farming and horticulture;
•Indigenous people's cultural survival lay knowledge; and,
•colonies that reproduce biological infrastructure for all economic systems.

Ecofeminists argue that the subsistence economy bridges human and natural cycles, because this work is universal, integral, and in touch with the ecosystem. It rarely uses up more matter/energy in resources than is needed for bodily provisioning. All of these members can meet their needs for heath care and nutrition through their resources and knowledge. Nevertheless their work is enforced by repression. Most times, they are reduced to cheap labour or DNA.

The *Natural Economy* is reproduced through ecological processes: the water cycle, the oxygen cycle, and the nitrogen cycle that together sustain all life forms. The ecosystem provides life through its free environmental services and it is a perfect recycling body where nothing goes to waste. The *water cycle* is organized around the heat of the sun that causes water in rivers, lakes, and seas to evaporate. The moist air rises and condensates into particles of water or ice. Water precipitates out of the clouds and back to the earth's surface. The *oxygen cycle* is produced around the plants that produce oxygen. Animals breathe the oxygen and exhale carbon dioxide (CO_2), and plants absorb CO_2 and produce more oxygen. The *nitrogen cycle* is organized by bacteria that help the nitrogen change between states, through decomposing bodies, until plants and trees absorb them through their roots. Human and non-human animals get nitrogen from plants. Nature produces free environmental services and does not produce waste, however nature's work is disrupted by ecocide or destruction of nature.

In sum, ecofeminists maintain that the market economy is a small island surrounded by an ocean of unpaid, caring, domestic work, and free environmental services.

Ecosocialists also contest the environmental management school. They claim that in the ecosystem, time and space modalities in the market economy are in conflict with the time and space of the natural economy. They maintain that the planet is materially finite, meaning there are biological limits to the volume of economic activity the biosphere can support. Elmar Altvater, for example, argues that the first law of thermodynamics (i.e., the total amount of energy in the universe is constant), and the second law of thermodynamics (entropy always increases in the universe) are natural laws that the economic system cannot overcome (84-85). He contends that ecological modalities of time and space are irreversible and inevitable as disorder increases in the universe. Jean-Paul Deléage contends that capital cannot control the reproduction and modification of the natural conditions of reproduction in the same way it aims to regulate industrial commodity production (50). Despite this reality, Michael Goldman maintains that "Global Resource Managers continue to serve the institution of development, whose *raison d'être* is restructuring Third World capacities and social-natural relations to accommodate transnational capital expansion" (47). According to John B. Foster, Brett Clarck, and Richard York, Marx's analysis highlights the *metabolic rift*, a rift that clashes with the natural cycles of life under capitalism. Capitalism robs and undermines the original sources of all wealth—soil and workers—by reducing nature's intrinsic and use values to goods and services for capital accumulation. The expropriation of producers from their conditions of labour constitutes the common basis of class struggle in the capitalist system. In sum, ecosocialists challenge the political ecology of neoliberalism, that is, the transforming of global nature into "market natures," but they still deny that capital accumulation is tied to the experience of the unwaged (women and Indigenous people) or poorly waged (peasants and colonies), being exploited, and their natural work condition being degraded.

What follows exposes how the use value of nature is damaged and how the unwaged and poorly waged are incorporated into capital accumulation by the Kyoto Protocol implemented in Costa Rica (see Isla) and the REDD+ programs implemented in Brazil. I argue that giving monetary value to the natural economy requires damaging the soil and devaluing other forms of peasant and Indigenous peoples' social existence.

THE "GREENING" OF THE UNITED NATIONS FRAMEWORK CONVENTION ON CLIMATE CHANGE (UNFCCC) AND ENVIRONMENTAL RACISM

Governments first agreed to tackle climate change at the 1992 Rio Earth Summit. Scientific research has highlighted the fact that forest vegetation absorbs and stores carbon that might otherwise trap heat in the atmosphere, driving up temperatures and speeding up climate change. At the Climate Change Convention held in Kyoto in 1997, industrial countries proposed the creation of mechanisms to reduce greenhouse gas emissions. Absorbing carbon dioxide (CO_2) from the overflowing waste of industrial countries to reduce the greenhouse effect has become part of the sustainable development agenda. According to the UNFCCC, countries or industries that manage to reduce carbon emissions to levels below their designated amount would be able to sell their credits to other countries or industries that exceed their emission levels.

The Kyoto Protocol was the aftermath to the UNFCCC, which set a non-binding goal of stabilizing emissions at 1990 levels by the year 2000. Among the six kinds of targeted gases is CO_2, which is discharged disproportionately by the industrial world. However, reducing emissions implies high costs for industries. So the major emitting corporations, with the backing of their governments, proposed a self-interested solution: the creation of a global market in carbon dioxide, focused on the forests of indebted countries. The Kyoto Protocol's Clean Development Fund evolved into the Clean Development Mechanism that allows a country to implement an emission-reduction project in developing countries, while giving industrialized countries some flexibility in how they meet their emission reduction. Since the industrial world is not held responsible for mitigating its own level of emissions, this type of solution allows the industrial world to continue polluting as long as it can purchase carbon credits from indebted rainforest-dense countries. Meanwhile, energy-related emissions produced by the increase in the amounts of fossil fuels, cement, and gas flaring proceed unimpeded.

Since Kyoto, rainforests have been valued economically in terms of the amount of carbon they sequester. As carbon became subject to trading on the open market, the rainforests of the world became valued as carbon sinks, with predictably disastrous results for the forest dwellers. An example of this is the widespread peasant land dispossession that took place in Costa Rica.

KYOTO PROTOCOL:
PEASANTS' EXPROPRIATION AND CRISES OF WOMEN AND NATURE

Costa Rica was the first country to take part in the Joint Implementation Program (JIP) organized by the United Nations (UNFCCC). JIP allows an industrialized country to earn emission reduction credits by buying them from another country. Costa Rica was one of the first countries to voluntarily sell carbon credits to the industrial world to help them achieve emission reductions. It was presented as the international model for Kyoto's Clean Development Mechanisms and guidelines were outlined by the World Wild Life Fund (WWF) and the International

Union for the Conservation of Nature (IUCN), in collaboration with the national government (Isla).

One of the worst effects of "greening" is the crisis of nature. The Costa Rican government, through its Ministry of Environment and Energy (MINAE), appraises the ability of private forest farms to sell carbon credits. However, selling carbon credits is particularly promoted among large-scale agricultural entrepreneurs in association with international capital. Lands categorized as forest reserves, which receive payments for environmental services, are exempted from property taxes. This tax relief, under a scheme called Fiscal Forestry Incentives (FFI), subsidizes plantations owned by international capital to promote foreign forest species of high yield and great market acceptance, such as gmelina (*Gmelina arborea* used by Stone Forestall, a U.S. corporation) and teak (*Tectona grand* used by Bosques Puerto Carrillo, a U.S. corporation and Maderas de Costa Rica S.A., or Macori, now Precious Wood Ltd., a Swiss corporation). These trees are native to South and Southeast Asia. Mono-arboriculture has been defined in this system as "reforestation" even though these plantations constitute artificial ecosystems, and corporations are allowed to cut the trees down after fifteen years of growth and transform them into wood for floors or paper, boxes for fruit export, or furniture. With credits provided by the World Bank, the Costa Rican government enthusiastically promoted the conversion of forest ecosystems into sterile monocultures by planting homogeneous forests (Baltodano "Monocultivos arbóreos," "Bosques en reservas"; Figuerola "Nativos y exóticos").

The monoculture of tree species has become a time bomb for biodiversity in Costa Rica. The natural forest of the humid tropics is a highly productive ecosystem. For instance, a hectare of tropical forest has on average more than three hundred species of trees. Biodiversity means that a forest will have a great number of *leguminosae* (trees, shrubs, plants) with leaves of different sizes, which lessen the impact of rainfall and prevent erosion. Sonia Torres, a forestry engineer, explains how teak plantations have resulted in the erosion of flatlands:

> Since the planting of these foreign species, I have observed that teak has a root system that grows deep into the soil, but in the rainforest the systems of nutrient and water absorption are at the surface. In general, nutrients and water are concentrated at a depth of between seventy and one hundred centimeters. As a result, teak trees are encircled by flaked soil. In addition, when it rains, the large-sized leaf accumulates great amounts of water

that then pours violently onto the soil. A drop of water, at a microscopic level, forms a crater; when water falls from fifteen meters or more it forms holes. Water descending on soft soil destroys the soil. The far-reaching spread of the roots and the shade produced by the leaves obstruct the vegetative growth on the lower forest layer, which could prevent the soil damage from the violent cascades. (Personal interview, August 2000)

Ecologists from Costa Rica oppose the payment of environmental services for arboreal monoculture. A monoculture is not a forest because it does not reproduce itself but rather needs external inputs such as agrochemicals to grow to maturity (Shiva). Many of the ecologists are not against selling so-called environmental services in general, but instead they promote reforestation through the natural and simple regeneration of secondary forests, which conserve biodiversity and regulate hydrology (Figuerola "Pago de servicios ambientales"; Franceschi). They argue that the conservation of forests with native wood species and associated plants and fauna should be a priority, that restoration and natural regeneration, with its own ecological complexity, is a legitimate goal, and that local peasants must be taken into consideration to avoid irreconcilable conflicts.

As the ecosystem disintegrates, it has powerful effects on the degree of oppression endured by peasant women and children. For them, the disappearance of forests is an issue of survival, forcing them to migrate to San José, the capital of Costa Rica, and/or to other ecotourist areas, in the hope of earning an income for themselves and their dispossessed families. Introduced into the cash economy, some impoverished women have little option but to earn all or part of their living as prostitutes. Ecotourism links conservation areas and promises a risk-free world of leisure and freedom for those with money to pay. At the same time, sex tourism offers women's and children's feminized bodies as commodities that are pure, exotic, and erotic. This image of Costa Rica entangles two aspects of capitalist patriarchal economics: the domination of creditors (core countries) over debtors (the periphery); and the psychology of the patriarchy in which men develop their "masculation." Masculation is the exploitative masculine identity created by the alienated world of patriarchal capitalism through compliant bodies (Vaughan). As Costa Rican people are increasingly impoverished, the enclosure of the commons, the mark of international power relations, is stamped on the bodies of its children and women.

Jacobo Schifter estimates that there are between 10,000 and 20,000 sex workers in the country, and between 25,000 and 50,000 sex tourists—he calls them "whoremongers," meaning regular clients—who visit each year. Eighty percent are U.S. citizens (43). Schifter concludes:

> Obviously, globalization has linked us to an international economy in which each country finds their specialization. In the Latin countries, it is increasingly concentrated in our genitalia. If in agriculture and industry our hands and feet had given us food before, now penises and vaginas do. In the case of Costa Rica, whether we like it or not, sex tourism is a strong component of our Gross National Product. (265)

Tim Rogers reports that the U.S. has become Costa Rica's pimp, as crack cocaine and sex with prostitutes helps narcissistic male tourists and old retirees affirm their masculinity and "escape reality" from their dissatisfied financial and social decline back home.

As Costa Rica slides into a subordinated position internationally, the country becomes a paradise for sex trafficking, paedophilia, and child pornography.

The Kyoto Protocol has not reduced greenhouse gas emissions, but instead has allowed capital to stake a total victory for a market-based approach to climate change. Thus, during the UNFCCC Conference of Parties (COP), new programs for payment for ecosystem services, such as Reduction of Emissions from Deforestation and Forest (REDD and REDD+) and the European Emissions Trading Systems (ETS) emissions certificates were sanctioned.

REDD+: EXPROPRIATION OF INDIGENOUS PEOPLES' TERRITORIES AND CRISES OF NATURE AND PEOPLE

The crisis of nature in Brazil lies in the Amazon rainforest. The Amazon Basin contains over 60 percent of the world's remaining rainforest. Through transpiration, the Amazon creates between 50 to 75 percent of its own precipitation, and its impact extends well beyond the Amazon Basin by feeding the largest river on the planet (the Amazonas), suppressing the risk of fire as well as creating moisture that travels through the canopy to Central and North America (Medvigy, Walko, Otte and Avissar). The canopy refers to the dense ceiling of leaves and tree branches formed by closely spaced forest trees that make up the level known as the overstory. The moisture created in the Atlantic

Ocean, in combination with the constant rainfall in the Amazon Basin, travels through tree canopies that have now been broken. Deforestation reduces local transpiration. As a result, increasing droughts in South, Central, and North America should be expected.

At COP 21 in France, in December 2015, the increase of greenhouse gas levels in the atmosphere, from around 360 parts per million (ppm) to over 400 ppm, was acknowledged. As a result, the international carbon markets and carbon pricing that achieved international recognition with the Kyoto Protocol, was central in the UNFCCC agenda. REDD+ is one type of payment for ecosystem services that ultimately translates into compensation for destruction. This initiative, however, claims that no damage is done because destruction of biodiversity in one place will be compensated by restoring a location elsewhere.

Who buys these ecosystem services? Corporations involved in extractive industries (e.g., oil, mining), industrial agriculture, the entertainment industry, airlines, the construction of large-scale infrastructure, as well as the World Bank, industrial countries, and international conservation NGOs buy these services. Jutta Kill and Giulia Franchi, referring to mining corporations, argues that the basic principal behind payment for ecosystem services is that a mining company that destroys 4000 ha. of forest for its open pit mining activities can "protect" another 4000 ha. of forest elsewhere. The communities affected, however, are not provided with any information about the industries financing or buying "offset credits" from their lands; further, they do not know why there are new restrictions on the way they have traditionally used the forest or why they are now being forced into hunger and food insecurity.

Those opposed to REDD+ implementation believe the initiative is highly questionable. At COP 20 in Lima, Peru, in 2014, the World Rainforest Movement (WRM) maintained that REDD+ is the largest land grab in history and that it is a false solution to climate change. Instead, they state, these instruments threaten to extinguish Indigenous people. WRM argues that REDD+ speculates with Indigenous peoples' territory; robs communities of their autonomy by creating restrictions and prohibitions for communities that depend on the forest for their subsistence; violates culture and tradition by integrating them into the international market; and, in the process, destroys and divides communities, threatens subsistence and food sovereignty, creates conflict, and exacerbates inequality, all while producing huge profits for corporations (WRM; "10 Alertas"; Bonilha).

At COP 20, the Heinrich Böll Stiftung Foundation supported a full-day seminar on *Financialization of Nature and Extractivism* organized by Latin American and Caribbean Friends of the Earth, and several other organizations. Several cases of the implementation of REDD+ in Indigenous territories were discussed and publications by the foundation were made available. For reasons of space, here I refer only to REDD+ in Brazil where divergences of opinion over the use of market-based mechanisms between the federal and state levels have been taking place.

At the gathering, Jutta Kill, a biologist and activist, argued that the logic of REDD+ is problematic:

> In order for the REDD offset project to generate carbon credits, the users of the land have to describe their activities as a threat to the forest. If the activities are not a threat to the forest, there is no risk of deforestation and therefore no credits that can be generated from avoiding deforestation!... Without such a story—that the forest would have been destroyed—there is no carbon to be saved, and no carbon credits to be sold. This necessity by design to describe the land use of forest dependent communities as a risk to the forest is already reinforcing the dangerous myth that forest dependent communities and small-scale farmers are among the most important agents of deforestation. ("REDD: Una coleccion de conflictos" 10)

Jutta Kill also argues that REDD+ deepens injustice and historical inequalities ("REDD in Brazil" ; "REDD: Una coleccion de conflictos"). She explains that since 1999, several individual forest carbon projects have been formulated in Brazil, among them the Guaraqueçaba Climate Action Project in the state of Parana. This was a joint initiative between the U.S. Nature Conservancy and the Brazilian Society for Wildlife Research and Environmental Education, and was funded by General Motors, American Electric Power, and Chevron. Kill shows how this project was presented by its proponents as an international model for REDD+. However, the locals involved and affected by its implementation consider it a failed project.[1]

Furthermore, REDD+ credits constitute a form of property title. Those who possess the credit do not need to be owners of the land, water, or trees that are on the ground, but they have the right to decide how these will be used. Cristiane Faustino and Fabrina Furtado researched two cases of REDD+ in Acre, Brazil. Both cases had been implemented under international guidelines since 2010, and were outlined by the World Wild Life Fund (WWF), the International Union for the Conservation of Nature (IUCN), the Federal University of Acre (UFAC), IPAM, The

Woods Hole Research Center, Embrapa, and the German Agency for International Cooperation (GTZ) in collaboration with the local governments (Herbert; WWF). Faustino and Furtado show that the organizations and social groups of Acre have denounced REDD+ for: 1) violations to land and territory rights; and 2) violations of the rights of the peoples in REDD+ occupied territories. Relatoria del Derecho Humano al Medio Ambiente (RDHMA) (Rapporteurship of the Human Right to the Environment) investigated the Acre case and found several problems. Among them are:

a) Indigenous people in Amazonia have no land title, deepening territorial conflicts. The Acre government has put Indigenous land on the REDD+ market without prearranging land title to the owners of the land. This situation violates ILO Convention 169 as well as the Federal Constitution;
b) territory for subsistence and traditional activities, such as family agriculture or fishing, has been reduced or eliminated;
c) blockage of rubber paths that constrains the main activity of the rubber tappers;
d) failure to generate sufficient income for Indigenous peoples' livelihood as they have lost their subsistence economy;
e) Indigenous land speculation has forced entire Indigenous families to move to the periphery of cities, such as the Jaminavá Indigenous people whose children are sexually exploited and/ or are forced into prostitution;
f) broken promises by the government and those that promote REDD+.

Faustino and Furtado cite Dercy Teles, Union President of Rural Workers of Xapuri, as she summarizes her outrage:

The impact of the green bag (REDD+) is that we lose all the rights that people have as citizens. [We] lost control of the territory. No longer can we plant. [We] can no longer do any daily activity. [We] receive some money, but only to be no more than observers of the forest, without being able to touch it. Thus, the true meaning of life as a human being is taken away. (5)

These authors conclude that REDD+ in Brazil occurs in a context of extreme inequality in which environmental NGOs such as WWF-Brazil, Comisión Pro Indio (Pro Indian Commission), Forest Trends, and Centro de Trabajadores de la Amazonia (The Centre for Amazonia Workers) are profiting while Indigenous people are dispossessed. REDD+ in this context has been "transferring responsibility for environmental degradation onto subjects that historically have maintained an environmental equilibrium throughout their traditional activities of sustenance. In this way, [people] are devalued and their different modes of land use and occupations practiced by traditional communities and Indigenous peoples are placed at risk" (Faustino and Furtado 22). Further, they submitted the following question: How is it possible, on the one hand, to meet the objectives of social and environmental recovery when, on the other hand, violations of rights occur?

La Carta de Belém Group maintain that other questions not answered by REDD proponents include: how to integrate the forest into the financialization framework; how environmental damage or mitigation payments can be calculated; what to include in the calculation; and who assesses the "true" value of an ecosystem? Belém's Group of Brazil declared that REDD+ programs are highly political:

> We stand together in rejecting mechanisms that commodify and financialize nature and market-based solutions to the climate crisis, because their impacts on territories, local residents, and workers cause the violation of social and territorial rights.... We believe that Brazil's proposal to estimate the historical contribution of each country, using a concentric differentiation, is a relevant approach.... We reaffirm our opposition to the introduction of forests into carbon markets.... We instead see the solution in mechanisms to build a just transition, which do not repeat or enhance the same forms of production and consumption that caused and continue to cause global warming and the loss of biodiversity.

The murder of Indigenous women and men, denounced at the U.N. (Conselho Indigenista Missionário) points to the dispossession of land and extermination of these peoples. These communities occupy strategic territories that transnational capital requires for REDD+ and other megaprojects.

CONCLUSION

The Kyoto Protocol enabled the replacement of the natural forests with artificial forest farms in Costa Rica. REDD+ has been murdering and displacing the guardians of the forest—Indigenous people—in Brazil. In this chapter, I have demonstrated that "greening" is a new imperialist stage of capital accumulation organized under the umbrella of the United Nations that entails four aspects:

First, credit instruments were expanded by financial capital to create new avenues for economic growth. Kyoto and REDD+ have been incorporated into Wall Street financial markets.

Second, the World Bank licensed big environmental non-governmental organizations (ENGOs) to broker the indebted countries' resources with large corporations involved in economic restructuring and globalization. The role of ENGOs is to establish the monetary values of the "global commons" of the indebted periphery, such as the forest, and to export these values into stock exchanges. These new experts, most of them biologists grouped in ENGOs, have emerged as new models of modernization and environmental protection by using the discourse of "protecting" the global commons in ecologically sensitive areas.

Third, there are new types of markets created—such as biodiversity for biotechnology and Intellectual Property Rights, scenery for ecotourism and forests for carbon credits—located in conservation areas. A conservation area is a designated domain where private and public activities are interrelated in order to manage and conserve a region's nature for capital accumulation.

Fourth, the "greening" process, in this case forests for carbon credits, results in peasants and Indigenous people losing their land and territories and acquiring in some cases new roles as service providers in new industries such as ecotourism; in other cases, they are forced into prostitution, hunger, and food insecurity.

This analysis shows that the new imperialism led by the United Nations, through the Kyoto Protocol and REDD+, and organized by the environmental management of the World Bank, ignores ecosystems, genders, and species, and promotes poverty and unsustainability in the indebted periphery. The imposition of monetary value onto the commons of peasants and Indigenous territories requires the destruction of nature as living grounds and devalues other forms of social existence, such as transforming agriculture skills into deficiencies; commons (scenery, forest, mountains) into resources; knowledge of biodiversity into ignorance; peasants and Indigenous people's autonomy into dependency; self-sufficiency of men and women into loss of dignity for women's and children's bodies.

We must undo patriarchy, colonialism, capitalism, and imperialism to stop the plundering of the forests, the earth, and its inhabitants in all parts of the world. Ecofeminists propose a subsistence perspective for the entire world to transform the nature of our economy. Subsistence orientation means achieving another relationship with our fellow humans and the non-human world. Starting points for another economy exist in the work done by women every day without pay, and poorly paid work done by peasants, Indigenous people, and peripheral countries. A manageable sized economy will allow us to live from our land, from our climate, and from our resources in that part of the earth we call home (Bennholdt-Thompsen). Ecosocialists warn us to create ecocentric social and political frontiers before capitalist accumulation poisons the atmosphere—the last ecological limit.

ENDNOTES

[1]For more information, please refer to the following three films: "Disputed Territory: The Green Economy Versus Community-Based Economies" (WRM 2012); "The Carbon Hunters" (Shapiro); "Suffering Here to Help Them Over There" (FERN).

REFERENCES

"10 Alertas sobre REDD para Comunidades," 11 October 2012. Web.

Altvater, E. "Ecological and Economic Modalities of Time and Space." *Is Capitalism Sustainable? Political Economy and the Politics of Ecology.* Ed. M. O'Connor. New York: Guilford Press, 1994. 76-89.

Baltodano, J. "Monocultivos arbóreos no merecen pagos de servicios ambientales." *Ambien-Tico* 123 (2003): 3-5.

Baltodano, J. "Bosques en reservas del Ida: biodiversidad y manejo." *Ambien-Tico* 133 (2004): 14-17.

Bennholdt-Thomsen, Vernonika. "Money or Life: What Makes Us Really Rich." Trans. Sabine Dentler and Anna Gyorgy. Bonn, Germany, 12 August 2011. Web.

Bennholdt-Thomsen, Veronika and Maria Mies. *The Subsistence Perspective: Beyond the Globalized Economy.* London: Zed Books, 1999.

Bonilha, Patricía. "A última fronteira do capital?" Caitalismo Verde = Neocolonialismo, 8 December 2014. Web.

Conselho Indigenista Missionário. "La Gran Muerte: Genocidio e violações de direitos são denun-ciados por indígenas e Cimi em Fórum da ONU," 19 May 2016. Web.

Carta de Belém Group. "Declaration by the Carta Belém Group at COP

20." Lima, 10, 2014. Web.

Deléage, J.-P. "Eco-Marxist Critique of Political Economy." *Is Capitalism Sustainable? Political Economy and the Politics of Ecology.* Ed. M. O'Connor. New York: Guilford Press, 1994. 37-51.

Fatheuer, T. *Nueva Economía de la Naturaleza. Una Introducción Crítica.* Berlin: Heinrich Böll Foundation, 2014.

Faustino, Cristiane and Fabrina Furtado. "Economia Verde, Pueblos de los Bosques y Territorios: Violaciones de derechos en el estado de Acre." Informe Preliminar de la Misión de Investigación e Incidencia. Rio Branco, Brazil, 2014. Dhesca: Plataforma De Dereitos Humanos. Economicos, Sociales, Culturais e Ambientas.

FERN. "Suffering Here to Help Them Over There." Video. 6 June 2012. Web.

Figuerola, J. "Pago de servicios ambientales a la tala rasa." *Ambien-Tico* 123 (2003): 10-11.

Figuerola, J. "Nativos y exóticos pero conservando la biodiversidad." *Ambien-Tico* 141 (2005): 16-17.

Foster, John Bellamy, Brett Clarck, and Richard York. *The Ecological Rift: Capitalism's War on the Earth.* New York: Monthly Review Press, 2010.

Franceschi, H. "Conflictos socio ambientales intercampesinos por los recursos naturales." *Revista de Ciencias Sociales* 111.12 (2006): 37-56.

Goldman, M. "Inventing the Commons: Theories and Practices of the Commons' Professional." *Privatizing Nature: Political Struggles for the Global Commons.* Ed. M. Goldman. New Brunswick, NJ: Rutgers University Press, 1998. 20-53.

Hamilton, K. "Genuine Savings, Population Growth and Sustaining Economic Welfare." Paper presented at the Conference on Natural Capital, Poverty and Development, Toronto, Ontario, September 2001.

Heede, Richard. "Tracing Anthropogenic Carbon Dioxide and Methane Emissions to Fossil Fuel and Cement Producers, 1854-2010." *Climatic Change* 122.1 (January 2014): 229-241. Springer Link, 22 November 2013. Web.

Herbert, Tommie. "Setting up Nest: Acre, Brazil, and the Future of REDD." Ecosystem Marketplace, 22 July 2010. Web.

Intergovernmental Panel on Climate Change (IPCC). "Climate Change 2007: Working Group II: Impacts, Adaptation and Vulnerability." N.d. Web.

Intergovernmental Panel on Climate Change (IPCC). "Fifth Assessment

Report (AR5)." November 2014. Web.

Isla, Ana. *The "Greening" of Costa Rica: Women, Peasants, Indigenous People, and the Remaking of Nature.* Toronto: University of Toronto Press, 2015.

Kill, Jutta. "REDD in Brazil: Two Cases on Early Forest Carbon Offset Projects." Rio de Janeiro: Heinrich Böll Foundation, 4 December 2014. Web.

Kill, Jutta. "REDD: Una coleccion de conflictos, contradicciones y mentiras." Montevideo, Uruguay: World Rainforest Movement, 4 December 2014. Web.

Kill, Jutta and Giulia Franchi. "Rio Tinto's Biodiversity Offset in Madagascar: Double Landgrab in the Name of Biodiversity?" A Field Report by WRM [World Rainforest Movement] and Re:Common. March 2016. Web.

"Let them eat pollution." *The Economist* 7745 (8 Feb. 1992): 82. Web.

Medvigy, D., R. L. Walko, M. J. Otte, and R. Avissar. "Simulated Changes in Northwest U.S. Climate in Response to Amazon Deforestation." *Journal of Climate* 26 (2013): 9115-9136. Web.

Roach, John. "We're Kidding Ourselves on 2-Degree Global Warming Limit: Experts." *NBC News*, 28 Nov. 2014. Web.

Rogers, Tim. *Costa Rica's Sex-Tourism Is Growing.* 16 October 2009. Web. Accessed 15 October 2015.

Pearce, D.W and J.J. Warford. *World Without End: Economics, Environment And Sustainable Development.* New York: Oxford University Press, 1993.

Salleh, Ariel. "Climate, Water, and Livelihood Skills: A Post-Development Reading of the SDGs." *Globalizations* 13.6 (2016): 952-959.

Salleh, Ariel. "Global Alternatives and the Meta-industrial Class." *New Socialisms: Futures Beyond Globalization.* Ed. R. Albritton, S. Bell, J. R. Bell, and R. Westra. London: Routledge, 2004. 201-211.

Schapiro, Mark. "The Carbon Hunters: On the Trail of the Climate's Hottest Commodity." Produced by Andres Cediel & co-produced by Daniela Broitman. Video. 11 May 2010. Web.

Schifter, J. *Viejos Verdes en el Paraiso: Turismo Sexual en Costa Rica.* San José, Costa Rica: Editorial Universidad Estatal a Distancia, 2007.

Shiva, V. *Staying Alive: Women, Ecology and Development.* London: Zed Books, 1989.

United Nations Framework Convention on Climate Change (UNFCCC). Joint Implementation (JI) [The Kyoto Protocol]. 15 November 2005. Web.

Vaughan, G. "Introduction: A Radically Different Worldview Is

Possible." *Women and the Gift Economy: A Radically Different Worldview is Possible*. Ed. G. Vaughan. Toronto: Inanna Publications and Education, 2007. 1-38.

Von Werlhof, C. *Madre Tierra o Muerte! Reflexiones para una Teoria Critica del Patriarcado*. Mexico: Cooperativa El Rebozo, 2015.

Waring, M. *If Women Counted: A New Feminist Economics*. San Francisco: Harper and Row, 1988.

World Commission on Environment and Development (WCED). *Our Common Future*. New York: Oxford University Press, 1987.

World Rainforest Movement (WRM) [Movimiento Mundial por los Bosques Tropical]. "El PSA convertido en permiso de destruccion ambiental." *Boletín* 198, 5 February, 2014. Web.

World Rainforest Movement (WRM). "Disputed Territory: The Green Economy Versus Community-Based Economies." Video. 11 December 2012. Web.

World Wildlife Fund (WWF). "Environmental Service Incentives in the State of Acre, Brazil: Lessons for Policies, Programmes and Strategies for Jurisdiction-wide REDD+." WWF Report, 21, October 2013. Web.

9.
Biotechnology and Biopiracy

Plant-based Contraceptives in the Americas
and the (Mis)management of Nature

RACHEL O'DONNELL

IN MAY OF 2015, A SCIENTIST at the Jamaican Scientific Research Council was granted a new U.S. patent on *Petiveria alliacea* called "Composition and Method for Treating Cancer" (Brooks 1). The "invention" of the plant compound in the laboratory relates to the treatment of various disease states with Dibenzyl Trisulfide (DTS), a phytochemical that can be extracted from Petiveria in the laboratory. As its name denotes, the patent relates to the use of DTS for therapeutic treatment for cancer. According to the patent application, the drug developed will have the potential to treat a wide range of cancers and other diseases. Some of the cancer cells that this drug has been found to be effective against include neuroblastomas and sarcomas, brain and skin cancers. One of the major findings, according to the patent, is that this drug made from the DTS compound does not appear to affect healthy cells, whereas most contemporary cancer treatments also damage healthy cells in the body.

The Jamaican Scientific Research Council reports that the cancer-fighting compound that was developed into this pharmaceutical was extracted from a local plant called Guinea Hen Weed, which is a common name in Jamaica for *Petiveria alliacea*. Many initial reports announcing the "discovery" also comment on how plants like Guinea Hen Weed have long been used by rural Jamaicans to cure a wide range of illnesses. Plants used by rural communities in local medical practice have been pursued by scientists with interest in "discovery," classification, research, and the global market. Indeed, in the first publications on Petiveria's curative properties for cancer in 2007, the scientist with this most recent patent reported,

> The data compiled in the present review on dibenzyl trisulphide (DTS) isolated from Petiveria alliacea L (the guinea hen weed or

anamu) revealed that the compound and its derivatives could be of tremendous pharmaceutical interest. (Williams et al. 17)

In the title of the article itself, the researcher includes local names Guinea Hen Weed and anamu for reference. How did the Jamaican scientist discover the activity of this plant compound? Why is the research understood as necessarily connected to pharmaceuticals? How did these plants end up being tested in the laboratory? The science of botany often disregards these questions of historical and cultural process, as well as the social, political, and economic relationships involved in the creation of a drug and its connection to plant-based knowledge.

Mainstream media has since reported on the scientists' findings and drummed up "global" excitement about a possible cancer cure. Many media reports include reference to the offer to the scientists from the pharmaceutical companies to patent the development; the total offered has already exceeded 100 million dollars (Lowe et al. 94; Williams et al. 23).

On the website of Cornell University, a graduate student has compiled five pages of information on *Petiveria alliacea* as part of a medicinal database. Thirty-one medicinal uses are listed there, including treatments for intestinal parasites in livestock and hysteria, rheumaticism and rabies in humans. In addition to its traditional medicinal use, Petiveria has gained the attention of a number of organic chemists. At least twelve different biologically active compounds have been assayed from root, leaf and seed of the plant. A study published in 2001 reported that the presence of two diastereomers of S-benzyl-L-cysteine, isolated from fresh roots. This is reportedly the first evidence of these compounds occurring in nature.

Apacina is the Guatemalan word I am familiar with as a researcher in rural Guatemala for *Petiveria alliacea*, a plant made into a tea that some women drink to prevent a pregnancy. It has a very strong acrid smell and an almost unpalatable taste. Mayan women sometimes boil it for use after intercourse and save the pot of tea to drink over three days. Interestingly, women have maintained this contraceptive usage of the plant, and the ongoing use of the plant and the development of the local knowledge surrounding it can be seen to represent efforts to resist twentieth century changes in their communities.

The maintenance of traditional plant-based forms of contraception coincides with efforts made to resist the family planning initiatives imposed on Mayan communities from development organizations operating in Western medical traditions. This case study of a particular region's knowledge about Petiveria focuses on resistance to family planning at the highest levels, starting in the 1960s, which persisted for more than three decades and impeded the spread of family planning. Multiple doctors and nurses in aprofam (Asociacion Pro Bienestar de la Familia or the Association for Family Well-being), the major family planning agency in Guatemala, partially funded and supported by usaid (United States Aid for International Development), in clinics throughout highland Guatemala spoke of the *"resistencia"* to family planning in Mayan communities. Despite the establishment of a dynamic private family planning association in the mid-1960s, 40 years later, Guatemala ranks last in contraceptive use in Latin America (Population Reference Bureau 2014). This resistance to change and Western forms of health care and pharmaceutical application deserves more attention than can be provided in these pages.

Recent studies have revealed that of the top 150 propriety drugs used in the Western hemisphere, 57 percent contained at least one major active compound derived from natural sources (Setzer et al. 21). One of the major aims of the pharmaceutical industry is to find small molecules that regulate the biochemistry of disease cells via "signal transduction" modes of action (Cohen 309). Dibenzyl trisulphide (dts) is one such molecule and was first coded in the laboratory when its insecticidal/ repellent activities were discovered. dts was first isolated from the root of Petiveria alliacea at a university laboratory in Brazil in 1990 (Sousa et al. 6353).1

Guinea Hen Weed is also used as a tonic by the Caribs on Dominica and Jamaica. In Grenada, it is commonly used by traditional healers for coughs, colds and as an effective 'cleaner' for the intestines. Its active compounds are often cited in scientific papers:

Cudjoe root [a local name for Petiveria alliacea, see Appendix 1] is well established, if only half-remembered, in Grenadian folk botany and medicine, used predominantly for sinusitis and colds, but also for diarrhea/dysentery and gynecological complaints. It has been shown to contain unique sulfur compounds (e.g. dibenzyl trisulphide) with immunostimulant and antiviral activities, both of which have value in diarrheal disease (e.g. Rotavirus). (Whittaker et al. 490)

It also features prominently into the medical traditions of Obeah in St. Kitts (Urueña et al. 30) and Grenada/Carriacou.

At the same time we see plants growing in use in pharmaceuticals, in our present political condition under contemporary neoliberal globalization, we are seeing an intensification of control over women's bodies. Much of this control manifests itself as reproductive control. In Latin America, especially, contraception and abortion have been contentious issues, and 'population control' remains an important component of international development policy. The United States has a long history of promoting sterilization and Western forms of pharmaceutical contraception globally; international organizations are making inroads in international family planning programs and birth control efforts. Struggles for justice over these issues have emerged in recent decades, especially surrounding forced sterilization and international adoption. We also see current debates over the right to control of fertility in Latin America, including the right to control birth and bodies and maintain access to contraception. Recent press over why most Brazilian women get c-sections highlight Brazil's attempts at pregnancy surveillance, including recent legislation to maintain state records of all pregnancies in an attempt to control population increase. With the stated aim of meeting the UN Millennium Development Goal of reducing maternal mortality, enacted a law in 2012 to establish a national system of registration, surveillance, and monitoring of pregnant and postpartum women. Under PM 557, every pregnant woman who enrolls is entitled to a small payment to assist with prenatal care, and the law intends to "improved access, coverage, and quality of maternal health care, especially in high-risk pregnancies" (Wilson 24).

This type of surveillance is directly related to the ways in which women's bodies are considered in modern science and medicine. Women are not protected from state control of their bodies, and a genealogy of contraceptive plant knowledge easily highlights the relationship between women and empire. The critique of the embeddedness of gender relations in both the practice of science and scientific knowledge itself has been one of the most important contributions of feminist science studies (Harding, *Is Science Multicultural?* 303; Subramanian 956). Feminists have also tracked how science and scientific knowledge exist in markets, capital, and the economy. Many have called modern science a scientific-industrial complex to point to the links between scientific knowledge, post-colonial peoples, and market application. The development of Petiveria as a cancer-fighting treatment must have come directly from the knowledge of lay people in the Caribbean, but we have no understanding of this development as a cultural process, or even a reference point to figure out how this knowledge traveled from local people to the scientific community in the development of a cancer-fighting drug. Feminists and critical science scholars have also connected laboratory funding and practice to global circuits of capital (Wilson 94), noting the false divide between 'science' and 'industry,' with universities supporting science and funding coming from industry. The flows of global science mirror the flows of capital, especially with the creation and marketing of drugs, and commodification of natural resources such as seeds, soil, and water. Therefore, bioprospecting and stories of pharmaceutical development are really at the heart of feminist science studies. Sandra Harding (2008) writes that it is difficult to produce any 'objectivity' in science and the myth of value-free knowledge. Situated knowledge (Haraway, "Situated Knowledges" 183), and strong objectivity (Harding, *Whose Science?* 138) have been particularly valuable in imagining a new "science" that would incorporate women's and lay people's knowledge. The postcolonial focus on Western science sees its developments as one category of many, and also considers how indigenous knowledge has been appropriated, and how science has been implicated in violent forms of colonialism. Alternative knowledge systems, practices, and sciences have been viewed more recently as equally worthy of inquiry (Shiva, *Staying Alive*; Harding, *Sciences from Below*), arguing that sciences must be understood in the plural (Subramanian 966).

Ultimately, the way we use Petiveria and the way it is used in the laboratory reflects our understanding of science, women's bodies, and narrow scientific study. In fact, so called 'folk medicine' has long made use of aqueous and alcoholic extracts from Petiveria alliacea, so this 'invention' was made first by the people doing this work in the Caribbean, Central and South America (Hernandez et al.) and thought it may have been modified in the laboratory, it is impossible to suggest that this knowledge arrived in the laboratory without local knowledge.

Knowledge of the properties of medicinal plants is seen as a local common resource in much of the world, not something to be held privately and made use of for individual profit-seeking. It is clear that the use of traditional knowledge pinpoints plant medicinal uses efficiently; without the use and abuse of Indigenous communities, the biotechnology industry would have no place to turn. Transnational corporations have defended their intellectual property rights and often won their freedom to purchase and patent biological materials, to the detriment of indigenous communities in the Global South. It has been clear that much of this biological knowledge gained and patented by transnational corporations is knowledge that comes from indigenous communities themselves, as problems have arisen where one indigenous community is contracted for their knowledge, but another may use it. *Anthurium tessmannii* is contraceptive in Colombia by three different indigenous nations, for example. The same medicinal compound may be used in various Indigenous communities (either for the same or different medicinal uses), but corporations have been able to claim that they acquired their biomaterials from whatever country and community they choose—under terms and conditions most favourable to them.[2] Knowledge of local geographies, geologies, animals, plants, classification schemes, medicines, pharmacologies, agriculture, navigational techniques, and local cultures that formed significant parts of European sciences' picture of nature were provided in part by the knowledge traditions of non-Europeans.

Such work, then, reminds us that plant collecting and knowledge about medicinal plants have been globally extensive and systematic before the colonialism took hold and that not only Europeans continued to learn about and classify the plant knowledge of the Americas. Genealogies of contraceptive plants correspond nicely with contemporary discussions about the patenting of natural knowledge and what has come to be termed "biopiracy," the exploitation of traditional knowledge through political means.

Scientists invested in plant biotechnology argue that the intensification of agriculture is necessary because of natural limitations that require enhanced and more efficient plant breeding to inspire "the release of economical, high-return and patentable plant-derived products" (Meiri and Altman 41), stressing the importance of this research to the pharmaceutical and agricultural industries. They argue that corporate funds must support advanced research and development in biochemistry, physiology, genomics and biotechnology of agricultural and medicinal plants (Borlaug), citing the population explosion as a global crisis and biotechnology as necessary solution (Meiri and Altman 3). The new plant biotechnology centers around three major areas: first, as an aid to "classical" breeding of plants, including ongoing genome mapping projects in global food staples, such as rice, maize and tomato, in an effort to shorten the time required for breeding cycles (Meiri and Altman 10-12). Second, the generation of engineered organisms: in view of the 'limitations' of naturally-occurring genes, a more 'efficient' engineering of plants has resulted in improved plants that grow faster or taller, or are able to withstand less water. Scientists value "the creation of novel, and otherwise impossible genetic recombinations" (Meiri and Altman) or the creation of plants in the laboratory that would not exist in nature without the hand of the scientist. Further work is being done to integrate microorganisms into plant production systems, that is, to develop plants that have genetic material to resist microorganisms that may harm them, such as fungi, bacteria, and insects using both laboratory-engineered plants and microorganisms.[3]

During the last two decades, new biotechnologies have been adapted to agricultural practices, meaning plants are being used in ways and in scientific disciplines where they never were before, and this will continue and intensify in the next decade. Plant biotechnology (especially in vitro regeneration and genetic modification) is changing the way we understand and know plants, as it affects everything about them, from their growth and characteristics to the ability to reproduce themselves. Plants are being specifically redesigned to produce specialty foods, biochemicals and pharmaceuticals.

In contemporary biotechnological development discourse, then, a complex ecosystem is reduced to individual plant properties, pieces of a system are taken to be a knowledge of its whole, particular properties are placed hierarchically above others, depending on the values of a particular population and time-period, to allow for the manipulation of the ones deemed significant. In particular time periods, including our own, contraceptive plant properties were not deemed significant or important and a plant is commonly reduced to its commercial value, manufactured into a commodity, and reduced again to a profit for pharmaceutical industry. Women's knowledge of plants and nature is manipulated to increase the production and distribution of these commodities, which are in turn legitimized scientifically as a productivity increase, even though its destruction decreases the reduction of diversity and the power or use of any medicinal property, recreating life and violating the ecosystem in the name of progress, science, and development. Profit is then the only gauge of a plant's value, and life as nature's organizing principle disappears. In science, the value is in the application, not in knowledge for other reasons, such as knowing for its own sake or for the betterment of human situation. The epitome of scientific understanding, the controlled experiment, is what Vandana Shiva calls "a political tool for exclusion such that people's experimentation in their daily lives was denied access to the scientific" (*Staying Alive* 31). Contemporary scientists are bound by such controlled experiments and as a result, disseminate research findings that correspond to a directly observable natural world.

In International Development Studies, population statistics are often used to measure poverty and development statistics, including measures of family planning and reproductive health. The World

Health Organization (WHO) estimates that 500,000 women die from pregnancy-related causes worldwide, and that perhaps one-third of those stem from unsafe abortion (WHO, *The Global Family Planning Revolution: 35 Years*). The WHO also ranks Guatemala as the most difficult to implement family planning initiatives, citing the ethnic makeup of the population as reason for it (WHO, *The Global Family Planning Revolution: 30 Years*). The scholars who write on the history of abortifacients often suggest that although plants were popular contraceptives, they were never particularly reliable (Himes; Riddle). Political Scientists likewise argue that these preparations were passed down prior to implementing Western medicine and disappeared because they were found to be unreliable and cause severe side effects. In fact, some say, plant-based contraception only disappeared when surgical abortion became much more accessible (Siedlecky 105).[4] Instead, plant knowledge and contraceptive recipes are still widely used and remain outside the knowledge realm of modern science, which would explain why the results (contemporary child spacing) do not appear to justify their reputation. Still, it is necessary that medical science record their efficacy? Many ethnobotanical studies mention plants used for menstrual regulation, but little information is provided, such as preparation dosage, reported effectiveness, or analysis of how women seek this care. In addition, individual women circulate information about abortifacients mainly by word of mouth and maintain their efficacy by developing this knowledge locally.

Indeed, the assumption that plant-based abortifacients have disappeared because of their lack of effectiveness remains prominent in the Western medical tradition. In contemporary scholarly work, it is common to depict the contemporary Western world as automatically and without question more 'advanced' than in previous historical periods. This is incorrect more often than not, and may cause us to ignore or dismiss the knowledge and understanding that people have gained in earlier times. The historical record demonstrates that women throughout many cultures and time periods have been managing their fertility without Western scientific methods of doing so, and have been long adept at understanding precise uses and misuses of nature. They have likewise made use of ambiguity in language in order to allow space for a variety of reproductive options, including plant-based ones. It is also common to assume that earlier times were more inflexible in comparison with our more "enlightened" age, but many feminists have demonstrated how women have historically had more freedoms regarding their bodies across times and in cultures, not less. Indeed, the twentieth century can in part be categorized as an era marked by the intensification of control over women's bodies (Bush 242; Ehrenreich and English 5-15).

The maintenance of fertility control remains part of this cultural maintenance and resistance to neoliberal globalization and corporate control of local knowledge and life forms. In the twentieth century, the solution to environmental and political problems often fell on the backs of Third World women, who were faulted for overpopulation and accused of ignorance to Western birth control, amid a political agenda that pressed for the sterilization of poor women worldwide. Indeed, instead of further investigating plant-based contraceptives and making them worthy of scientific investigation, recent efforts have been to patent particular life-forms and develop them under corporate ownership for profit in laboratories.

In the Guatemalan countryside, erosion is a continuing problem from slash and burn agriculture and monoculture that pushed indigenous communities further and further into the highlands. As a result, much natural plant life is dying out. We might also cite capitalist development as part of the destruction of this natural knowledge. A midwife I interviewed in the highlands of Guatemala was very angered to tell me that that the younger midwives do not have as much knowledge of plants "and do not know where to find them" (Personal communication, 2014). She cited that they were too urban and that only midwives "further out" (that is, in more remote locations) had more knowledge of where to find Apacina.

Pharmaceutical companies are well aware of the plant-based remedies that have spurred drug development, and indeed, their marketing campaigns for "natural" medicines and nature's remedies are vast. The Native American Ethnobotany database at the University of Michigan-Dearborn brings up 160 plant species that can produce abortion. Still, scholarly treatment of folk methods of birth control remains focused on its absurdity, and some catalog magic and superstition along with herbal and barrier methods. Indeed, comments in many botanical texts refer in passing to plants that may have "some antifertility activity," reminding us that to much of the scholarly world, at least, this information remains hidden. Further, we cannot assume that because an herbal drug has not been tested, it has no contraceptive effect.

Some of the most popular contraceptives, such as birth control pills and intrauterine devices (IUDs) have had adverse health consequences for women. Depo-Provera, an injectible hormonal birth control, it has been argued, has been deemed dangerous by women in the developed

West and has been outsourced to the Global South. The introduction of the birth control pill appears to have actually increased pregnancies in some areas of the Global South, in part because indigenous methods of child spacing, such as extended breastfeeding or abstinence, have been displaced (McClaren 6). Breastfeeding does not provide a completely effective contraceptive method, but it has no known side effects. Depo-Provera has been linked to cardiovascular and uterine problems. Latin American feminists have attempted to make this known, but the powerful campaigns of pharmaceutical corporations have squashed their efforts. Many women in rural Guatemala who use Western forms of birth control choose Depo-Provera because this method is easy to hide from family members, as it only requires one injection every three months. Another problem with many of these pharmaceutical contraceptives is that their continual use must be maintained, whereas plant-based contraceptives are only used when needed. In my interviews, many women cited the importance of only drinking the tea after being intimate with a man, a type of control women cannot find in contemporary pharmaceutical forms of birth control. Contraceptive plants offer women a sense of privacy and control because no one else has to know they are using them, including local health providers and families.

As Donna Haraway has written, it is almost impossible to separate nature from ourselves, but we do so consistently, especially in the context of the developing world when the carbon dioxide production of industrial cultures is absorbed by plant materials and the plants themselves become service providers for the industrial economy, providing a clear example of her concept of "naturecultures": we cannot divide a view of nature from the cultural (Haraway, *How Like a Leaf* 25). The botanical history and contemporary biopiracy of Apacina demonstrates how knowledge and power are intimately linked, and this is nowhere more obvious than in the present global political economy. The base of the global economic system shifted in the twentieth century from heavy industry to information technologies and service industries, and scientific innovation has moved decisively to the base of the contemporary economy. Those who own nature and are able to access its product as well as the global knowledge of it are able to decide how to make use of nature's resources and profit from contemporary scientific innovation and technological change. The majority of the world's people, and especially women, have few of these resources, no ownership of pieces of nature nor the resources to access it, and are in fact systematically denied the knowledge of how to gain access to nature's abundant resources.

214

Ultimately, what is demonstrated by this non-Western worldview among the Maya K'iche' is that medicine and food are not separate life pieces, but part of a holistic understanding of health and bodily care. Market-based herb sellers often asked if I wanted Apacín as a food condiment or a remedy; either way, one woman said, the plant has excellent effects, suggesting that plant-based abortifacients and emmenagogues are most often used for a wide variety of women's health complaints. In rural Guatemala, food and medicine are one and the same, maintained and valued as part of women's knowledge base. In Western contexts, we rarely recognize the possibilities of plant-based remedies and their political function. Yet for other parts of the world, these understandings, long missing from Western scientific study, are very much part of women's knowledge bases and everyday lives.

ENDNOTES

[1]A substance found in Petiveria, dibenzyl trisulphide (DTS), exhibits antitumor and immunomodulatory activities (Williams et al. 17). The extract displayed several mechanisms of action that may explain its antitumor activity, such induction of cytoskeletal reorganization and DNA fragmentation (Urueña et al. 1). Several compounds isolated from Petiveria alliacea compounds have antibacterial and antifungal activities (Benevides et al. 744) and another showed promise as a wound treatment (Schmidt et al. 5223). Anti-inflammatory and analgesic effects have also been studied and reported (Lopes-Martins et al. 245). A 2009 publication supported the molecule's possible role in the treatment of inflammatory ageing diseases (Williams et al. 57). A product that includes the patented dibenzyl trisulphide compound was indicated for the treatment of cancer in 2015 (Patent 20080070839 A1).
[2]See "A Closer Look at the Royalty Payment Agreement Negotiated by Monsanto Corporation and Washington University ICBG Bioprospecting Agreement for Collection of Peruvian Medicinal Plant."
[3]From Proceedings of "Plant Biotechnology and In Vitro Biology in the 21st Century," conference held in Jerusalem, June 1998.
[4]A future project will look at the ways in much migrant women maintain access to this knowledge, even though access to particular plants may become limited.

REFERENCES

"A Closer Look at the Royalty Payment Agreement Negotiated

by Monsanto Corporation and Washington University icbg Bioprospecting Agreement for Collection of Peruvian Medicinal Plants." Published by the etc group in response to License Option Agreement between G. D. Searle & Co. (licensee) and Washington University (Licensor) for Peruvian Plant Extract Collection under the International Cooperative Biodiversity Group Agreement, 1994.

Meiri, H. and A. Altman. "Agriculture and Agricultural Biotechnology: Development Trends Towards the Twenty-First Century." *Agricultural Biotechnology*. Ed. A. Altman. New York: Marcel Dekker, Inc. and Information Systems for Biotechnology (ISB) News Reports,1998. 1-17.

Bajaj, Y. P. S. "Biotechnology in Agriculture and Forestry." *Medicinal and Aromatic Plants* 7.28 (1994): 36-41.

Benevides, P. J. C., M. C. M. Young, A. M. Giesbrecht, N. F. Roque, and V. S. da Bolzani. "Antifungal Polysulphides from Petiveria alliacea L." *Phytochemistry* 57.5 (2001): 743-747.

Borlaug, N. E. "Feeding a World of 10 Billion People: The Miracle Ahead." *Plant Tissue Culture and Biotechnology* 3 (1997): 119- 127.

Brooks, Floyd. "Jamaican Scientists Offered $100 Million for Rights to Patent." Accuteach.com. Web. May 29, 2015. Accessed June 24, 2015.

Bush, Barbara. *Slave Women in Caribbean Society: 1650-1838*. Bloomington: Indiana University Press, 1990.

Cohen P. "Protein Kinase—The Major Drug Targets of the Twenty-first Century." *Nature Reviews Drug Disovery* 1 (2002): 309.

Ehrenreich, Barbara and Deirdre English. *Complaints and Disorders: The Sexual Politics of Sickness*. New York: Feminist Press, 1973.

Haraway, Donna. "Situated Know-ledges." *Simians, Cyborgs, and Women*. New York: Routledge, 1991.

Haraway, Donna. *How Like a Leaf*. New York: Routledge, 2000.

Harding, Sandra. *Is Science Multicultural? Postcolonialisms, Feminisms, and Epistemologies*. Bloomington: Indiana University Press, 1997.

Harding, Sandra. *Whose Science? Whose Knowledge?* Ithaca, NY: Cornell University Press, 1991.

Harding, Sandra. *Sciences from Below: Feminisms, Postcolonialities, and Modernities*. Durham: Duke University Press, 2008.

Hernández J. F., C. P. Urueña, M. C. Cifuentes, et al. "A Petiveria alliacea Standardized Fraction Induces Breast Adenocarci-noma Cell Death by Modulating Glycolytic Metabolism. "*Journal of Ethnopharmacology* 153 (2014): 641-649.

Himes, Norman E. "Medical History of Contraception." *New England Journal of Medicine* 210 (1934): 576-581.

Lopes-Martins, R. A., B. D. H. Pegoraro, R. Woisky, S. C. Penna and J. A. A. Sertié, "The Anti-in-flammatory and Analgesic Effects of a Crude Extract of Petiveria alliacea L. (Phytolaccaceae)." *Phyto-=medicine* 9.3 (2002): 245-248.

Lowe, Henry I. C. et al. "Petiveria alliacea L (Guinea Hen Weed) and Its Major Metabolite Dibenzyl Trisulfide Demonstrate HIV-1 Reverse Transcriptase Inhibitory Activity." *European Journal of Medicinal Plants.* 5.1 (2014): 88-94.

McClaren, Angus. *A History of Contraception from Antiquity to the Present Day.* Oxford: Basil Blackwell, 1990.

Meiri, H. and A. Altman. "Agriculture and Agricultural Biotechnology: Development Trends Towards the 21st Century." *Agricultural Biotechnology.* Ed. A. Altman. New York: Marcel Dekker, Inc., and Information Systems for Biotechnology (ISB) NEWS REPORTS, 1998. 1-17.

Riddle, John. *Eve's Herbs: A History of Contraception and Abortion in the West.* Harvard University Press, 1997.

Rice, Susan E. "Poverty Breeds Insecurity." *Too Poor for Peace?* Eds. Lael Brainard and Derek Chollet. Washington, DC: Brookings Institution, 2007. 44-46.

Schmidt, C., M. Fronza, M. Goettert, et al. "Biological Studies on Brazilian Plants used in Wound Healing." *Journal of Ethnopharmacology* 122.3 (2009): 523–532.

Setzer C. M. et al. "Biological Activity of Rainforest Plant Extracts from Far North Queensland, Australia." *Biologically Active Natural Products for the 21st Century.* Ed. L. A. D. Williams. Delhi: Research Signpost, 2006. 21-46.

Shiva, Vandana. *Staying Alive: Women, Ecology and Development.* New York: South End Press, 2010.

Shiva, Vandana. "Bioprospecting as Sophisticated Biopiracy." *Signs: Journal of Women in Culture and Society* 32.2 (2007): 307-313.

Shiva, Vandana. *Biopiracy: The Plunder of Nature and Knowledge.* New York: South End Press, 1997.

Siedlecky, Stefania. "Pharmacological Properties of Emmenagogues: A Biomedical View." *Regulating Menstruation: Beliefs, Practices, Interpretations.* Eds. Etienne Van De Walle, and Elisha P. Renne. Chicago: University of Chicago Press, 2001. 93-113.

Sousa, Jose R. et al. "Dibenzyl Trisulphide and Trans-n-Methyl 4 Methoxyproline from Petiveria Alliacea." *Phytochemistry* 29.11

(1990): 3653-3655.

Subramanian, Banu. "Moored Meta-morphoses: A Retrospective Essay on Feminist Science Studies." *Signs: Journal of Women in Culture and Society* 34.4 (June 2009): 951-980.

Urueña, C., C. Cifuentes, D. Castañeda, et al. "Petiveria alliacea Extracts Uses Multiple Mechanisms to Inhibit Growth of Human and Mouse Tumoral Cells." *BMC Complement Alternative Medicine* (Nov. 18, 2008): 8-60.

Whittaker, J. A., M. Smikle, E. N. Barton, and Y. A. Bailey-Shaw. "Immunological Evidence Supporting the Use of Extracts from Boehmeria Jamaicensis Urb for Treating the Common Cold and Sinus Infection." *West Indian Medical Journal* 56 (2007): 487-90.

Williams, Lawrence, et al. "A Critical Review of the Therapeutic Potential of Dibenzyl Trisulphide Isolated from Petiveria alliacea L (Guinea hen weed, anamu)." *West Indian Medical Journal* 56.1 (2007): 17-23.

Wilson, M. R. *A Highland Maya People and their Habitat; The Natural History, Demography, and Economy of the K'ekchi*. Doctoral dissertation, University of Oregon, 1972.

World Health Organization (WHO). *The Global Family Planning Revolution: 30 Years*. Geneva: WHO, 2010.

World Health Organization (WHO). *The Global Family Planning Revolution: 35 Years*. Geneva: WHO, 2015.

World Health Organization (WHO). *WHO Expanded Program of Research, Development and Research Training in Human Reproduction*. Task Force on Indigenous Plants for Fertility Regulation. Geneva: WHO, 1977.

10.
Building Food Sovereignty through Ecofeminism in Kenya

From Capitalist to Commoners' Agricultural Value Chains

LEIGH BROWNHILL, WAHU M. KAARA AND TERISA E. TURNER

I̱N 1997, WE CO-AUTHORED AN ANALYSIS of a women-initiated movement of Kenyan farmers to break free from long-standing coffee commodity value chains in which they had been entangled for a century. The article described how coffee was first grown in the temperate Central Kenyan highlands on white settler plantations in the early part of the twentieth century. By 1940, coffee was one of the most lucrative commodity markets in the world. After protest against discriminatory coffee licensing practices, large land-holding African farmers were allowed to grow coffee. Smallholders were later provisionally given license to grow coffee, but restrictions still barred the majority, whose under-four acre farms were deemed in policy as uneconomical for coffee production.

By the 1980s, the World Bank was actively advocating for farmers to grow more coffee, and for governments of peasant nations to adopt policies that would facilitate increase of export production, in part as a means for those governments to repay international loans. Corporate coffee value chains came to benefit auctioneers and middle men, as well as foreign bankers, far more than the average farm husband, whose name was on the land title, the production contract, and the bank account (Brownhill, Kaara and Turner 43). That piece described the gendered class dynamics within which peasant women in coffee growing districts like Maragua in Central Kenya uprooted coffee trees, and in tandem planted food crops on their small farms.

Our 1997 analysis emphasized the ways that women collectively asserted their usufruct rights to grow food on their own family land. In asserting this claim, women broke the "male deals"[1] that had bound husbands (and wives) into unfair production and trade relationships with men of the state and of international capital, relationships which were becoming harmful to the environment through chemical application, and to producers' health and well-being through exposure to chemicals but also significantly to loss of food gardens when the coffee economy expanded.

The article also examined the "ethnicized gendered class alliance" that emerged to support women's return to cultivation that prioritized local food value chains, starting with feeding farm families and supplying basic nutritious foods, like organic bananas, to urban markets. Women farmers found allies among young men who worked in transport and trade, and who built links between rural farmers and urban eaters. We held up the case as an example of the contemporary state of global struggle that saw the enclosures of neo-liberal corporate rule being resisted by peasant women, whose efforts in alliance with young, dispossessed men were aimed at enabling the re-emergence of subsistence-oriented agriculture and local trade. Twenty years later, how far have women and men in places like Maragua succeeded in re-establishing what we can call in contemporary parlance, local food sovereignty?

THE COFFEE ECONOMY
TRANSFORMED: MARAGUA IN 2016

In March 2016, I, Leigh Brownhill, returned to Kenya to visit my co-author, sister, and friend, Wahu Kaara. During that stay, we made a daytrip to the collective farm of a youth group in Maragua to learn about the progress of the agricultural transformation there, in the heart of a region that spurned coffee some thirty years ago. What had happened to the farm women's efforts to transition to the cultivation of food crops? What follows is our collective reflection on this question, with attention to the gender dynamics and ecological outcomes in the Maragua case.

It is one thing to uproot coffee trees and plant bananas and vegetables to supply household subsistence, as women in Maragua and surrounding areas did beginning in the mid-1980s. But it is quite another thing to continue to maintain a food-first focus year after year in a market and policy context that is hostile to subsistence agricultural production and trade.

Farmers who have abandoned a cash crop, for reasons varying from world market forces to plant disease and cost burdens, need not only to grow their own food, but also to find alternative sources of income to cover purchases of goods they cannot produce, and for their other cash needs. To produce surpluses of vegetables for sale takes time, not only to convert the farm, but also to find the customers and build the market links. Hence in rural areas of Kenya today, there are large surpluses of fruits and vegetables that never reach markets or paying customers, due to lack of affordable transport.

The Maragua ex-coffee farmers were not given much chance, or any official support, to re-establish local agricultural skills, indigenous seed stocks, and local produce markets. The dominant view guiding policy and development planning in Kenya remains entrenched in a neo-classical economic paradigm whose proponents assert that subsistence means poverty, and global markets mean development (Seavoy 252). International advisors have designed and imported development interventions that sought (and continue to seek) to fully commercialize Kenyan farming (as elsewhere in the world) by linking smallholder farmers into global value chains.

In this neo-liberal policy milieu, instead of supporting the coffee farmers' initiatives with research, extension, and education focused on the crops that farmers do prefer, agricultural scientists and international advisors simply sought to recapture the coffee labour force and channel it into alternative export crop production. Canadian and U.S. development agencies, for example, in alliance with biotech and big agribusiness firms, promoted ex-coffee farmers' adoption of other mass-market-oriented crops, especially perishable horticultural products, such as strawberries, macadamia nuts, and French beans (Komu).

Maragua, as it happens, was the site of one such project to promote tissue culture bananas (Karembu 2). "Tissue culture" is a method of propagation of plant material, done in laboratories in a specially formulated nutrient medium, using the tissue of disease-free plants. Tissue culture bananas produce uniform, disease-free plants and fruits, but require more labour, more inputs, and more expenditures, as planting material must be purchased each season rather than propagated on-farm from suckers from mature banana plants (Indimuli 7). Many of these crops were touted as "women friendly" and marketed to appeal to female-headed households (Kabunga et al 22). Now the old coffee "male deal" was being repackaged as a new kind of deal for both women and men to independently contract their labour for supply to global horticulture value chains.

About six years ago (2010), when the Shiriki youth group began a collective organic farm in Maragua, they were confronted with the reality of a local political economy still heavily slanted towards highly chemicalized commercial production. Tissue culture bananas were only one of the new cash crops that farmers had adopted to replace coffee.

Some neighbours grow cucumbers for urban wholesalers. Some have tried French beans, destined for export. But in Maragua, even for those who adopted the production of these alternative cash crops, many have found similar problems, including high costs, variable incomes, unequal burden of labour in the household, and negative ecological implications. Farmers' low adoption and discontinuance of those crops has become a problem for the bio-tech firms operating in Maragua (Indimuli 5; Kabunga et al 2). The alternative cash crops have turned into false solutions to critical problems of rural hunger and youth unemployment.

The Shiriki group is comprised of young Rastafari, most of whom were originally living in urban slums. Rastafari in East Africa is rooted in the region's nineteenth-century Nyabingi militant matriarchy and the global Rastafari movement (Turner 9). "Shiriki" means "participation" in Kiswahili, and signified the youths' commitment to self-sufficiency and self-determination. The group organized for the purpose of raising money from craft manufacture and fundraising cultural events to enable them to move to the rural area, find land to farm, and build houses, storage sheds, and a communal hall. They did so after forging an agreement with a widow who could not farm the land due to her employment in a distant urban centre. Now a core and fluctuating group of some five to fifteen youth live in Maragua and operate a two-acre collective farm, with the long-term aim of establishing a rural cultural education and media centre.

The group's goals include food self-sufficiency through organic ecological farming, and establishment of rural food processing, crafts, and trade. Members also seek to build a larger community of "fossil fuel-free" farmers (Shiva), and so educate and disseminate information in the weekly village market, where they sell surpluses of their local organic foods and handmade arts and crafts. They do outreach among neighbours, school children, visitors, and international volunteers, along with demonstrations of the use of their powerful solar cooker. The land is richly endowed with spring water and a high water table, allowing for readily available ditch irrigation. They have a local as well as a continental view of their initiative for "cultural independence" (Shiriki website).

The Shiriki group has gained ground since 2010 in becoming largely food self-sufficient. The initiative has grown as group members have integrated with the community. The seed varieties they have planted have come from local women farmers, especially elder women seed savers, who encourage the Shiriki youth's effort to revitalize local farming with Indigenous agricultural knowledge and seeds. The biggest challenge they reported was that while most of their neighbours want to "go organic" and to focus on local food crops, the markets for these crops are not developed, and the need for increased labour (e.g., for weeding and pest control) acts as a barrier, due to the migration of youth away from rural areas and farming professions. To counter these challenges, the Shiriki youth have worked diligently to build marketing links, both in local village markets and central urban areas. They have even managed to represent their group at an international trade fair in Tanzania (Shiriki).

Rural-urban migration is a highly gendered phenomenon. Shiriki's group members, the majority of whom are young men, have developed a position on the gender dynamics of contemporary agricultural communities, where both women and men are prone to leave for the town or urban areas. They ask,

> ...where are the young men and women of this community? And the answer is quite apparent for the young men are [taking work in the] unintelligently designed industry of motorcycle transport [taxis]. And if the men are not in the farm, who shall attract the young women to till the land? So the sisters have found themselves in this exodus to the urban [areas] where they offer themselves for pure exploitation. (Shiriki)

Their analysis relies on an understanding of land relations in the country where women rarely get independent access to land. Without up-turning the whole colonial and cultural history of bias against women's land access, the only viable option for most young women is to seek employment in towns or cities to meet their subsistence needs. This gendered dynamic is firmly at the root of the emptying out of rural areas and the filling up of urban slums—the cumulative socio-cultural, economic, and geographic impacts of waves of colonial and post-colonial imperial enclosures. Ecological degradation and deep social injustice in both rural and urban areas have also been the persistent, disastrous results of these ongoing enclosures.

The Shiriki youth do not *advocate* women's access to land only through relationships with men. They rather *acknowledge* that the huge project of establishing food sovereignty and transforming the capitalist food system must be started—*is being started*—even before all of the perfect conditions are present in the society. The construction of a food sovereign future requires concrete work now. And the Shiriki youth have shown themselves prepared to work within, against, and around the existing policy and market barriers, and discriminatory social norms around land access, in order to get a footing from which to replace the old hierarchical and exclusionary order with an inclusive, gendered, and generational alliance for food sovereignty.

As much as farming is a seasonal and cyclical activity, it is also a generational matter. So the willingness and ability of the Shiriki youth to ground themselves on land can only go so far, and for so long, if there are no young women around. Both in the day-to-day and seasonal cultivation of the farm, and as importantly in the founding of successor generations, women's presence and labour can be a deciding factor in the success of the organic transition in the long-term, at Shiriki farm and more widely.

Given the double impact of a history of women's exclusion from land ownership and decades of youth outmigration from the countryside, young women's return to rural areas in large part depends on the availability of land. Access to that land may come both through marriage (which then requires securing of potential spouses' land rights) and through independent means (owning, leasing, or borrowing). Very few women, particularly young women residing in urban slums, can ever afford to buy or lease land and make an income to cover all the associated costs of farming. And most do not have any direct relative from whom to borrow even a small plot. So how will the landless, men or women, be able to return to farming in Kenya, and elsewhere for that matter?

Barring massive state-sponsored land re-distribution, which is unlikely, the transformation of the rural economy from its present doldrums and decline to the emergence of vibrant, youthful, agroecological communities may well depend upon a re-invigoration of the customary practices of overlapping entitlements. In particular, the critical return of young women to rural areas calls for the re-institution and re-invention of some long-lost Indigenous forms of land tenure that applied specifically to women (e.g., variations on old Kikuyu customs of some daughters' and widows' rights to occupy land singly). Farmers who cannot utilize all of their land or who need extra help on their farms can find their own ways of following the example of land-sharing at Shiriki, to bring young women and young men back to the land, perhaps as farm apprentices in exchange for periods of usufruct rights. Whatever path the Maragua farmers choose, their decisions will shape the future of farming on their own land and within the wider community, including within the urban diaspora, for generations to come.

The transformation of the food system in Kenya remains in progress in 2016. This progress is accompanied by diverse efforts to retain and rebuild Indigenous knowledge, seed systems and soil fertility, not only at Shiriki's farm, but in hundreds of farmer groups mobilizing for food sovereignty across Kenya. One of the biggest challenges to the food sovereignty movement is the need to transform the policy context into one that supports local trade of local goods, and spurs youth self-employment and urban-rural migration. In tandem with farmers' own efforts and initiatives towards the building of movements and communities of organic farmers, public budget support and international development funding could go a long way towards advancing the transformation of rural economies globally from food insecure to food sovereign.

When going into organic farming for local markets, while farmers' incomes may begin small, they come with other unpriced values (health, biomass, soil fertility, social capital) which can be re-invested in the farm to enable the small income to grow. The main lesson gleaned from the visit to Shiriki was that an ecologically-conscious new generation of Kenyan farmers has emerged with the energy, skills, willingness, and orientation to realize an agricultural transformation that has eluded previous generations. The younger generations' potential stems in part from their cosmopolitan urban experiences and from their exclusion from their elders' lifelong entanglement with export crop production regimes, commercialization policies, and agro-chemical applications.

Because youth have for decades been taught to "rise above" mere farming, and so have fled rural areas for elusive urban educational and job opportunities, most youth have also never been enmeshed in corporate agricultural value chains. There is great willingness among Kenyan farmers and capacity among the youth, in Shiriki and in other food sovereignty initiatives, to switch from corporate control over oil-dependent, global agricultural value chains to commoners' value chains that enable organic and Indigenous methods, seeds and culinary preferences, and are characterized in particular by farmers' sovereign decision-making over land, seed, knowledge, and on-farm energy sources.

But besides a willingness to learn and capacity to work hard, the wider transformation of the farming system towards agroecology requires people who can change the social relations of rural areas from hierarchical and extractive towards horizontal and harmonious. We have characterized these relations as "male deals" and "ethnicized gendered class alliances" (Turner and Brownhill 1046). The most important change characterizing the Shiriki farming initiative was the establishment and elaboration of ethnicized, gendered class alliances, first between the young men who wanted to return to farming, and the elder women who had the advantage of land, and Indigenous knowledge and wisdom to share. The extension of this alliance to wider constituencies depends on the increased involvement of young women who share an interest in returning to the land. Such alliances, rooted in making access to land more equitable, remain key to the transformational power of ecofeminist food sovereignty initiatives.

Eric Holt-Gimenez, the Executive Director of Food First and the Institute for Food and Development Policy in Oakland, California, recently blogged that researchers need to go beyond studies of and strategies for "scaling-up" agroecology and "changing the food system," to inquire more deeply into "how agroecology [can] help us transform capitalism itself" (Holt-Gimenez). In reviewing what is needed to make agroecology "the norm rather than the alternative," Holt-Gimenez also partially answered his own question, when he noted that, "agroecology requires extensive human labor coupled with place-specific knowledges—both of which are incompatible with the current system's need for vast, cheap inputs" (ibid).

Extensive human labour and place-specific knowledge? It sounds like the emerging youth agricultural initiatives, urban-rural migration and Indigenous knowledge systems that the Shiriki youth and the wider food sovereignty movement are championing. It is precisely in the process of new generations actually reversing the erosion of farming as a livelihood, and learning and sharing Indigenous knowledge and seed systems, that agroecology transforms the capitalist system (Giacomini). Agroecology provides grounds for the more widespread replacement of capitalism by offering the opportunity for household and community food self-sufficiency and "cultural independence" outside of corporate value chains.

CONCLUSION

The decades-long commercialization drive of international development through trade and aid has failed to overcome the chronic humanitarian disaster of hunger and malnutrition, especially in East Africa (Brownhill, "A Climate for Change" 225). Neo-liberal policies have only served to scuttle, divide, and weaken the alternative answers to the hunger question, including local food self-sufficiency initiatives, crop biodiversity, and Indigenous knowledge. Market-driven agricultural development has meant that public resources (whether Kenyan scientific organizations or publicly funded international development and research funding) are failing to realize or to mobilize the knowledge and power of local agricultural systems. They only deem it worthy to do so if there is something that can be patented and commercialized, whether by a local or foreign firm that will privatize the knowledge and exclusively reap the profits.

Instead, through direct household food self-sufficiency and the re-institution of local marketing of local goods, the Shiriki farmers and others linked (often through mobile technology) into wider Kenyan, East African, and global food sovereignty networks, are tackling hunger and malnutrition by refusing to participate in the corporate market, and by implication, by strengthening Indigenous agricultural technologies and practices and recreating peoples' markets. They do so by changing with whom they do business (urban and rural commoners rather than local or foreign capital) and under what terms they carry out their farming initiatives (cooperative, organic and life-centred rather than competitive, chemical-intensive and profit-driven).

In an age of extreme extractivism, post-peak oil, climate chaos and political and economic uncertainty, the Shiriki case highlights the critical relation of youth to the coming transition to post-capitalist (and therefore more labour-intensive) organic food systems, globally. It also suggests that the reversal of rural-urban migration requires, to start, a wider recognition of the many kinds of opportunities that could entice jobless youth into self-employment in services and trades supporting farmers' food sovereignty initiatives. Everywhere in the world are urban dwellers, especially youth, who are returning (or wishing to return) to rural livelihoods, learning to farm without expensive and damaging chemical inputs, and grappling with the imbalance between the labour required and the labour available. Youth unemployment and farmers' on-farm labour shortages could both be addressed through further urban-to-rural migration, and this likely requires support through the establishment of new types or re-constitution of old customs of land-sharing, such as that reached by Shiriki youth and the widow on whose land they farm.

Youth are a powerful catalyst for rural transformation, insofar as they can inject energetic labour, skills, and creativity into labour-intensive agriculture-related fields, and whose self-employment in rural areas could facilitate farmers' existing efforts. Supportive industries range from supply of local construction materials and carpentry services (e.g., for irrigation, fences, granaries, coops and barns), to ox-ploughing, composting and post-harvest value-addition in crop processing, packaging, branding, and distribution.

This kind of transformation of the farming system is not the effort of one season; but the slow, steady work of rebuilding self-sufficient, subsistence-oriented rural political economies. In Maragua, that work is well advanced. Elaborating on the transformation begun by a women's coffee protest in Maragua twenty years ago, the enthusiastic initiative of young Rastafari women and men in Shiriki sets a stellar example of the vast untapped potential for a generational succession from extractive corporate agri-business to a new era of climate-adapted, youth-mobilized agroecology.

ENDNOTES

[1]Male deals are cross-class and cross-ethnic collaborations between capitalists and colonized or working class men, for the purpose of channeling the land, labour, fertility and other resources of women, poor men and whole communities into the commodified, cash economy and global markets (see Brownhill, *Land, Food, Freedom* 24).

REFERENCES

Brownhill, Leigh. "A Climate for Change: Humanitarian Disaster and the Movement for the Com-mons in Kenya." *Women, Gender and Disaster: Global Issues and Initiatives*. Eds. Enarson E. and P. G. Dhar Chakrabarti. Delhi: Sage International, 2009. 224-232.

Brownhill, Leigh. *Land, Food, Freedom: Struggles for the Gendered Commons in Kenya, 1870-2010*. Trenton: Africa World Press, 2009.

Brownhill, Leigh, Wahu M. Kaara and Terisa E. Turner. "Gender Relations and Sustainable Agriculture: Rural Women's Resistance to Structural Adjustment in Kenya." *Canadian Woman Studies/les cahiers de la femme* 17.2 (Spring 1997): 40-44.

Giacomini, Terran. "System Change Grounded in Food Sovereignty at the 2015 Climate Talks in Paris." *La Via Campesina, International Peasant's Movement*, 27 January 2016. Web. Accessed 21 July 2016.

Holt-Gimenez, Eric. "If Agroecology Is So Great, Why Aren't All Farmers Doing It?" *Huffington Post Blog*, 8 July 2016. Web. 10 July 2016.

Indimuli, Raphael. "Factors Influencing the Discontinuance in Adoption of Tissue Culture Banana Technology: A Study of Smallholder Farmers in Maragwa District." Master of Arts, Development Studies,

University of Nairobi, October 2013.

Kabunga, Nassul S., Thomas Dubois and Matin Qaim. "Information Asymmetries and Technology Adoption: The Case of Tissue Cul-ture Bananas in Kenya." Courant Research Centre Discussion Paper 74, Georg-August-Universität Göttingen, March 2011.

Karembu, Margaret. "Enhancing the Diffusion of Tissue Culture Banana to Small-Scale Farmers in Kenya." Tissue Culture Banana Policy Brief. *The International Service for the Acquisition of Agri-Biotech Applications*, May 2007.

Komu, Nicholas. "Nyeri coffee farmers switch to macadamia citing frustrations." *Daily Nation*, May 4, 2016. Web. Accessed 10 May 2016.

Seavoy, Ronald E. *Subsistence and Economic Development*. Westport: Praeger, 2000. *Shiriki: Maragua Project*. n.d. Web. Accessed 1 July 2016.

Shiva, Vandana. *Soil Not Oil: En-vironmental Justice in an Age of Climate Change*. Boston: South End Press, 2008.

Turner, Terisa E. "Women, Rastafari and the New Society: Caribbean and East African Roots of a Popular Movement to Reclaim the Earthly Commons." *Arise Ye Mighty People! Gender, Class and Race in Popular Struggles*. Ed. Terisa E. Turner.Trenton: Africa World Press, 1994. 9-55.

Turner, Terisa E. and Leigh Brownhill. "African Jubilee: Mau Mau Resurgence and the Fight for Fertility in Kenya," *Canadian Journal of Development Studies* 22 (2001): 1037-1088.

11.
Monsanto and the Patenting of Life

JENNIFER BONATO

ECOFEMINISM IS A FEMINIST THEORY that aligns the oppression of women with the oppression of nature based on their reproductive labour. In a visual depiction of the Capitalist-Patriarchal economy, "The Iceberg Model," splits the economy into two parts (Bennholdt-Thomsen and Mies 32). The visible economy includes only capital and waged-labour, represented by Gross Domestic Product (GDP) and Gross National Product (GNP). The invisible, reproductive work of nature, women, the unpaid and poorly paid, and Indigenous people is therefore valued as free. The devaluation of women and nature is essential to the maintenance of a capitalist-patriarchal social order. Reproductive labour is viewed as value-less—feminized, primitivized—and exploited as surplus.

Land is the mother of wealth, and labour its father (Salleh *Ecofeminism as Politics* 13). By making the connection between the exploitation of women's bodies and the body of nature, ecofeminists challenge the way that production, rather than reproduction, has been centralized under capitalist-patriarchy. The capitalist-patriarchal economy characterizes the world as resource, while the generation of wealth is based on turning resource to commodity (113-115). It is, according to ecofeminists, systems of reproduction that create an economic "surplus," for instance, life-giving work—that of water, rocks, air, animals, women and care-takers—that are not factored into a nation's GDP or GNP. The word re-production implies a secondary action, but reproductive systems create the conditions for production to take place. The capitalist-patriarchal economy places reproductive labour somewhere between "natural resource" and "condition of production" (108).

Ecofeminists argue that the market economy is not based on reality, but on the fragmentation of natural resources. Market growth is not based on human need, but on capital accumulation and increased profits. This system is destructive, and "kills off its own future options as it goes" (121). The delicate balance of the natural ecosystem is ignored, and nature is broken down into commodity goods in the name of progress and technological advancement (Salleh "Eco-Sufficiency" 4).

PRIMITIVE ACCUMULATION:
THE OPPRESSION OF REPRODUCTIVE BODIES

Ecofeminists have long declared Marx's theory of primitive accumulation to be inadequate, and view primitive accumulation as an ongoing process and permanent condition of capitalist-patriarchy. Silvia Federici, in her book *Caliban and the Witch* (2004), sustains that the European witch-hunts of the twelfth to seventeenth centuries were the first form of primitive accumulation (Federici 12). The witch-hunts meant that women, cast as witches, were reduced to bodies for use by the church and state. Having her property seized, a woman accused of witchcraft was tried and tortured, her body fragmented through a process that generated significant profits for the church and state, and lead to the development of modern day legal and medical systems.

According to Vandana Shiva, twenty-first-century primitive accumulation is enacted through biotechnology, genetic engineering, and patenting of genetic material (*Staying Alive*). Using biotechnology, private companies reduce organisms from whole beings to fragmented genes, and genes are characterized as blueprints for the living object (Shiva "Biotechnological Development" 209-210). Shiva explains the privatization of genetic material as a continuation of the private property laws established during the fifteenth and sixteenth centuries (194). Seeds possess a dual character, they are, as seed, means of production and, as grain or plant, final product (199). This is a problem under capitalist-patriarchy, as the natural reproductive capacity of a seed means that it will freely and continuously reproduce.

Genetic engineering involves extracting DNA from plants or animals and inputting them into the DNA of *different* plants or animals so that future generations will show traits desired by private companies. Using these processes to alter the genes of a seed, companies can patent and add "value" to a seed. However, this view of "value" is not neutral (199), it is a capitalist-patriarchal perspective (as depicted in the Iceberg Model) that centralizes uniformity, control, and profits while ignoring the "feminine principle" that views the world as life-giving, diverse, interconnected, and whole (Shiva, *Staying Alive* 14). Quite often, the

231

modifications are resistance to herbicide or pesticide that use chemicals like glyphosate (a likely carcinogen according to a 2017 report by the World Health Organization's *International Agency for Research on Cancer* [75-65, 112]). Companies then patent their creations in order to contract and sell their use.

A profitable seed is one that has been *produced* through scientific engineering, not *reproduced* by its innate reproductive ability. Thus, in order to become profitable, a seed must be patented and its reproductive capacity must be eliminated. Patenting the seed ensures that the individual seed *and* its subsequent generations remain private property. Seeds that have historically been cultivated by those in the global South, Indigenous people, or subsistence farmers, then, are considered to be "primitive cultivars," while those created through biotechnology are considered "advanced" and "elite" (Shiva "Biotechnological Developent" 209). This characterization (again, as depicted in the Iceberg Model), upholds a global hierarchy where private companies like Monsanto have free use of "primitive," "value-less" seed, while resources stolen or cheaply bought from the global South are accessible as raw material. Under capitalist-patriarchy, seeds only become valuable once they have been subject to investment of time, research, money, and patents (209).

The distinction between "primitive" and "advanced" seed provides a link to Shiva's assertion that this process is a form of primitive accumulation. It also links to von Werlhof's theory: justifications of modern progress sustain that all life has to be fragmented and appropriated in order to be "civilized." "Primitive" seed is thus "advanced" through laboratory research, analysis, and engineering. Adapting the seed from "primitive" to "advanced," Monsanto and other biotechnology companies engage in a political process whereby the location of power is shifted from within the seed, as "use value," to patent-owners, for "exchange value." Using biotechnology, patent-owners appropriate the seeds for profit, and the reproductive capacity of the seed becomes private property.

Below, I describe biotechnology and align definitions from the United Nations Convention on Biological Diversity (1993), the United States Food and Drug Administration, and Health Canada. I discuss the history of Monsanto, and their use of biotechnology to manufacture, patent, and sell seeds and chemicals. Finally, I provide an overview of the findings from a Major Research Project, during which I explored the U.S. Supreme Court case *OSGATA et al. v. Monsanto* using an ecofeminist framework.

BIOTECHNOLOGY, MONSANTO, AND THE COURTS

Biotechnology
The United Nations Convention on Biological Diversity defines biotechnology as "...any technological application that uses biological systems, living organisms, or derivatives thereof, to make or modify products or processes for specific use" (United Nations Convention on Biological Diversity 3). Biotechnology is explained as a natural and necessary progression of science and technology that is intended to "feed the world" and address challenges associated with climate change. Scientifically, biotechnology involves extracting DNA from one organism and inputting it into another, or altering the DNAsequence of an organism in order to produce the traits desired by private companies. This process turns them from "nature's seeds"—accessible resource— to "corporate seeds"—a new commodity for consumption (Shiva "Biotechnological Development" 200).
The process of biotechnology creates genetically modified organisms (GMOs). According to the United States Federal Drug Administration, GMOs are "originally used by the molecular biology scientific community to denote a living organism that has been genetically modified by inserting a gene from an unrelated species" (Vecchione, Feldman, & Wunderlich 329). Health Canada states that "a GM [Genetically Modified] food is one derived from an organism that has had some of its heritable traits changed ... using chemicals or radiation to alter the genetic make-up of the organism's cells called mutagenesis," or "introducing a gene from one species into another species" (Health Canada "Genetically Modified").
GMOs are categorized by Health Canada as "novel foods," defined as:

(a) a substance, including a microorganism, that does not have a history of safe use as a food;
(b) a food that has been manufactured, prepared, preserved, or packed by a process that
has not been previously applied to that food, and
causes the food to undergo a major change; and
(c) a food that is derived from a plant, animal, or microorganism that has been genetically modified such that
the plant, animal, or microorganism exhibits characteristics that were not previously observed in that plant, animal, or microorganism,
the plant, animal, or microorganism no longer exhibits

characteristics that were previously observed in that plant, animal, or microorganism, or one or more characteristics of the plant, animal, or microorganism no longer fall within the anticipated range for that plant, animal, or microorganism. (Health Canada "Genetically Modified")

After engineering a seed, the product is patented and then requires an annual fee for its use. Every year, a number of novel foods receive Health Canada approval and enter the Canadian food system. Many of these products are altered strains of corn, soy, or canola to be used for bio-fuel and animal feed, with built-in resistance to chemical pesticide or herbicide to be used in the growth cycle (Health Canada "Genetically Modified").

Since 2014, twenty-one "novel foods" have been approved in Canada (Health Canada "Genetically Modified"). Some companies, such as Okanagan Specialty Fruits and Aqua Bounty, have products that are standalone and sold whole. Past whole food approvals include pink pineapple, potatoes, yogurt, margarine, birdseed, alfalfa, and orange juice products (U.S. Food and Drug Administration; Health Canada "Genetically Modified").

Monsanto

The multi-national corporation Monsanto is a major player in the global biotechnology industry. Founded in 1902 by John F. Queeny, Monsanto's first products included saccharine (an artificial sweetener) and herbicide 2,4-D (used alongside 2,4,5-T to make Agent Orange, part of the U.S. military's herbicidal warfare program against Vietnam, produced by Monsanto from 1965-1969) (Monsanto, n.d.). Despite Monsanto's origin in war products such as Agent Orange, Round-Up and Glyphosate (*OSGATA et al. v. Monsanto*, 2011), strategic public relations have positioned Monsanto as a leader in industrial agriculture, on a quest to "feed the world." Monsanto describes its biotechnology research program as established in 1981, with the set up of its first molecular biology group ("Original Monsanto Company"). Biotechnology is, according to Monsanto, the solution to climate change, drought in Africa, the decline of honeybees, and sustainable agriculture.

Monsanto's vision of sustainable agriculture is based on increased output with decreased input. This means less land, less water, and less energy per unit produced while managing weeds, insects and environmental changes. In 2015, Monsanto boasted to shareholders about their annual progress: expansion of soybean crops to fifteen million hectares in South America (with expectations to double that coverage in the following year) and holding share in every major corn market (Monsanto *Annual Report* 3). Monsanto projects to double yields in core crops by 2030, enabling "farmers to get more from every acre of farmland" through "advanced plant breeding, biotechnology and improved farm-management practices" (Monsanto "Improving Agriculture").

OSGATA ET AL. V. MONSANTO

In a Major Research Project entitled *Monsanto and the Patenting of Life: Is Biotechnology a New Form of Primitive Accumulation in the 21ˢᵗ Century?*, I wrote about biotechnology using an ecofeminist framework. I examined the U.S. Supreme Court case, *Organic Seed Growers and Trade Association et al. v. Monsanto* in order to critique the role of the legal system in favouring private ownership over public concerns.

My research found that new avenues of profit require new legal frameworks crafted by the United Nations Convention on Biological Diversity and North American Free Trade Agreement, and applied by the Supreme Court. Patent law, sanctioned by the U.S. Supreme Court, allows Monsanto to appropriate the bodies of seeds, and subjects small organic farmers in the United States and Canada to risk of contamination of their crops. The right of Monsanto—the colonizer— to patent and own genetic material was entrenched by the United States Supreme Court ruling, while the rights of those represented by the Organic Seed Growers and Trade Association, seeds, peasants and subsistence farmers, Indigenous people, and those in the global South— the colonized—are eroded in favour of ongoing Primitive Accumulation (Bonato 56-57).

International and North American legal and regulatory systems (United Nations Convention of Biological Diversity, North American Free Trade Agreement, the Supreme Court, the U.S. Food and Drug Administration and Health Canada) can thus be considered instruments of primitive accumulation; their functions are based on upholding the economic power structure made visible by the Iceberg Model of the capitalist-patriarchal economy.

Below, I discuss recent developments in the North American GMO and biotechnology industries. The ongoing debate around labels for genetically engineered products has been largely won by lobbyists of the food and biotechnology industries who have funded campaigns against GMO labels and who supported then-U.S. President Obama's 2016 Bill S.764. Also in 2016, multi-national biotechnology companies Monsanto and Bayer announced that they would merge in a $66 billion deal that will establish their monopoly over global seed and pesticide markets (Philpott).

RECENT DEVELOPMENTS IN NORTH AMERICAN GMO INDUSTRIES

The Label Debate

The fight for labels on foods that contain genetically engineered products is fierce. Companies have argued that labels might scare consumers away from their products, which could impact their profits and result in lost business. A 2016 report from the Environmental Working Group, citing lobbying disclosure forms from the U.S. House of Representatives and U.S. Senate, totaled U.S. anti-label lobbying contributions by food, farming, and biotechnology industries at $192.8 million since 2013 (Coleman).

In Canada, Health Canada is responsible for setting food labelling policies. While 78 percent of Canadian citizens say that they would like genetically engineered products to be labeled (Harris), to date, labels are only required in cases where there are health and safety concerns about allergens or changes in nutritional value. The Canadian General Standards Board established these standards in the form of CAN/CGSB – 32.315-2004 in 2004 under Prime Minister Martin, and reaffirmed them in May 2016 under Prime Minister Trudeau (Canadian General Standards Board). While this standard is presented as an act of transparency, it has not prompted GMO producers to add labels. Rather, programs such as the Non-GMO Project verification, established in 2007, test and label products to ensure that GE content remains below 0.9 percent (Bain and Dandachi).

During the Stephen Harper era (2006-2015), Bill C-474 was introduced in Canadian Parliament following a 2009 crisis of genetically engineered flax contamination that impacted Canada's organic flax exports (CBAN). The Bill was meant to support farmers by requiring an "analysis of potential harm to export markets be conducted before the sale of any new genetically engineered seed is permitted," and was the first real action in Parliament to interrogate the negative impact of genetically engineered crops (Sharratt). A motion to extend the debate was voted down on October 28, 2010 in a four-vote victory (Sharratt).

In July 2016, then-U.S. President Obama signed bill S.764 into law. Renamed by critics as the DARK (Deny Americans the Right to Know) act, the bill allows companies to use QR codes in the place of clear, plain-language labels about the presence of genetically engineered crops. S.764 was signed into law one week after the state of Vermont's labelling law was enacted, and effectively nullified any state law that required labels. It received the support of Monsanto, the U.S. Grocery Manufacturers Association, and the Organic Trade Association (OTA). The Organic Seed Growers and Trade Association announced its withdrawal from the OTA following its endorsement, and stated that the OTA had betrayed them by supporting S.764 and creating partnerships with Monsanto. Even the FDA criticized the bill for a narrow definition of bioengineering that exempts many genetically engineered foods from its label requirements (Chow).

Understanding the debate about labelling genetically engineered products is important in engaging in an ecofeminist critique. According to the Bennholdt-Thomsen and Mies' Iceberg Model (32), the legal and regulatory institutions that structure and govern the economy were established to uphold a capitalist-patriarchal power structure. Thus, North American governments and their agencies (Health Canada, U.S. Food and Drug Administration, etc.) function in a manner that supports ongoing primitive accumulation. Despite public concern and efforts to seek labels on genetically engineered products, North American governments have chosen to reject labelling the products of biotechnology and industrial agriculture, in the name of unhindered profits.

Monsanto and Bayer: A Merger Under Trump

In September 2016, Monsanto announced that they were in talks to forge a $66 billion merger with Bayer, a multinational chemical and pharmaceutical company (Bayer "Bayer and Monsanto"). Monsanto's shareholders approved the offer in December 2016. The following month, company CEO's Werner Baumann (Bayer) and Hugh Grant (Monsanto), met with then U.S. President-elect Trump to discuss the future of the company and its role in U.S. agriculture (Monsanto "Joint Statement"). In a joint statement on January 17, 2017, the meeting was described as a way to share their view on "increasing and accelerating innovation to help growers around the world address challenges like climate change and food security," with the intention to ensure that the U.S. "remains anchors of the industry" (Bayer "Joint Statement") as a global agriculture leader.

U.S. President Trump's support of biotechnology and industrial agriculture can be gauged by his nomination of former Georgia Governor, Sonny Perdue for Secretary of the United States Department of Agriculture (USDA) (Hirschfeld Davis and Haberman). Perdue's most recent role was head of a global agribusiness trading company, and he was awarded Governor of the Year in 2009 by the Biotechnology Industry Association. Accepting his award, Perdue stated, "we'll continue to support the [biotechnology] industry because their research improves our lives and brings jobs and investment to the state." Below, I discuss two new products, the Arctic® Apple and AquAdvantage® Salmon, whose presence in the North American food market has been met with controversy and concern ("Genetically Modified apple"; "Federal Court"; 2016 Resolution Status; Dehaas). The Arctic® Apple's use of a QR code as a label represents former President Obama's Bill S.764 in action, and AquAdvantage® Salmon is the first genetically engineered animal to be allowed into the global food system. Both examples have largely been considered "game-changers" by industry critics, and, as I argue, signify entry into an era I call "late-biotechnology."

NEW PRODUCTS IN THE NORTH AMERICAN MARKET

The Arctic® Apple

The Arctic® Apple was developed by Okanagan Specialty Fruits in order to prevent browning. When an apple is cut, Polyphenol Oxidase (PPO) cells mix with other polyphenolics to produce a brown melanin. By genetically engineering four genes within the apple's DNA sequence, scientists used "gene silencing" to stop the apples from turning brown after they've been cut (Arctic® Apple), and created Arctic® Apple trees with low PPO production. The Arctic® Apple was approved by the U.S. FDA in February 2015, and Health Canada in March 2015 after years in regulatory limbo, with hopes that they would hit national markets in 2016 and 2017 respectively.

The apple has been met with concerns by a number of organizations, including Society for GE-Free B.C., who objected based on the risks of cross-pollination of genetically-modified with non genetically-modified apples, and the impact on future apple generations. In addition, at its 2016 annual general meeting in Kelowna, the B.C. Fruit Growers Association passed a resolution to ask the Canadian government to deregister the product (2016 Resolution Status). The Arctic Apple was not de-registered and, as of February 1, 2017, ten grocery stores in the mid-western U.S. will carry the Arctic® Apple, which will be sold sliced and packaged.

According to Arctic® Apple grower Neal Carter, the "apple will do great things for the industry by preserving more fruit throughout the production system" ("Genetically modified apple"). The Arctic® Apple will not be labeled as a GMO, but will include a QR code for customers to scan if they want to learn more about the product (Arctic® Apple). While Arctic Apple's website is generally open about its process and use of genetic engineering, their choice to include a QR code instead of a plain-language GMO label is significant. First, one has to be in the financial position to have a smart phone and access to the internet. Second, one has to be interested in learning about the product enough to engage in the process of scanning a code with a smartphone application, and taking the time to investigate a fruit as common as an apple. The use of a QR code is a decision that is intended to appear as an act of transparency (it is in compliance with U.S. Bill S.467) while not actually being transparent.

AquAdvantage® Salmon

AquAdvantage® salmon was approved by the U.S. FDA in November 2015, and Health Canada in May 2016. Genetically engineered to grow faster, require 25 percent less feed than non-GMO salmon, and reach the market in less time, Aqua Bounty makes no mention of its use of biotechnology or genetic engineering on its website. Instead, it simply states that AquAdvantage® salmon will help to produce more fish in less time with minimal environmental impact (Aqua Bounty Technology).

Critics have fought its approval, referring to AquAdvantage® salmon—an atlantic salmon that is modified with genes from "chinook salmon" and a fish called "ocean pout"—as frankenfish (Dehaas). Despite its approval, some retailers have declined to sell the fish, including Trader Joe's and Whole Foods (Dehaas). Organizations The Ecology Action Centre and Living Oceans Society appealed to Canadian Federal Courts to reconsider Environment Canada's approval of AquAdvantage® salmon, citing concerns about the adequacy of health and safety assessments for what is the "world's first GMO animal to be approved for farming and sale," but the appeal was not granted ("Federal Court").

Aqua Bounty's purpose as a company is to grow salmon quickly, with less feed, getting the fish to market faster to generate profits more rapidly. While there are many concerns about the fish escaping and breeding with non-GMO salmon, AquAdvantage® fish are all female and all sterile, thus possessing no reproductive capacity (Aqua Bounty Technology). This action ensures that AquAdvantage® salmon "... shift from the ecological processes of reproduction to the technological processes of production" (Shiva "Biotechnological Development" 201). Rendering the salmon unable to reproduce on its own, Aqua Bounty engages in primitive accumulation.

CONCLUSION: A NEW ERA OF BIOTECHNOLOGY

This paper has examined the work of Monsanto and other biotechnology companies in ongoing primitive accumulation. Using an ecofeminist perspective that connects women and nature based on their reproductive labour, and according to the framework in the Iceberg Model of the capitalist-patriarchal economy, legal and regulatory systems were established as an extension of the property rights that took shape between the fifteenth and sixteenth centuries. The first form of primitive accumulation took place during the European witch-hunts, when women's bodies were used to generate profits for the church and state. According to Vandana Shiva, primitive accumulation takes place today by commodifying the bodies of seeds using biotechnology and patent law. In order to support new avenues of profit, "[h]uman rights, including the right to a livelihood, must ... be sacrificed for property rights that give protection to the innovation processes" (193). Under capitalist-patriarchy, "primitive" seed is only made valuable after it has been subject to investment of research and patents (209).

Industrial agriculture and biotechnology have shifted the concept of private property and engage in primitive accumulation by genetically altering and patenting seeds and the pesticides to support them. Selling their products as "feeding the world," they have expanded the reach of industrial agriculture across the globe (Monsanto *Annual Report* 5). Genetically modified products are sold as logical and necessary in order to face impending ecological challenges such as global food shortage, population growth, drought, poverty, and pollution (Moser 1). Instead, this process has resulted in massive dispossession in the global South *and* global North in order to establish corporate control over the reproductive capacities of seeds, plants, and animals.

Just as the Supreme Court sided with Monsanto in *OSGATA et al. v. Monsanto*, courts, governments, and their regulatory bodies such as Health Canada, the Canadian General Standards Board, and the United States' Food and Drug Administration and Department of Agriculture, continue to be tasked with issues associated with biotechnology.

However, "the master's tools will never dismantle the master's house" (Lorde), and these bodies, established under capitalist-patriarchy, vote down debate (CBAN Sharratt), create technologically advanced and inaccessible label laws (S.764; Chow), nominate biotechnology industry-insiders to lead national departments (Hirschfeld Davis and Haberman), and reject public concerns about genetically engineered animals in the global food market ("Federal Court"; Dehaas).

We have entered a new era of biotechnology, one that includes monopolies over seed and pesticide markets, QR codes instead of labels, and genetically engineered animals in our food system. Creating ever-new avenues of profit, Monsanto and other biotechnology and industrial agriculture companies engage in primitive accumulation, and sell their "creations" as innovative solutions to the problems of capitalism. In the twenty-first century, biotechnology has expanded primitive accumulation by creating a surplus of raw material to be scientifically engineered and patented in the name of adding "value" to free, genetic resources. Ecofeminists must ask: what next?

REFERENCES

2016 Resolutions Status. Kelowna, B.C. Fruit Growers Association, 11 Feb. 2016.

Arctic® apples. "Distinctly nonbrowning apples." 2017. Web. 30 Jan. 2017.

Aqua Bounty Technology. 2017. Web. January 30, 2017.

Bain, C. and T. Dandachi. "Governing GMOs: The (Counter) Movement for Mandatory and Voluntary Non-GMO Labels." *Sustainability* 6.12 (2014): 9456-9476.

Bayer AG. "Advancing Together." No date. Web. 30 Jan. 2017.

Bayer. "Bayer and Monsanto to Create a Global Leader in Agriculture." 2016. Web. 30 Jan. 2017.

Bayer. "Joint Statement: Monsanto, Bayer CEOs Meet with New Administration." 2016. Web. 30 Jan. 2017.

Bennholdt-Thomsen, V. and Maria Mies. *The Subsistence Perspective: Beyond the Globalised Economy*. London: Zed Books, 1999.

Bonato, J. "Monsanto and the Patenting of Life: Is Biotechnology a New Form of Primitive Accumulation in the 21st Century?" Major Research Paper, St. Catharines: Brock University, 2016.

Canadian General Standards Board, Standards Council of Canada. *Voluntary Labeling and Advertising of Foods that Are and Are Not Products of Genetic Engineering*. Gatineau: Canadian General

Standards Board, 2004. Reaffirmed May 2016.

Canadian Biotechnology Action Network (CBAN). No date. Web. January 30, 2017.

Chow, L. "Obama Signs Industry-Backed GMO Label Bill Into Law." 2016. Web. 30 Jan. 2017.

Coleman, R. "Food Lobby Spends $101 Million in 2015 to Avert GMO Labeling." 25 Feb. 2016. 14 March 2017.

Dehaas, J. "Health Canada approves salmon genetically modified to grow more quickly." 19 May 2016. Web. 30 Jan. 2017.

"Federal Court to hear case against genetically-modified salmon." The Canadian Press 16 Nov. 2015. Web. 30 Jan. 2017.

Federici, S. *Caliban and the Witch.* New York: Autonomedia, 2004.

"Genetically modified apple that won't brown under fire." *CBC News* 4 Nov. 2013. Web. 30 Jan. 2017.

Georgia Governor Sonny Perdue. "Governor Perdue Named BIO Governor of the Year." 23 April 2009. Web. Retrieved 30 Jan. 2017.

Goldenberg, S. "Prop 37: Food companies spend $45m to defeat California GM label bill." 5 Nov. 2012. Web. 14 March 2017.

Harris, K. "Consumers opposed to genetically modified foods, but don't know what they are." 9 Oct. 2016. Web. 30 Jan. 2017.

Health Canada. "Genetically Modified (GM) Foods and Other Novel Foods." Health Canada, 2016. Web. 30 Jan. 2017.

Health Canada. "AquAdvantage Salmon." 2016. Web. 30 Jan. 2017.

Hirschfeld Davis, J. and M. Haberman. "Sonny Perdue Is Trump's Choice for Agriculture Secretary." 18 Jan. 2017. Web. 14 March 2017.

International Agency for Research on Cancer. "Some Organophosphate Insecticides and Herbicides: Glyphosate." *Iarc Monographcs on the Evaluation of Carcinogenic Risks to Humans* 112.10 (2017): 1-92. World Health Organization. Web. 1 Feb. 1 2017.

Lorde, Audre. "Master's tools will never dismantle the master's house," comments at "The Personal and the Political" panel, Second Sex Conference, October 29, 1979.

Monsanto. "Improving Agriculture." 2015. Web. 30 Jan. 2017.

Monsanto. "Original Monsanto Company." 2017. Web. 30 Jan. 2017.

Monsanto. "Who We Are. Our Commitment to Sustainable Agriculture." No date. Web. 31 August 2015.

Monsanto. *Annual Report: Ideas Big Enough for a Growing World.* St. Louis, MO: Monsanto, 2015. Web. 31 August 2015.

Moser, I. "Critical Communities and Discourses on Modern Biotechnology." *Biopolitics: A feminist and ecological reader on biotechnology.* Eds. V. Shiva and I. Moser. London: Zed Books, 1995. 1-24.

OSGATA *et al. v. Monsanto and Monsanto Technology LLC.* United States District Court Southern District of New York. 29 Mar. 2011. Web. 31 August 2015.

Philpott, T. "Trump is ready to bless Monsanto and Bayer's massive merger." *Mother Jones* 19 Jan. 2017. Web. Retrieved 30 Jan. 2017.

S. 764, 114th Congress: A bill to reauthorize and amend the National Sea Grant College Program Act, and for other purposes. 4 January 2016. Web. January 30, 2017.

Salleh, A. *Ecofeminism as Politics: Nature, Marx, and the Postmodern.* 1997. London: Zed Books, 2017.

Salleh, A. "From Eco-Sufficiency to Global Justice." *Eco-Sufficiency and Global Justice: Women Write Political Ecology.* Ed. A Salleh. London, England: Pluto Press, 2009. 291-312.

Salleh, A. "Organized Irresponsibility: Contradictions in the Australian Government's Strategy for GM Regulation." *Environmental Politics* 15.2 (2006): 388-416.

Sharratt, L. "CBAN – Canadian Biotechnology Action Network." 28 Oct. 2010. Web. 30 Jan. 2017.

Shiva, V. *Staying Alive: Women, Ecology, and Development.* London, England: Zed Books, 2010.

Shiva, V. "Biotechnological Development and the Conservation of Biodiversity." *Biopolitics: A Feminist and Ecological Reader on Biotechnology.* Eds. V. Shiva and I. Moser. London: Zed Books, 1995. 193-213.

United Nations Convention on Biological Diversity. 1993. Web. U.S. Food and Drug Administration. "AquAdvantage Salmon Approval Letter and Appendix." 2015. Web. 30 Jan. 2017.

Vecchione, M., C. Feldman and S. Wunderlich. "Consumer Knowledge and Attitudes About Genetically Modified Food Products and Labelling Policy." *International Journal of Food Sciences and Nutrition* 66.3 (2014): 329-335.

12.
"I Know About My Own Body ... They Lied"

Race, Knowledge, and Environmental Sexism
in Institute, WV and Old Bhopal, India

REENA SHADAAN

*I had come back from my [in-law's] house that day. My daughter
was eight days old.... [M]y eyes were tearing, and my daughter
was coughing. I thought somebody was burning chillies.... [E]
verybody was coughing and vomiting.... My daughter died
about two and a half years after the disaster.*
—Noorjahan, Old Bhopal, India (Personal communication,
Jan. 2012)

*They lied about that emission.... I don't care how long it takes.
We're going to know one day that they lied to us.... I know
about my own body.... They lied to us.... They come out
smelling like a rose, while the community goes on with this
stench.*
—Sue Davis, Institute, West Virginia (Personal communication,
Apr. 23, 2015)

*For the sense of smell, almost more than any other, has the
power to recall memories and it's a pity we use it so little.*
—Rachel Carson (*Silent Spring* 83)

IN 1962, RACHEL CARSON PUBLISHED her ground-breaking work,
Silent Spring. With that, Carson introduced the American public to the
harmful ecological and health impact of pesticide-use, and particularly
DDT.[1] She noted, "Residues of these chemicals linger in soil.... They
have entered and lodged in the bodies of fish, birds, reptiles, and
domestic and wild animals.... [They] are now stored in the bodies of
the vast majority of human beings..." (Carson, *Silent Spring* 15-16).
As a result of *Silent Spring*, Carson faced a barrage of industry-led
attacks, questioning her knowledge claims. Notably, much of Carson's

critiques were rooted in her gender identity. Specifically, she was touted as a "hysterical" and "uninformed woman" (Hess). Decades later, little has changed. Women-activists in environmental justice struggles are similarly belittled by "experts" in industry, the state, and the scientific community. In reference to her work with white working-class women-activists in the U.S. environmental justice movement, Celene Krauss observed, "Male officials ... exacerbated [the] intimidation by ignoring the women, by criticizing them for being overemotional, and by delegitimizing their authority by labelling them 'hysterical housewives'" (139-140).

Early scholarship often highlighted the central role of women in the environmental justice movement, but also noted the dismissal of their knowledge and lived experiences. As Phil Brown and Faith Ferguson aptly summarized,

> The women activists transform their everyday experiences, most typically their own and their neighbors' children's illness, into knowledge that they can use in the struggle against toxic waste, and they insist on its validity as knowledge. Such validity is contested by scientific experts and professionals, whose cultural beliefs about women and science lead them to refuse to accept the women activists' claims about the consequences of toxic exposure. (151-152)

Recent environmental justice scholarship calls for a renewed feminist lens—one which highlights multiple layers of oppression and broader systems of exploitation that produce disproportionate environmental harms (Malin and Ryder; Pellow; Scott). This chapter addresses this call, but also demonstrates the continued dismissal of toxic-impacted women's knowledge, which was first identified in early scholarship. Specifically, the chapter will consider the dismissal of women's health-related knowledge and lived experiences within the context of two different but historically connected environmental justice struggles. The first is located in Institute, West Virginia, within the wider region of Kanawha Valley—known as "Chemical Valley." The second is in Old Bhopal, India, an area severely impacted by the 1984 Bhopal Gas Disaster—known as the "world's worst industrial disaster" (Hanna, Morehouse and Sarangi). In both areas, women discuss the dismissal of their health experiences, and specifically doctors' refusal to draw connections between their health struggles and their exposure to industrial pollution—which many identify as the root cause of their illnesses. Diverse factors, such as the absence of comprehensive health studies and industry-touted misinformation contribute to this denial. Consider Institute resident, Donna Willis's observation: "So, when we heard that [the chemical company] maintained that they contained [the chemical release] into the fence area, you either have to be an idiot or a stone-cold fool to think that the chemical didn't get outside that chain-linked fence that's got big holes in it. They actually could con our legislatures into believing that crap" (Personal communication, Apr. 23, 2015). This kind misinformation contests residents' claims of toxic exposure and subsequent illness. Or consider that following the Bhopal disaster, Union Carbide's Bhopal-based medical officer informed hospital staff that methyl isocyanate (MIC) is non-poisonous and that applying a wet towel on the eyes is a sufficient remedy (Agarwal, Merrifild and Tandon). While these issues contribute to the denial of toxic-impacted women's health experiences, the dismissal is further linked to race and class oppression, which influences these women's particular experiences of environmental sexism. Traci Brynne Voyles defines environmental sexism as,

...when women's roles as caretakers compound the burden of environmental problems in their lives: it is women who take up the labor of care when family members become sick; it is women who often assume doubled financial responsibilities when their husbands or partners die and women who undertake a large amount of the labour of family care; and it is women who are at the front lines of the reproductive havoc that many modern toxins ... wreak on human bodies, including increased risks of miscarriage, stillbirth, and birth defects. Moreover women, particularly women of color, are often the most economically vulnerable and politically powerless members of a community, making them less likely to have been consulted when toxic industries move into their communities. (142)

This paper extends this definition to include the dismissal of racialized women's health-related experiences in relation to toxic exposure.

INSTITUTE, WEST VIRGINIA IN "CHEMICAL VALLEY"

Institute is located within the region of Kanawha Valley (West Virginia). Kanawha Valley has been site to a number of major chemical facilities, which have wreaked havoc on workers and residents. According to Maya

Nye, former spokesperson of the Institute-based environmental justice group, People Concerned About Chemical Safety (People Concerned),[2] "Institute is primarily an African American community... Other surrounding communities ... are mostly poor or working class white. The entire area is Appalachian ... a marginalized culture stereotyped as being ignorant and poor" (Personal communication, Sept. 27, 2013).

Institute is a mixed-income African American community that has historically faced much of the brunt of "Chemical Valley." The community was profiled in Robert Bullard's *Dumping in Dixie: Race, Class, and Environmental Quality*, which is a foundation text in the framing of environmental racism. Pam Nixon—spokesperson for People Concerned and former Environmental Advocate at the Department of Environmental Protection—notes, "[O]ther than the ammonia tank up at the DuPont plant [in Belle] ... the Institute area had the most dangerous chemicals. They had 1,3 butadiene, they had the phosgene, they had ... MIC [methyl isocyanate]" (Personal communication, Apr. 21, 2015). That environmental racism is killing Institute's predominantly Black residents led People Concerned member, Donna Willis to reflect, "We'll hold up our hands and say Black Lives Matter" (Personal communication, Apr. 23, 2015).

Parts of Kanawha Valley are known for high cancer rates. This became apparent when driving through Institute with Donna Willis and another life-long Institute resident and member of People Concerned, Sue Davis. The following excerpt aptly described our two-hour drive in and around Institute: "Mr. Pruitt over there, he had cancer.... Billy had cancer.... Jerome James, he died of cancer. He lived right there ... [and] his widow died of cancer.... Diane Carter was raised right here.... She died of cancer" (Donna Willis, Personal communication, Apr. 23, 2015). However, cancer is just one of many health issues plaguing residents. In 1987, People Concerned, with support from allies at several U.S. universities, carried out a comprehensive health survey of Institute[3]—the only one to date. When compared to national statistics, Institute residents have "significantly higher" rates of bronchitis, cataracts, hay fever, itching skin, tinnitus, indigestion, psoriasis, constipation, goiter/thyroid issues, bladder problems, hearing impairments, ulcers, and tachycardia (Hall and Wagner). Interestingly and as if anticipating mistrust, the study notes: "The respondent-assessed health status is generally in line with national estimates. Respondents also reported less experience of stress in their lives than is the case nationally.... [This] would tend to indicate ... that *respondents in this survey were not overemphasizing their health problems*" [emphasis added] (Hall and Wagner 6).

"THE WORLD'S WORST INDUSTRIAL DISASTER" – BHOPAL, INDIA

Bhopal, India is the site of the 1984 Bhopal Gas Disaster—the "world's worst industrial disaster" (Hanna et al.). In 1969, the American-owned Union Carbide Corporation (UCC) sited a facility in Old Bhopal, the poorest subsection of the city of Bhopal. According to Satinath Sarangi of the Bhopal Group for Information and Action (BGIA),[4]

> [O]ld Bhopal is [largely] composed of ... immigrants, driven out of their [villages] ... as a result of ... mechanized agriculture ... and other "development" projects. Over 75% ... earned their livelihood through daily wage labor and petty business....
> [T]he Muslim poor ... formed over 35% of the population....
> ("The Movement" 101)

Utilizing untested technology and amidst a myriad of safety hazards, the UCC-Bhopal plant formulated and later manufactured pesticides to be used in India's Green Revolution (Hanna et al.).

On December 3, 1984, 40 tonnes of methyl isocyanate (MIC) leaked, killing workers and residents of the predominantly poor communities that surrounded the plant. Within three days, up to 10,000 were killed (Amnesty International *30 Years Too Long* ...) and to date, approximately 25,000 have died (Sarangi "Compensation"). Additionally, 150,000 people face a myriad of chronic health issues, including respiratory illnesses, eye diseases, immune system impairments, neurological damage, neuromuscular damage, endocrine system disruption, reproductive health issues, gynecological disorders, mental health issues (Amnesty International *Clouds of Injustice*), as well as cancers (Sarangi, Personal communication, Jun. 28, 2014). Notably, many of the illnesses experienced by the gas-affected population are dismissed, as will be discussed further in greater detail. In fact, an early health study felt it necessary to clarify, "Each symptom was described in such graphic detail that it was obviously based on the patient's own experience and *could not be malingering or wild imaginations as some are apt to allege*" [emphasis added] (MFC 6).

THE HISTORIC CONNECTION BETWEEN INSTITUTE AND BHOPAL

While the Bhopal Gas Disaster is an important reference point for the U.S. environmental justice movement (see Pariyadath and Shadaan), it has particular significance in Institute. Union Carbide's Institute facility also produced MIC. UCC claimed that the facility was safe; however, the company's records indicated that the plant had leaked MIC 28 times between 1979 and 1984. UCC later admitted to 62 MIC leaks (Agarwal, Merrifield and Tandon).

The Bhopal disaster, coupled with decades of toxic emissions and industrial pollution, led to the emergence of People Concerned, which has been at the forefront of environmental justice in Kanawha Valley. Sue Davis expressed the deep connection between Institute residents and Bhopal gas survivors. She notes, "I share their grief. I share their heartache and heartbreak. When it happened to them, it happened to me" (Personal communication, Apr. 23, 2015). Davis also discussed her brother, Warne Ferguson's pivotal role in forming People Concerned:

> He got very upset after the Bhopal incident... He virtually founded [People Concerned], even though you had a lot of people who worked on that original committee... He went to Don Wilson, this young man that lives in West Dunbar... and said, "Don, we've got to do something." They, in turn, went to Ed Hoffman who was at [West Virginia State University]... So, that's when it started.... (Ibid)

In effect, while Institute and Old Bhopal are distinct sites of environmental racism, they are also sites of an intertwining history, and shared struggle. This deep connection is aptly summarized in the following statement by People Concerned in 1985:

> We are residents, professors, and college students[5] who oppose MIC production in our community. We do so not only because a disaster similar to Bhopal could happen here, but also out of respect for the victims and survivors in your city.... We see Union Carbide's haste to make profits again from methyl isocyanate as an indication of little concern for what happened to the Indian people, and little concern for the predominantly black community that lives just downwind from the Institute plant. The lesson of the Bhopal disaster for us is that Union Carbide cannot be trusted to insure our safety.... We hold our hands in brotherhood to you. May our common concern for safety and health bond your community and ours for many years to come. (Agarwal, Merrifield and Tandon 31)

The ongoing relationship of solidarity is further indicated in the statement released by the International Campaign for Justice in Bhopal (ICJB) following the 2014 Elk River spill:

[ICJB] ... expresses solidarity with the communities of West Virginia that are facing a toxic nightmare.... The contamination of our water ... is a heinous crime. Like you, Bhopalis have faced widespread groundwater contamination.... Toxic facilities are routinely situated in areas populated by the poor, working-class and/or racial minorities and, left to self-regulate, chemical industries will continue to pose a threat to the lives and environments of such communities. ("The West Virginia Chemical Spill")

Old Bhopal and Institute are connected by history and their fight against environmental racism. They converge to (informally) form a transnational environmental justice struggle that contests the disproportionate burden of industrial pollution and the resultant health impacts.

METHODOLOGY

A feminist and phenomenological methodology underlies this study. Qualitative methodologies are "flexible, fluid and better suited to understand the meanings, interpretations and subjective experiences of those groups who may be marginalized, 'hard to reach' or remain silenced" (Bhopal 189). Feminist methodologies, in particular, aim to "capture women's lived experiences in a respectful manner that legitimates women's voices as a source of knowledge" (Campbell and Wasco 783). This is pertinent, as environmental justice women-activists are routinely excluded from knowledge production (being viewed as hysterical, ignorant, and suspicious) largely due to the intersections of environmental racism, environmental sexism, and class oppression. Moreover, a phenomenological approach "argue[s] ... that the patient's self-understanding and experience of illness ... offers a legitimate source of relevant medical knowledge," making this framework particularly apt (Goldenberg 2628).

As an observer of the struggle for justice in Bhopal, as well as a participant in the International Campaign for Justice in Bhopal, North America (ICJB-NA), I draw on historical data, as well as narratives gained from my various interactions with women-activists impacted by the ongoing Bhopal gas disaster.[6]

250

It is in the capacity of ICJB-NA that I learned about and connected with People Concerned. In April 2015, I travelled to Institute in order to meet and learn from the women who are part of the core of People Concerned. While I had initially planned to explore issues surrounding women's motivations for activism in People Concerned, health and healthcare emerged as prime issues in each of my interactions with members of People Concerned. A similar trend was apparent in my interactions with Bhopali women-activists. The struggle for environmental justice is, after all, also a struggle for health.

MAKING THE CONNECTION

In both Old Bhopal and Institute, women find their knowledge and experiences of illness (particularly the causation) dismissed by medical professionals. As Phil Brown and Edwin J. Mikkelsen note, "Science is ... limited in its conceptualization of what problems are legitimate and how they should be addressed.... [P]hysicians are largely untrained in environmental and occupational health matters, and even when they observe environmentally caused disease, they are unlikely to blame the disease on the environment" (132). In the context of Institute, and in relation to an autoimmune disorder she developed after a chemical emission, Pam Nixon notes,

> *[The doctor] didn't want to... give a causation of it.... None of the doctors here in the Valley ... would ever say what was causing your problem.... I know [the chemical release] affected me.... [T]he reason I say that is [because] every time I would almost go into remission, and they would have a release at a particular unit ... I'd have symptoms again.... I knew it was coming from the plant, but I couldn't get a doctor to say it was coming from the plant.* (Personal communication, Apr. 21, 2015)

Although not in Institute, Stephanie Tyree, a Board Member of People Concerned and a long-time environmental justice advocate in West Virginia, relates a similar experience. Following the 2014 Elk River spill in the region,

> *What they told you to do was turn on your hot water at full force, and just flush all the water out. The chemical that was spilled, when it was heated up, it turned into ... a neurotoxin....*

When we were doing this … [the] whole apartment filled up with gas. You could smell it. It was really intense…. I got a really bad migraine from it that lasted for like a week…. [E] ventually I went to … a MedExpress Center…. I told them that I got it because of the [contaminated] water, and the doctor didn't believe me…. He was like, "I don't think that's what it's about… I think you just have a bad headache…." I was like, I think I know what it's about…. (Personal communication, Apr. 24, 2015)

Maya Nye concurs. "A lot of stories that I heard after the Elk River chemical spill was that doctors were refusing to make any sort of connection between their symptoms and the exposure to the MCHM [crude 4-methylcyclohexane methanol]." However, "[A]fter the Elk River chemical spill … I'll bet doctors were more likely to make the connection versus a one-off release … because more people were impacted [and] … people of affluence were also affected … white folks, not … people of colour" (Personal communication, Jun. 8, 2015).

The experiences of Nixon and Tyree are consistent with those of women in Old Bhopal. Shortly after the Bhopal disaster, a journalist observed,

There are shocking tales of mothers who have lost their offspring or who are bringing up deformed infants, the shocking accounts given by the junior staff of the hospitals, midwives and nurses who insist they have never seen any birth-and-death cycle of this kind before…. Against this we have the official version of bureaucrats and senior doctors who are under instruction not to talk. An attempt is being made to cover up the deformities and abnormalities being recorded…. Nobody knows if the trauma will end with this generation, or the next. (Sarin)[7]

While Sarin's account points to the silencing of reproductive health concerns, the notion that infants' bodies are damaged via "deformities and abnormalities" reproduces problematic notions of normative bodies (Clare; Di Chiro).

A 2004 study re-iterates the hesitancy to relate illnesses to MIC exposure. Following visits to various hospitals, it was noted, "Doctors were refusing to admit to the fact that MIC … had affected major organs" (We for Bhopal 37). Further, "The doctors were not willing to admit that the high incidence of cancer can be related to the affect of the gas. There was simply not enough research to back up any such theory..." (We for Bhopal 56). Here, it is imperative to ask: Why are there so few studies on the health impacts of toxic exposure in Old Bhopal and Institute? Consider that over three decades after the Bhopal disaster, survivors' organizations continue to have to ask for, "[T]reatment protocols specific to exposure related health problems and ... medical research that benefits the gas victims and those exposed to contaminated ground water" (ICJB 2014a). Or that when travelling though Institute, Donna Willis noted, "I just had an aunt die of cancer. I just had a girlfriend who ... was raised with me that just died of cancer. Her mother died of cancer," to which Sue Davis responded, "It's unbelievable, and we can't get a study" (Personal communication, Apr. 23, 2015). Maya Nye provides a useful response in the context of Institute, but with lessons that can be applied in Old Bhopal and elsewhere:

> [T]here have been [more] studies done on the Elk River chemical spill, [than] any of the spills that happen ... in Institute or in Belle, [which is] more of an economically depressed area.... [M]ore affluent people were impacted [which] is the reason why more studies happen. [emphasis added] (Personal communication, Jun. 8, 2015)

RACE, CLASS, AND ENVIRONMENTAL SEXISM

The denial of illness is influenced by gender, race, and class oppression. Prior to delving into the particular context of Institute and Old Bhopal, it is imperative to note the knowledge/power hierarchy that can underlie the relationship between medical professionals and lay persons. Martha Balshem provides a useful summary from her in-depth case study of a working-class community in Philadelphia. She notes the strained relationship between the medical establishment and residents, who attribute their illnesses to industrial pollution—an assertion that the medical community is less willing to make.

> For many lay people, contact with the medical-care system has at some point involved the felt experience of a loss of personal authority. These experiences are often dramatic and terrifying.... Medical social scientists have described in elaborate detail the physician's power to confer or deny legitimacy to particular interpretations of patient sign, symptom, and behaviour; charged that through the distinction between scientific knowledge and

folk knowledge, lay interpretations are cast as illegitimate and inconvenient counterpoints to real medical knowledge.... (6-7)

Likewise, C. Sathyamala, who has worked with Bhopal gas survivors, notes, "In the doctor-patient relationship, generally the patient is considered a malingerer unless clinical and laboratory tests prove that she/he show some changes.... This is exaggerated when the complainant belongs to an "inferior category"... either in terms of class or sex" ("The Medical Profession" 40). The role of gender has warranted scholarly attention. As Marci R. Culley and Holly L. Angelique have summarized, "Science ('rational/masculine') has typically rejected women's 'ways of knowing' in antitoxic efforts ('informal,' 'experience based,' 'housewife surveys') as unscientific, unobjective, and irrational" (447). This analysis is pertinent in the context of Old Bhopal where "[t]he belief that women are emotional and hysterical creatures, led researchers to conclude that the effect on pregnancies was due to the enormous stress these women underwent.... Stress, of course, may have taken its toll, but the tendency was to put the entire blame on the emotional state of the women" (Sathyamala, "Condition" 130). This is also pertinent in the context of Institute. Sue Davis notes,

[They] say, "Oh, she suffers from paranoia. She's paranoid regarding the chemical plant"…. Both hospitals said it…. [T]hen you look at their descriptions, and … you see where they created all this stuff that they lied about…. They were so rude, and so … non-caring, and they don't know what we go through…. [T]hey said, "She has fears"…. Who am I gonna fear? Who am I gonna fear? I don't fear them. I don't fear their chemicals. I live so that if I die tomorrow, I know where I'm going. (Personal communication, Apr. 23, 2015)

Allegations of "paranoia" and "hysteria" are not limited to women. In the context of a working-class community in Southern West Virginia, Stephanie Tyree shared,

They were having [coal slurry] through their water systems, and having a lot of health impacts from that, and cancer clusters…. [T]hey [both men and women] really struggled to get doctors to recognize the health impacts…. A lot of doctors telling them that they were imagining things…. They were hysterical…." [emphasis added] (Personal communication, Apr. 24, 2015)

Donna Willis shared the particular ways in which Black women's knowledges, lived experiences, and resistance is discounted:

Black women aren't afraid of speaking out against injustice.... Society is quick to place a label on anyone who isn't snowed in by their hypocrisy.... There are elements in our environment that are killing us ... If a Black women returns an insult in kind, then up pops the race card. She's loud, trying to intimidate us, threatening.... (Personal communication, Feb. 17, 2016)

Andrea Simpson writes that Black women are routinely denied credibility. She refers to the environmental justice work of Doris Bradshaw in Memphis, Tennessee, and notes that not only are Black women silenced when advocating for themselves and their communities; they are vilified for it, as their concerns are misrepresented in media and otherwise as self-interested.

In effect, reducing the discussion to gender serves to ignore other other facets of oppression that deny legitimacy. In the context of Old Bhopal, race, gender, and class oppression serve to deny legitimacy, in particular, to gas-affected women. To illustrate, a gynecological health study conducted in 1985 found a correlation between MIC exposure and gynecological illnesses. The findings were widely contested by Bhopal's medical establishment in a manner that indicates gender-based discrimination, as well as anti-poor and anti-Muslim sentiments. Gynecologists affiliated with India's leading, state-sponsored medical body, the Indian Council of Medical Research (ICMR), said, "Oh, these poor women live in such filthy conditions. All of them have pelvic infection. It is very frequent amongst Muslim women" (Kishwar 38). Moreover, three Bhopal-based gynecologists, and one Bhopal-based obstetrics and gynecology professor identified "...gynaecological symptoms as 'usual,' 'psychological,' or 'fake' and the gynaecological diseases ... as 'usual,' 'tuberculour,' or 'due to poverty and poor hygiene'" (Sathyamala, "Medical" 50).

The perception that gynecological illnesses are "fake" warrants further discussion. It is rooted in the oft-touted allegations of "compensation neurosis" (feigning illness in order to gain larger sums of compensation), which is linked to anti-poor attitudes. Sathyamala observes, "The gas victims were poor and a larger proportion were women ... and it was easy for the medical community to dismiss their complaints out of their ... suspicion of such people" ("Medical" 40). These perceptions persist. In a 2004 interview with the then Director of the Bhopal Gas Tragedy

Relief and Rehabilitation Department, Bhopal's Chief Medical Officer, and several high-ranking Bhopal-based hospital officials, it was noted, "Look don't be taken in by what people tell you.... It is all gimmicks.... When the question of giving them compensation money came up they all lined up.... The fact of the matter is that 95% of these people are not gas victims" (We for Bhopal 57). Notably, the suspicion of poor, working-class, and racialized communities is pertinent in the context of Kanawha Valley as well. Maya Nye notes,

> I would say that there's ... a stereotype of "working the system," the healthcare system.... "Working the system" ... would probably be someone who is low-income, on disability potentially, who has MedicAid.... Those are people who are considered people who "work the system," people who are living off the system.... I would say that, with the class issue, you're less likely to be taken seriously if you're on disability, or if you have some sort of public assistance with healthcare.... You're less likely to have quality healthcare, I would say, or to be taken seriously.... [A]s far as race goes, I would say that there are probably similar barriers.... I would say that the barriers are the same regardless of [income] when it comes to race.... (Personal communication, Jun. 8, 2015)

In effect, in both Institute and Old Bhopal, toxic-impacted communities have been labelled by some of the medical community as "paranoid," "hysterical," "ignorant," and "fake."

However, both communities have led a relentless struggle for health and justice. Aware of Union Carbide's attempt to silence the health impacts of MIC exposure, Bhopali survivors and solidarity activists opened the People's Health Centre in 1985, noting:

> For the past six months, politicians have hidden the problems of gas victims; withheld effective cures and blindly pumped people full of random drugs, playing havoc with their lives. We have fundamental human rights to health and proper medical care.... This clinic was made for the people by the people, it is for the benefit of gas victims. We intend to make it a model for public struggle against the merchants of death... Down with the murderer Union Carbide! The fight for medical care is a fight for our rights! (Bhopal Medical Appeal and Bhopal Group for Information & Action 27)

Although the People's Health Centre was raided by police, and shut down after twenty days, it is an early and notable example of the struggle for healthcare in Bhopal that addresses the needs of residents impacted by MIC exposure. Today, the Sambhavna Trust Clinic carries on the tradition of the People's Health Centre. Established in 1995, and staffed primarily by survivors, this clinic provides free ayurvedic and allopathic care to those impacted by the Bhopal gas disaster. In addition, medical research and public health education is carried out to the benefit of survivors, their children, and those living in the communities impacted by groundwater contamination.

These efforts are in addition to the ongoing work of the International Campaign for Justice in Bhopal—led by survivors' groups and a grassroots support group. Under the guiding framework of the precautionary principle, the polluter pays principle, right-to-know, international liability, and environmental justice, ICJB focuses on the short-term goals of relief and rehabilitation and the long-term goals of justice and accountability. This includes adequate medical care and research on the long-term impacts of MIC exposure, social and economic rehabilitation, environmental remediation, adequate compensation, and justice and accountability within the Indian and U.S. legal system. The Bhopal campaign's over three decades of work has led to a number of significant victories, although the struggle for justice is ongoing.[8]

In Institute, the only comprehensive health study to have taken place was sponsored by People Concerned (Hall and Wagner). The study aimed to identify recurring health issues, develop a community health profile, demonstrate the need for improvements in healthcare responses, and demonstrate the need for epidemiological studies (Holt). A group newsletter during the time noted, "Was it deliberate oversight or mere negligence that led to little effort to obtain systematic information about the health status of communities living in close proximity to chemical plants?" (Holt 1).

Like ICJB in Bhopal, People Concerned has been at the forefront of both prevention and remediation efforts in West Virginia for more than three decades. This includes the development of a Pollution Prevention Program in the 1990s. A notable part of this initiative was an "Odor Patrol" in which community members would monitor, identify, and report odors stemming from the chemical facilities. In addition, a community air-monitoring program was put into place, following the 2008 explosion at Institute's Bayer CropScience facility. Finally, People Concerned has advocated for effective emergency response plans, third-party safety audits, greater transparency, and has been a key voice in calling for chemical safety and environmental legislation in West Virginia (Maya Nye, Personal communication, Jul. 30, 2016).[9]

CONCLUSION

Race, class, and gender oppression confer to produce racialized women's unique experiences of environmental sexism. In both Institute and Old Bhopal, residents find their knowledge claims dismissed as they are perceived as "hysterical," "paranoid," "fake," and "suspicious." This is significant as it means that medical professionals do not link the myriad of health issues to toxic exposure—a refusal that is at odds with residents' claims, which, in turn, impacts residents' ability to gain recourse.

However, these communities possess a knowledge that is integral to understanding environmental harm. Underlying their knowledge claims is a historical awareness, an understanding of the larger framework (Brown), as well as sensory perceptions. As Phil Brown and Edwin J. Mikkelsen have noted,

[P]eople often have access to data about themselves and their environment that are inaccessible to scientists. In fact, public knowledge of community toxic hazards in the last two decades has largely stemmed from the observations of ordinary people.... Even before observable health problems crop up, lay observations may bring to light a wealth of important data.... Yellow Creek, Kentucky residents were the first to notice fish kills, disappearances of small animals...." (127)

Consider the importance of smell in Pam Nixon's narrative, "[W]here I grew up ... we were downstream from the Belle [DuPont] plant.... [W] hen they would have releases into the water, there'd be large fish kills, and of course the fish would float down river ... and so the smell of the dead fish would come up into [our home]" (Personal communication, Apr. 21, 2015). Or the significance of sound for Savannah Evans, following the 1985 leak of methylene chloride and the aldicarb oxide in Institute. At a town hall meeting, Evans noted, "The birds stopped singing Sunday morning, and they came back Wednesday" (qtd. in Pickering and Lewis). It is largely through the senses that these communities *know* the impacts of toxic exposure on their bodies and environments. It is this community-based, expert knowledge that must underlie and drive the comprehensive health studies that are so desperately needed in both Old Bhopal and Institute.

ENDNOTES

[1]It is crucial to note the foundational work of labour which was at the forefront of promoting industrial health and safety long before Carson's work brought these issues to mainstream attention.
[2]People Concerned About Chemical Safety was formerly called People Concerned About MIC. To avoid confusion, I refer to the organization as People Concerned.
[3]The study includes Institute, Pinewood and West Dunbar, which collectively make up the community of Institute.
[4]BGIA is one of five Bhopal-based groups that make up the leadership of the International Campaign for Justice in Bhopal (ICJB).
[5]Institute is the site of West Virginia State University, a historically Black college.
[6]ICJB-NA is the North American solidarity tier of the International Campaign for Justice in Bhopal. I have been involved in ICJB-NA peripherally since 2006, and more centrally since 2013.
[7]Here, it is worth noting the power dynamics (gendered, and otherwise) *within* healthcare institutions. In Sarin's observation, it is the "junior staff of the hospital, midwives and nurses" that note the prevalence of reproductive health issues, while the "bureaucrats and senior doctors" choose to silence this narrative.
[8]For more information and to support the ongoing work of the Sambhavna Trust Clinic and ICJB visit: bhopal.org (Sambhavna Trust Clinic) and bhopal.net (International Campaign for Justice in Bhopal).
[9]For more information and to support the work of People Concerned, visit: peopleconcernedaboutmic.com.

REFERENCES

Agarwal, A., J. Merrifield, and R. Tandon. *No Place To Run: Local Realities and Global Issues of the Bhopal Disaster*. New Market, TN: Highlander Center and New Delhi: Society for Participatory Research in Asia, 1985.
Amnesty International. *Clouds of Injustice: Bhopal Disaster 20 Years On*. Oxford: Alden Press, 2004.
Amnesty International. *30 Years Too Long ... To Get Justice*. London: Amnesty International, 2014.

Balshem, M. *Cancer in the Community: Class and Medical Authority.* Wash-ington: Smithsonian Institution Press, 1993.

Bhopal, K. "Gender, Identity and Experience: Researching Marginalized Groups." *Women's Studies International Forum* 33 (2010): 188-195.

Bhopal Medical Appeal and Bhopal Group for Information & Action. *The Bhopal Marathon.* Brighton: BMA/BGIA, 2012.

Brown, P. "Popular Epidemiology and Toxic Waste Contamination: Lay and Professional Ways of Knowing." *Journal of Health and Social Behaviour* 33 (1992): 267-281.

Brown, P. and F. Ferguson. "'Making a Big Stink' Women's Work, Women's Relationships, and Toxic Waste Activism." *Gender & Society* 9.2 (1995): 145-172.

Brown, P. and E. Mikkelsen. *No Safe Place: Toxic Waste, Leukemia, and Community Action.* Berkeley: University of California Press, 1990.

Bullard, R. D. *Dumping in Dixie: Race, Class, and Environmental Quality.* Boulder, CO: Westview Press, 1990.

Campbell, R. and S. Wasco. (2000): "Feminist Approaches to Social Sciences: Epistemological and Methodological Tenets." *American Journal of Community Psychology,* 28.6 (2000): 773-791.

Carson, R. *The Sense of Wonder.* New York: Harper & Row, 1956.

Carson, R. *Silent Spring.* Boston: Houghton Mifflin, 1962.

Clare, E. *Brilliant Imperfection: Grappling with Cure.* Durham: Duke University Press, 2017.

Culley, M. and H. Angelique. "Women's Gendered Experiences as Long-term Three Mile Island Activists." *Gender & Society* 17.3 (2003): 445-461.

Di Chiro, G. "Polluted Politics? Confronting Toxic Discourse, Sex Panic, and Eco-Normativity." *Queer Ecologies: Sex, Nature, Politics, Desire.* Eds. C. Mortimer Sandilands and B. Erickson. Bloomington: Indiana University Press, 2010. 199-230.

Goldenberg, M. J. "On Evidence and Evidence-based Medicine: Lessons from the Philosophy of Science." *Social Science & Medicine* 62 (2006): 2621-2632.

Hall, B. and G. Wagner. *Report on the Health Survey of Institute, Pinewood Park and West Dunbar.* People Concerned About MIC, 1988.

Hanna, B., W. Morehouse, and S. Sarangi. *The Bhopal Reader.* New York: The Apex Press, 2005.

Hess, A. "50 Years After *Silent Spring*: Sexism Persists in Science." *Slate* Sep. 28 2012. Web.

Holt, M. "Community Health Survey: An Historical Kanawha Valley

Event." *Downwind News* 7 Jun. 1988: 1-8.
International Campaign for Justice in Bhopal (ICJB). "Our Demands." 2014. Web.
International Campaign for Justice in Bhopal (ICJB). "The West Virginia Chemical Spill: Solidarity from the International Campaign for Justice in Bhopal." 2014. Web.
Kishwar, M. "People's Right to Know: The Struggle in Bhopal." *Manushi* 29 (1985): 38-40.
Krauss, C. "Challenging Power: Toxic Waste Protests and the Politicization of White, Working-Class Women." *Community Acti-vism and Feminist Politics.* Ed. N. Naples. New York: Routledge, 1998. 129-150.
Malin, S. A. and S. S. Ryder. "Developing Deeply Intersectional Environmental Justice Scholarship." *Environmental Sociology* 4.1 (2018): 1-7.
Medico Friend Circle (MFC). *The Bhopal Disaster Aftermath: An Epidemiological and Socio-medical Survey. A Summary of the Report.* 1985.
Pariyadath, R. and R. Shadaan. "Solidarity after Bhopal: Building a Transnational Environmental Justice Movement." *Environmental Justice* 7.5 (2014): 1-5.
Pellow, D. N. *What Is Critical Environmental Justice?* Cambridge: Polity Press, 2018.
Pickering, M. and A. Lewis. *Chemical Valley.* Dir. Mimi Pickering and Anne Lewis. Whitesburg, KY: Appalshop Films, 1991.
Rajagopal, A. "And the Poor Get Gassed." *The Bhopal Reader.* Eds. B. Hanna, W. Morehouse, and S. Sarangi. New York: The Apex Press, 1987. 24-27.
Sarangi, S. "The Movement in Bhopal and its Lessons." *Social Justice* 23.4 (1996): 100-109.
Sarangi, S. "Compensation to Bhopal Gas Victims: Will Justice Ever Be One?" *Indian Journal of Medical Ethics* 9.2 (2012): 118-120.
Sarin, R. "The Babies of Bhopal." *Sunday Magazine.* July 1985.
Sathyamala, C. "The Medical Profession and the Bhopal Tragedy." *Lokayan Bulletin* 6.1/2 (1988): 39-41.
Sathyamala, C. "The Condition of Bhopal's Women." *Bhopal: The Inside Story – Carbide Workers Speak Out on the World's Worst Industrial Disaster.* Ed. T. R. Chouhan. New York: The Apex Press, 1994. 130-131.
Scott, D. N. *Our Chemical Selves: Gender, Toxics, and Environmental Health.* Toronto: University of Toronto Press, 2015.

Simpson, A. "Who Hears Their Cry? African American Women and the Fight for Environmental Justice in Memphis, Tennessee." *The Environmental Justice Reader: Politics, Poetics and Pedagogy.* Eds. J. Adamson, M. Evans, and R. Stein. Tucson: University of Arizona Press, 2002. 82-104.

Voyles, T. B. *Wastelanding: Legacies of Uranium Mining in Navajo Country.* Minneapolis: University of Minnesota Press, 2015.

We for Bhopal (2004). *Closer to Reality: Reporting Bhopal Twenty Three Years After the Gas Tragedy.* 2004.

13.
"Water is Worth More Than Gold"

Ecofeminism and Gold Mining in the Dominican Republic

KLAIRE GAIN

"WATER IS WORTH MORE THAN GOLD," eight-year-old Juan[1] chanted along with members of his community in Las Piñitas, Dominican Republic (personal interview, 2013). He was born and raised here, neighbouring the largest foreign direct investment project the country has ever seen, the Pueblo Viejo gold mine. Pueblo Viejo, owned by Canadian companies Barrick Gold Corporation (BGC) (sixty percent) and Goldcorp Inc. (forty percent), began commercial production in 2014. Since then, community members of Las Piñitas, Las Lagunas, El Naranjo, and La Cerca have expressed great concern regarding environmental devastation, which they believe has directly impacted their health and livelihoods.

Unfortunately, the case of the Pueblo Viejo gold mine is not unique and Virginia Rodriguez notes that over sixty percent of all mining companies

View of Pueblo Viejo from a community members' home in Las Piñitas, 2015.
Photo: Klaire Gain.

263

operating abroad are Canadian. Therefore Canada is increasingly viewed internationally as the face of the mining industry. According to Todd Gordon and Jeffrey Webber, Latin America and the Caribbean (LAC) is home to twenty-five percent of the world's forests and forty percent of its biodiversity and due to its mineral rich environment is a prime destination for Canadian mining projects to explore, exploit and export non-renewable resources.

Throughout this article, open-pit gold mining within the Dominican Republic will be discussed with the purpose of portraying the intersections between gender and mining. First, I will present an ecofeminist analysis on the oppression of women and nature. It will be followed by a discussion of the impacts of open-pit mining and the complicity of Canadian foreign policy in the human and environmental exploitation within the mining industry. Next, a case study of the Pueblo Viejo gold mine in the Dominican Republic will present narratives from communities impacted by open-pit mining including personal findings gathered from 2014 to 2018. Finally, stories of historical and current resistance movements at the Pueblo Viejo gold mine are discussed, focusing on women-led community strength and resistance.

ECOFEMINISM: OPPRESSING WOMEN AND NATURE

Ecofeminism is considered a convergence of feminism and ecology and emphasizes the notion that both women and nature are oppressed and exploited. Through an ecofeminist lens, women and nature are interconnected, leading to mutual patriarchal domination. Ariel Salleh contends that an ecofeminist political analysis links the current day ecological crisis with a "Eurocentric capitalist patriarchal culture built on the domination of nature and the domination of women as nature" (13). Silvia Federici's book, *Caliban and the Witch* connects the rise of capitalism to a war on women, explaining, "women's power had to be destroyed for capitalism to be developed" (17). Federici relates the beginning of capitalism and women/nature oppression with the enclosures of the commons. She contends that women are disproportionately impacted by the enclosures since,

> as soon as land was privatized and monetary relations began to dominate economic life, they found it more difficult than men to support themselves, being increasingly confined to reproductive labor at the very time when this work was being

completely devalued. This phenomenon ... had accompanied the shift from a subsistence to a money-economy in every phase of capitalist development. (*Caliban* 74)

Silvia Federici explains that through the creation of capitalism, primitive accumulation and the enclosures, women became for men "the substitute for the land lost to the enclosures, their most basic means of reproduction, and a communal good anyone could appropriate and use at will" (*Caliban* 97). She links the exploitation of this "communal good" and the environment by explaining, "women's labour began to appear as a natural resource, available to all, no less than the air we breathe and the water we drink" (*Caliban* 97). Through this, women, like the environment are oppressed and marginalized within a capitalist system based on exploitation and perpetual growth. Federici ("Feminism") also notes the importance of traditional lands for community well being noting that "the commons too were for women the centre of social life, the place where they convened, exchanged news, took advice and where a women's viewpoint on communal events, autonomous from that of men, could form" ("Feminism" 72). Therefore, the enclosure of the commons and exploitation of the environment leads to a decline in the health and well being of women and communities.

The enclosure of the commons is a necessity in the creation of open-pit mining and a study conducted by the Labrador West Status of Women Council, in collaboration with Mining Watch Canada noted, "mining affects health of women at various levels, through environmental contamination of air, water and soil, through noise pollution, and through disasters and pit closures" (12). Mining also leads to "indirect" impacts for women including increased mental health concerns, community breakdown and physical, emotional and sexual violence. Within mining impacted communities, women are "drawing connections between the building of large-scale mines and the corresponding deterioration of social, cultural, economic and physical well being of local populations" and recognizing that the vast impacts of open-pit mining disproportionately affect women (International Women and Mining Network 4). An ecofeminist perspective provides evidence of the impacts of mining on both ecology and women and the following discussion of Canadian open-pit mining describes the sectors stance on Corporate Social Responsibility (CSR) and the government's complicity in environmental and human rights violations throughout LAC.

CANADIAN MINING INDUSTRY AND HUMAN RIGHTS VIOLATIONS

According to Barrick Gold Corporation, their company stives to "respect human rights ... and recognize the equality and dignity of the people we interact with everyday." However, Canadian mining corporations have a well-known history of environmental destruction and human rights violations. Gordon and Webber note, "mining investment in most instances simply cannot proceed without a community—often Indigenous—being dispossessed of their land, natural resources and livelihoods" (67). Katy Jenkins notes that this dispossession often comes from forced evictions, pollution of water sources, decrease in agriculture production, death of livestock and an array of negative health implications due to the toxic waste produced by open-pit mining. Despite the attack on communities, the Canadian government provides significant support for Canadian mining companies.

The failure of proposed Bill C-300, "Corporate Accountability for the Activities of Mining, Oil or Gas Corporations in Developing Countries" was a prime example of the Canadian government's complicity in human and environmental injustice committed by Canadian companies. Yves Engler explains, "under Bill C-300, companies that failed to adhere to, relatively lenient, standards of social responsibility would lose the support of Canadian embassy officials and taxpayer funded agencies" (45). However, an aggressive lobbying campaign was launched and BGC Gold stated that Bill C-300 would "risk the competitive position of Canadian companies, result in the reputational damage to Canadian companies and ignore existing regulatory frameworks for CSR" ("Barrick Gold Sets Out Position"). Furthermore, the Prospectors and Development Association of Canada (PDAC) director, Tony Andrews, stated "the development of a CSR framework specifically targeted at Canadian Mining Companies could disturb the "level playing field" and place them at a competitive disadvantage.... The voluntary nature of social responsibility must be maintained" (Keenan 4). Due to the fear that Bill C-300 passing would lead to strict regulations and therefore loss of profit, Canadian mining companies hired over 300 lobbyists to persuade parliament members and the Bill failed by a mere 6 votes (Labrador West Status of Women Council). This ruling told the world that the Canadian government is not only allowing mining companies to operate without regulations abroad, but will continue funding corporations even if they are found guilty of environmental or human rights violations.

With the growth of an extractive economy backed by government, the dangers of mining have drastically increased; Ana Isla explains, "open-pit mining... now uses the extremely toxic cyanide lixiviation technique" which poses huge risks to both the environment and community members (147). Gordon and Webber maintain that cyanide is an incredibly dangerous chemical compound and a rice size grain could kill a human being within five minutes. Rodriguez argues that cyanide residue left from the gold extraction process is often held in insecure, human-made tailings ponds where changes of pH levels frequently lead to toxic gas emissions. Furthermore, tailings ponds are historically known for overflow and cracks, allowing the poisonous substance to leak into local water sources. EU Issue Tracker ("Ban on Cyanide") has shown the devastating impact of cyanide in a spillage in Romania in 2000 where a tailing pond leaked cyanide contaminated water and heavy metals into local river systems. This one spill caused extensive pollution throughout Romania, Hungary, Serbia, Bulgaria and the Black Sea.

Although Glevy's Rondon explains that most mines have a life span of ten to thirty years, the environmental impacts often last decades after mining activities have ceased, causing problems for generations to come. To depict these findings, I will next present a case study of the Pueblo Viejo gold mine in the Dominican Republic.

THE PUEBLO VIEJO GOLD MINE

According to Barrick Gold Corporation (BGC), the Pueblo Viejo mine is its largest project with plans to extract "8,087,000 ounces of proven and probable gold reserves" throughout the mine's lifetime The Pueblo Viejo Gold mine is shared with Canadian company Goldcorp Inc. (forty percent) and it is located in the province of Cotui, Dominican Republic. Pueblo Viejo was acquired by BGC in 2008 and began open-pit commercial production in 2013, with a mine life of 25+ years (Tomayo) The zone in which Pueblo Viejo now sits was historically a small, state-run mining site, which operated from 1975 until 1999 by company Rosario Dominicana (RD) (Cardenas, Miranda and Krutzelmann). When BGC acquired the mining sitethe scope of the project increased and BGC now has more than "3.7 Billion dollars in mine construction capital" (Tomayo).

Virginia Rodriguez informs that BGC currently has twenty-six open-pit gold mining projects operating predominately in Latin America. She maintains that the open pit mining conducted at Pueblo Viejo is the most dangerous type of extraction with each ounce of gold producing an average of 79 tons of waste (11). Due to this, Rodriguez argues that both community members and researchers contend that the operations of the former Rosario Dominicana created environmental, social and financial harms. In 1999, the operation was shutdown due to mismanagement and lack of financial resources, however RD's operations exposed enough sulfide ore to initiate acid mine drainage which left a community water source, the Margarita river in an acidic state. However, Barrick's enlargement involved an enormous expansion of the mining site, which required land to be taken from local communities. She also sustains that although communities generally receive minimal compensation for their land, they are often misinformed and deceived into believing mining will provide positive changes for their families and communities.

Local newspaper, *Maimon Punto De Encuentro* noted, "the traditional perception was that the operation of a mine will create a booming economy. Logically then, the announcement of the mine created expectations of economic development, but not of pollution and sickness" ("El peligro"). Rodriguez argues that in order to gain control over communities and the local political economy, government officials are often bribed, given false promises of employment and poverty reduction, intimidated and threatened at the hands of BGC (31). For example, she writes that within the communities surrounding Pueblo Viejo, the mayor was a past adversary of BGC and was elected on the grounds that he would resist the mine. However, after his election, Barrick provided a large sum of money to the community and in exchange the city hall allowed the expansion of Pueblo Viejo open-pit mining to begin. He later admitted that he was threatened by BGC and "the whole issue had been a strategy for Barrick to get better control of the town" (Rodriguez 25).

The political and economic gains that BGC's has acquired through Pueblo Viejo have not come without controversy. Rodriguez notes that in 2005, the Dominican government consented to the mining under the agreement that they would receive twenty-five percent plus taxes on all gold exported. Yet Rodriguez explains that during a short visit to the Dominican Republic, former Canadian Prime Minister, Stephen Harper, quietly passed an amendment stating the Dominican Republic would only receive this after BGC recuperated the $3.5 billion they invested and were making a ten percent return on investment (23). Therefore, despite the Pueblo Viejo gold mine being the largest foreign direct investment project that the country has ever seen, the presence of BGC has not been of significant benefit to the Dominican economy nor its citizens. Instead, community members have experienced fear and uncertainty while living beneath BGCs "El Llegal" tailings pond.

The 2001 Romania tailings pond spill shown by EU issue tracker ("Ban on Cyanide") is also a frightening reality for communities surrounding the Pueblo Viejo gold mine. The threat of this is critical as the Dominican Republic is small enough to fit into Romania six times, making the danger of a tailings pond spill not only detrimental for surrounding communities, but the entire island. Furthermore, Rodriguez notes that there were four occasions in 2011 where "dozens of families of surrounding communities and thousands of Pueblo Viejo workers had to be temporarily relocated due to the risk of the tailing pools collapsing because of the record amount of sudden rain being registered in the area" (27).

The risks are not only for community members, but also for those working at the mine and Rodriguez explains in 2011, close to 1,000 workers at Pueblo Viejo fell ill to what the company defined as "food poisoning." However, workers claimed it was due to exposure to a chemical explosion and after an investigation conducted by the Dominican Academy of Science, it was concluded that the intoxication had a chemical origin (24). Furthermore, Pueblo Viejo security forces were reported to be "accompanying the employees to the health centers and trying to avoid any contact with them from the public" (24). Although BGC states "Barrick believes that transparency—whether through disclosing payments to governments, reporting on our energy and water use, voluntarily opening ourselves to third-party scrutiny, or otherwise- is integral to being an honest partner" ("Barrick Gold Sets Out Position"), this case portrays that the opposite may be true and creates scrutiny of Barrick's transparency, accountability and safety measurements when utilizing dangerous substances.

While Barrick Gold has signed the international code of practice for the handling of cyanide at their Pueblo Viejo mine and claims that they conduct community viewable tests on water and air, community leaders state the contrary and explain that the company has refused to provide results of these tests to the people ("Mining in the Dominican Republic"). Due to this, community leaders contacted the Ministry of the Environment, who found that the water in the Margarita River, located downstream from the mine was highly acidic. They also revealed that the river contained cyanide residue, sulphides, copper and mercury, all of which cause severe lesions, sicknesses and death, specifically for women, within the surrounding communities ("Mining in the Dominican Republic"). In the next segment, I present my findings and pictures regarding to effects of the Pueblo Viejo mine on the bodies of ecology and women.

IMPACTS OF MINING ON COMMUNITY MEMBERS
SURROUNDING PUEBLO VIEJO

Many community members note that they did not initially recognize the threats that an open-pit mine would bring to their community. Escarlin from Las Piñitas notes, "Barrick Gold caught us sleeping. It tricked us and the state tricked us" (personal interview, 2014). While BGC relocated a small number of families throughout the expansion of the mine, local newspaper *Maimon Punto De Encuentro* explains, "the construction of the neighborhood, where some of the evicted are relocated, has failed to comply with the contract" (El peligro") as the infrastructure is poor, the land is coarse and the water service is precarious, leaving families without access to subsistence living. Pablo, who was relocated by BGC, explains, "while I may have four sturdy walls I have nothing else and I still have to live with the sounds, smells and pollution from the mine" (personal interview, 2018). Juan expands upon the misconceptions of what communities were told, versus reality and states, "when Barrick first arrived they promised employment with more than eighty percent of workers being from our communities, but it has been the exact opposite" (personal interview, 2018).

In addition to a lack of economic gains, communities assert that their suffering has not ceased, but rather increased with the expansion from Rosario Dominicana to Pueblo Viejo (personal interview, 2014), despite the companies claim of using "state of the art technology" (Barrick "Barrick Gold Sets Out Position"). In an interview, carried out in 2015, community member Julia from Las Piñitas explains, "Barrick Gold says that the environmental impact that exists were left by the former Rosario Dominicana. But with Rosario Dominicana, at least we had water." These concerns are echoed by community members who believe that although BGC claims corporate social responsibility and strong community connections, the devastation communities experience is related to BGC's methods of operation within their Pueblo Viejo gold mine.

While the general rhetoric of mining corporations involves community collaboration and positive change, Omar, an environmentalist within the area, contradicts such claims, explaining, "A lot of the communities are still unreachable. Roads still haven't been built besides those going

The wall of BGC's El Llegal tailings pond behind homes in La Naranja, 2018.
Photo: Klaire Gain

to the mine. Infrastructure is completely lacking. So, where is this CSR [Corporate Social Responsibility]?" (personal interview, 2014). Also in 2014, community member Natalia living beside the mine furthered this notion; "They say that we are living well in this community yet they spend all of their money on publicity to tell those lies." Local ecologist Jorge notes that when BGC acquired the mining site in 2006 it updated the extraction method to a process called pressure oxidation using the four largest autoclaves in the world, which allows them to extract gold from low-grade ore that Rosario Dominicana could not (personal interview, 2014). This process is ideal for BGC as it is quick and inexpensive, making it economically viable for a capital driven company. Despite the economic gains, this new extraction procedure is extremely resource intensive and uses industrial quantities of cyanide, a hazardous chemical compound that creates lasting impacts on communities surrounding the mine (Jorge, personal interview, 2014).

Community members experienced another close call in October 2016 when they claim that BGC's El Llegal tailings pond rose to a dangerous level during a time of high rainfall. According to Juan, although there was a "red alert" sounding and mine employees were evacuating, BGC did not communicate with community members surrounding the mine, leading to fear, confusion and complete mistrust of the company (personal interview, 2016). Community members surrounding the mine believe that there is no positive outcome with the presence of BGC's El Llegal tailings pond above their communities. In a 2018 interview, Juan explained, "We are at the point of death, if they don't drain the tailings

Barrick Gold's El Llegal Tailings Pond, 2017. Photo: Jamie Foxton

pond it will overflow and we will die, but if they do drain the pond it will go into our river source and we will die."

While the community members fear overflow and collapse of the tailings pond, they believe that the sole presence of the mine is detrimental to their health. In an interview in 2014, community member Maria stated, "Barrick Gold could be compared to a cancer for the people. A cancer that grows bit by bit, destroying each one of their organs until it takes their life." Furthermore, community members argue that the Pueblo Viejo mining site disproportionately impacts women. Since Pueblo Viejo began operation, over one hundred members of the community, largely women and children, have rashes and lesions on their bodies (personal interviews, 2018). Additionally, in 2018, Maria explained that the presence of vaginal infections has drastically increased since the arrival of BGC. Community members believe that these lesions and infections are linked to environmental pollution as they state most appeared after having direct contact with local water sources while bathing (personal interview, 2014).

Community member Juan believes that the lesions are consistent with skin damage caused from cyanide exposure (personal interview, 2018).

Due to this, five members of surrounding communities who had fallen ill underwent blood testing in an attempt to portray the toxic impact of the mining process. When results were returned, all five shared that they tested positive for cyanide traces above safe levels (personal interview, 2014). Furthermore, community member Rubin alleges, "there have been twenty-three people in the community who had zinc, lead, and chromium in their blood. This was an analysis done by a doctor [name not noted] in Santo Domingo. But Barrick would never admit this" (personal interview, 2014). According to community members, seventeen out of twenty-three diagnosed were women, and communities identify that it is the women who most drastically experience the vast range of impacts caused by Pueblo Viejo (Escarlin, personal interview, 2014).

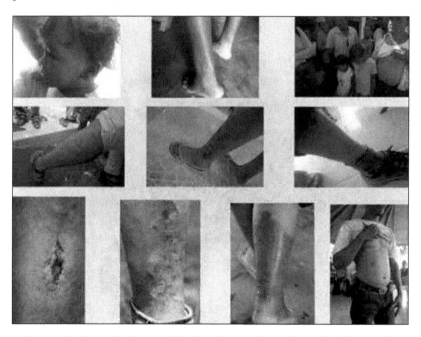

Images of lesions and rashes on the bodies of community members allegedly caused by mine pollution, 2016. Photos: Klaire Gain

Women in the surrounding communities note that they are more likely to be exposed to dangerous chemicals from the mine due to their matriarchal care roles within their families. Community member Marisella explains, "the women have to work with water many times a day, and it is not the men, but the women who are feeling the major decline in the community" (personal interview, 2014). The struggle for clean water has been ongoing since BGC began operation and after years of community complaints to both the company and the ministry of the environment regarding inadequate water sources, BGC began providing each family in the area with eight bottles of water per week. Community members share that this was in response to a so-called "drought" within the area and was part of its community development initiatives. The water service continued for a year, and although it was not sufficient or sustainable, community members shared that it was better than nothing at all. However, communities claim that on February 14, 2015, after creating community dependency, BGC suddenly ended the program, leaving families with no safe drinking water (personal interview, 2015). Community member Juanita stated that she believes that, "[Barrick] found that this public service was too expensive and they were damaging their reputation in the media.... When you went to the Internet and the first thing that came up was that they were giving out large supplies of water, it meant that they had accepted that the water was damaged. So they regretted it and they took the bottled water from us" (personal interview, 2015). Geologist Domingo Abreu noted the interconnections between water, health and poverty and states that since community members are no longer able to utilize their local water sources, "they are obligated to receive or buy water in bottles for all uses. They use bottled water to bathe, wash dishes, cook, clean ... for everything. This affects the economics of the families because before Barrick the families had free clean water, they would simply search for their water in the rivers. Now they are using money that they could use for food to buy water" (personal interview, 2015). Echoing Abreu's words, community members also express concern over food insecurity allegedly caused by the mining industry.

Since BGC began constructing the Pueblo Viejo mine community members shared that they have experienced drastic decreases in their agricultural production. In a 2014 interview, community member Carmen notes, "in the fields, we had good production and lived well and peaceful lives with our families. But now Barrick, like a curse, has come." With the creation of the Pueblo Viejo gold mine, community members claim that they have lost over 80 percent of the cacao production in the area (Marianna, personal interview, 2014). When questioned about the potential reasoning for this, Marianna from Las Piñitas, stated that she believes the rain is so polluted that it does not provide nourishment for crops to grow. She furthered, "rain used to be something we embraced, praised and cherished. It grew our crops, flourished our communities and nourished our children.

Now it is killing us" (personal communication, 2014). Additionally, local newspaper notes, "2000 livestock have died since the opening of Pueblo Viejo" ("Campesinos desalojados"). This does not only create food insecurity, but also impacts the livelihoods of the families as communities state that due to the severity of the situation, many of them no longer even bother to plant crops (Jorge, personal communication, 2018).

A mango from the community of Las Lagunas, 2014. Photo: Klaire Gain

Community member Marisella alleges that, "the mangoes used to be everywhere and our children would play and take them from the trees to eat, but now, they are dangerous" (personal interview, 2014). Communities explain that before the mine, the mangoes grew healthy and plentiful and claim that the black spots on the mangoes are caused by the rain and air pollution from the mining site (personal interview, 2014). The women in the community worry about the health of their families, particularly their children as they continue to consume the mangoes without a thorough understanding of the risks. Furthermore, community leader Eugenia explains "we, the mothers of these communities, are very worried for the future of our children, because here, there is no tomorrow" (personal interview, 2014). Rosalie further explains that within the past year, four women in the community have had miscarriages which they claim is a drastic increase from previous years. Luis notes that "the women are no longer fighting for them, but for their children and grandchildren, because without the fight they will have no future" (personal interview, 2018).

Community members surrounding the Pueblo Viejo gold mine claim the impacts run deeper than agricultural decline and water pollution, with sounds and odours from the mine creating poor living environments for their families. They allege that machinery at the mine operates twenty-four hours a day and children often wake up in the night in fear from loud bangs and rock explosions (personal interview, 2014). The odours that emit from the mine cause physical illness with youth. Daniella from Las Piñitas explains, "when we are at school and the odour arrives we vomit, we have nausea, we have headaches, we are dizzy ... everything. The teachers have to send us home and we can't even learn" (personal interview, 2014). Juan echoed this comment and noted that on February 28, 2018, an unbearable odour was emitted from the mining site which lasted the entire day and caused headaches, blurred vision, and vomiting. He continued, "everyday that passes becomes more difficult. There is more and more contamination everyday" (personal interview, 2018).

Women are also concerned about the fabrics of their families due to the dangers that have become part of their daily lives. In reference to the El Llegal tailings pond, a woman in the community notes, "every time it rains hard we live in fear that it will overflow" (Daniella, personal interview, 2014). Due to this fear, women state that they have created a system in which two members of the community stay awake each night to monitor the activity at the mine and the levels of the tailings ponds. Since they do not trust the company, this system allows them time to evacuate their communities if a disaster occurs (Marianna, personal interview, 2017). Communities explained that it is the men who have taken on this task, but shared that this has consequently intensified the roles of the women within the families since their husbands do not have sufficient energy or time to contribute after being "on watch" all night. In a 2014 interview, Eugenia furthered, "because of the mine women now have many roles, not just more around the household, but we have to worry about getting food, water and keeping everyone healthy." Despite the hardships that women face, their strength is evident and women are leading resistance movements for community justice within the Dominican Republic. Marianna from Las Piñitas explains, "we live day to day fighting for ourselves and our children just to survive" (personal communication, 2018).

WOMEN IN RESISTANCE

Community members throughout El Naranjo, Las Lagunas, La Cerca and Las Piñitas recognize the importance of mobilization in resistance against BGC. Local activist Carolina notes "As many gold companies as there are exploiting [in Latin America] there are groups resisting" (personal interview, 2018). This notion is depicted throughout Latin America and the Caribbean with women often at the front lines of organizing and resisting Canadian mining companies. This comes in many forms and often involves protesting, petitioning governments and organizing community movements. The strength of the resistance at Pueblo Viejo in the Dominican Republic is evident as community member Maria notes, "if we have to die of thirst, of hunger or of cancer we prefer to die with a bullet in our head defending our land" (personal interview, 2014). While resistance has been strong since the mine began construction, recent mobilization is calling for BGC to adhere to community demands.

On November 6, 2017 community members decided to take direct action in resistance to BGC's Pueblo Viejo mine. After years of silence from BGC, communities began occupying space outside of the mines

Community members spray-paint "NO MINE HERE" on the road leading to the Pueblo Viejo gold mine, 2014. Photo: Klaire Gain

Signs reading: "Welcome to the valley of death. Danger, area is contaminated by Barrick"
posted on electricity poles leading to the mining site, 2017. Photo: Klaire Gain

headquarters, chaining themselves to chairs and demanding relocation for the 600+ families surrounding the mine. Relocation has been an ongoing struggle since the opening of the mine and local newspaper notes, "the area has become a cemetery to people who must live in the pollution prevailing in the area due to the mining operation" ("Campesinos desalojados" 1). The current protest is gaining national attention and a sign at the protest explains that community members will not be moved until BGC agrees to "a dignified and immediate relocation of the endangered communities, payment for all the properties of the affected, investment in the environment, agriculture and livestock production and investment in the community" (personal interview, 2018).

This is not the first time that impacted communities have taken to the street. In February of 2015, when BGC suspended water supply to neighbouring communities, the communities were strong in their opposition. A local newspaper notes that community leaders organized a march to the community relations office of mining with over 300

Community members chained outside of BGCs Pueblo Viejo mine, 2018. Photo: Klaire Gain

participants carrying empty bottles and chanting, "we want water, without water there is no life" (Jorge, personal interview, 2015). After receiving little response from BGC, they later created a community road blockade beginning at 5:30 am prohibiting the mining vehicles from passing. Unfortunately, they shared that they were met with violent repression from BGC armed security and had to stand down. This militarization around the mining site is not uncommon and a community member claims, "even when we gather to go to church security guards come and ask us what we are doing" (Juan, personal interview, 2014).

Despite the constant repression, community members continue to mobilize in protection of their environment, identities and livelihoods. As Marisella explains, "the only good thing that Barrick Gold has brought to this community is that it has taught us to resist" (personal interview, 2014). Community relations have grown strong throughout the years of resistance against BGC and they are committed to continuing the struggle for reclamation of their health, land and lives. In a 2015 interview, Carolina from Las Piñitas passionately declared, "there is no secret that where there is a mining area there is pollution. Wherever there is an area of mining there is horrible poverty, because they take the riches, while we are left with holes... The people have to get up with one voice before they [BGC] damage the whole country. What will they leave us with? The dead to be buried? ...Barrick Gold, leave the country!"

CONCLUSION

An ecofeminist analysis of Canadian mining within the Dominican Republic demonstrates the impacts of the Pueblo Viejo mining project on the bodies of ecology and women within the communities of La Cerca, El Naranjo, Las Piñitas and Las Lagunas. My findings suggest that despite claims of corporate social responsibility within the mining sector, community narratives portray the realities of extraction and exploitation within predominately impoverished campesino communities. However, with strength and resilience, community members are resisting for their right to live their lives in health and peace. Both Canadian and international solidarity must accompany this direct community action to hold mining companies accountable and work towards justice for mining impacted communities on a global scale.

ENDNOTES

[1]All names have been changed to protect the identities of community members.

REFERENCES

"Ban on Cyanide in Gold Mines?" EU Issue Tracker, 2010 Web.

"Barrick Gold – Sustainability – Transparency Hub." Barrick.com (n.d.). Web. November 26, 2017.

"Barrick Gold Sets Out Position on Bill C-300 and Provides Facts." Press Release. (n.d.). Web. 28 October 2017.

"Barrick Gold Operations – Pueblo Viejo." Barrick.com, 2016. Web.

Cardenas, Rosmery, Hugo Miranda and Hogel Krutzelmann. (2018, March 19). *Technical Report on the Pueblo Viejo Mine, Sanchez Ramirez Province, Dominican Republic* (Rep. No. NI 43-101). Roscoe Postle Associates Inc. 19 March 2018. Web. 01 March 2019.

Bhanumathi, K. "The Status of Women Affected by Mining in India." *Tunnel Vision: Women, Mining and Communities.* Eds. Ingrid Macdonald and Claire Rowland. Victoria, Australia: Oxfam Community Aid Abroad, 2002. 20–24.

Engler, Yves. *The Ugly Canadian: Stephen Harpers Foreign Policy.* Vancouver: RED, 2012.

Labrador West Status of Women Council, in collaboration with Mining Watch and the Steelworkers Humanity Fund. *Effects of Mining on Women's health in Labrador West. Final Report.* Labrador City, NL,

7 Nov. 2004. Web.
"El peligro del uso del cianuro y la Mega Minería en Rep. Dom." *Maimon Punto De Encuentro* 20 June 2012. Web.
Federici, Silvia. *Caliban and the Witch*. Brooklyn: Autonomedia, 2004.
Federici, Silvia. "Feminism and the Politics of the Commons." *Uses of a WorldWind, Movement, Movements, and Contemporary Radical Currents in the United States*. Ed. Team Colors Collective. Oakland: AK Press, 2010.
Fordham, Maureen. "Gender, Disaster and Development: The Necessity for Integration." *Natural Disasters and Development in a Globalizing World*. Ed. Mark Pelling. London: Routledge, 2003. 73-90.
Gordon, Todd and Jeffrey R. "Webber. Imperialism and Resistance: Canadian Mining Companies in Latin America." *Third World Quarterly* 29.1(2008): 63-87.
Happel, Ellie. "Water is More Valuable than Gold." NACLA, 25 April 2016. Web. 03 December 2017.
International Women and Mining Network/Red Internacional Mujeres y Mineria (RIMM). *Women From Mining Impacted Communities Speak Out: Defending Land, Life & Dignity*. Ed. Tanya Roberts-Davis. Andhra Pradesh, India: International Women and Mining Network, 2010.
Isla, Ana. "A Struggle for Clean Canadian Mining in Costa Rica." *Canadian Women Studies/les cahiers de la femme* 21/22.4,1 (2002): 158-164.
Jenkins, Katy. "Women, Mining and Development: An Emerging Research Agenda." *The Extractive Industries and Society* 1.2 (2014): 329-339.
Keenan, Karyn. "Canadian Mining: Still Unaccountable." NACLA *Report on the Americas* 43.3 (2010): 29-34.
Macleod, Morna. "Development or Devastation? Epistemologies of Mayan Women's Resistance to an Open-Pit Goldmine in Guatemala." *AlterNative: An International Journal of Indigenous Peoples* 12.1 (2016): 86-100.
"Mining in the Dominican Republic, Sickness and Wealth." *The Economist* 21 Sept. 2013. Web. 01 November 2017.
Rodriguez, Virginia. "Natural Resource Exploitation in the Caribbean: From Colonialism to Neoliberalism: Case Study of Pueblo Viejo Gold Mine in Dominican Republic." Masters Thesis, Environment, Development and Policy, University of Sussex, 2011.
Rondon, Glevy. "Canadian Mining in Latin America: Corporate Social Responsibility and Women's Testimonies. *Canadian Woman Studies/*

les cahiers de la femme 27.1 (2009): 89-96.

Salleh, Ariel. "An Ecofeminism Bio-Ethic and What Post-Humanism Really Means." *New Left Review* 217.1 (1996): 138-147.

Tamayo, Carlos. "Dramatic Improvements at Margajita River at Pueblo Viejo Gold Mine." Barrick.com (n.d.). Web. March 01, 2019.

United Nations Office for Disaster Risk Reduction. *Disaster and Gender Statistics*, 2009. Web. January 20, 2018.

14.
Indigenous Andoas Uprising
Defending Territorial Integrity and Autonomy in Peru

ANA ISLA

THIS CHAPTER ADDRESSES THE SOCIAL and ecological catastrophes caused by fossil fuels extraction activities by examining a 2009 court case that was part of the aftermath of a 2008 uprising in Andoas town of three Indigenous groups living adjacent to three river basins: (a) the Achuar in Corrientes River; (b) the Quichua in Pastaza River; (c) the Urarina in Tigre River. Andoas is located in Loreto, Datem del Marañon province, Peru. This chapter continues the analysis presented in an earlier article in *Capitalism Nature Socialism* (Isla "Eco-Class-Race") that examined land privatization and regional Indigenous responses to it. It focused on the defence of the commons by the Awajun-Wampis Indigenous people in Bagua Province, Peru, and the bloody attack of President Alan Garcia's administration against Indigenous commoners. The 2009 article established a context of regional civil disobedience within which Indigenous populations from the rainforest rose up against oil production by Pluspetrol Norte S.A. (Pluspetrol) operations as they began to experience life-threatening conditions in their ecosystem and subsistence economies. This confrontation marks a significant moment in the battle for livelihood and the dispute over the Commons.

In this chapter, I argue that Indigenous labour, resources, and territories in the rainforest are the new sources of accumulation within the global state and "green" capitalism re-launched in the face of ecological crisis. I look at the role of the neoliberal extractivist Peruvian state, working under the banner of sustainable development, as a political instrument associated with violence. In particular, this instrument has involved the process of dispossessing Indigenous peoples from Corrientes, Pastaza, Tigre, and Maranon rivers of their homes and their hunting grounds, and destroying the livelihoods of the communities living near petroleum concessions known as Blocks

1AB and 8, as their territory and economy go unrecognized or invisibilized. I use the concept of territory (water, forest, traditions, cosmovision, etc.) as a place of dispute, which is linked to use value or livelihood provisions for Indigenous communities and to exchange value or profit for oil corporations. I argue that Indigenous people's existential requirements for subsistence and needs for reproducing their provisions has been the source of their empowerment and of their resistance. I also look at the ecological crisis introduced by oil production on the Indigenous peoples' territories, first by Occidental Petroleum Corporation, and later by Pluspetrol Norte S.A. Based on these interactions, I use the concept of ecocidal (in)action of the Peruvian state to include the violation of human and Earth rights of the land of marginal populations.

Through a detailed review of these gridlock situations, first I present ecofeminist critiques of capitalism. Critics have underlined "enclosure" (Federici) and "housewifization" (Mies) as part of the working of patriarchal capitalism, which is a system that maintains relationships of domination and subordination in general and men's domination of women in particular. I then apply these insights to analyze the Indigenous Amazonian human condition before oil production and the role of the neoliberal extractivist Peruvian state. Here I highlight the preparation of Peru as a member of the global state, after the Earth Summit in1992, and the use of the law as an instrument of dispossession. I document a war against subsistence, understood as a livable economy, by examining: (1)how oil fields have gone hand in hand with encroachment on Indigenous people's land, water, and health; (2) how subsistence confiscation resulted in the incorporation of Indigenous people into discriminatory and poorly paid jobs; (3) how a protest against poor wages placed Indigenous people into the Judicial Court in Loreto in 2009; and (4) how the Court Verdict recognized Indigenous people's collective rights in opposition to state criminalization. I conclude with an ecofeminist view of the court case, and provide grounds to characterize the ecocidal (in) action of the Peruvian state.

ECOFEMINIST CRITICS TO CAPITALISM AND SUSTAINABLE DEVELOPMENT

I have chosen the ecofeminist subsistence perspective because this perspective argues that it has been possible to sustain the illusion that economic growth is a positive and benign process because the costs have been borne primarily by what has been devalued: the bodies of women, peasants, Indigenous people, the so-called underdeveloped world, and nature.

Silvia Federici, in *Caliban and the Witch,* argues that capitalism produced a new sexual division of labour, in which women's position in society as providers was redefined in relation to men, to become wives, daughters, mothers, and widows, all of which "hid their status of workers, while giving men free access to women's bodies, their labour, and the bodies and labour of their children" (79). With the advent of capitalism, women's reproductive labour was placed at the service of an international system of accumulation, women's wombs became public property, and they were forced to function as an instrument for the reproduction of labour and the expansion of the workforce, the common territory of men and the state. In this sexual division of labour, women and those bodies that are feminized were forced into a state of reproductive labour—used as a resource, devalued, dependent, and deviant.

Maria Mies, in *Patriarchy and Accumulation on a World Scale,* argues that when women, peasants, and Indigenous people are described as "closer to nature," they are made exploitable—their bodies and labour power are free for the taking. Mies uses the term "housewifization" to capture the process whereby hitherto productive (life sustaining) work is captured, confined, devalued, and put to use in support of "real" (monetized) production, in effect as a free subsidy. The division of labour into reproductive (subsistence, supplemental, housewife) labour and productive (income-earning, breadwinning) labour is necessary for capitalism to function (102). Mies presents a detailed case for understanding this division as existing on a world scale via colonialism, in much the same way that it exists in families and households. The labour and bodies of racialized peasants and Indigenous people in the Global South are appropriated and objectified by white industrialists by means of violence and forced dependency.

Maria Mies and Vandana Shiva, in *Ecofeminism* argue that the "good life" and freedom are possible when people and nature are not separated, since it is the inhabitants' subsistence know-how that helps to conserve the *conditions* of life by valuing sufficiency and autonomy while recognizing the necessity of life being in harmony with the natural world: "It is this kind of materialism, this kind of immanence rooted in the everyday subsistence production of most of the world's women (and Indigenous people) which is the basis of our ecofeminist position" (19).

THE INDIGENOUS AMAZONIAN HUMAN CONDITION AND
THE ROLE OF SUSTAINABLE DEVELOPMENT

Before 1971, the authority of the Peruvian State in the rainforest was minimal, and for the Indigenous people and *bosquecinos* almost inexistent. Jorge Gasche and Napoleon Vela, researchers from Instituto de Investigaciones de la Amazonia Peruana (IIAP), produced a concept of *sociedad bosquecina* (bush people) for those living in the Amazon rainforest where society and nature are still considered one sociological unit. They argue that labour for *bosquecinos* (women and men) is a multitask activity that relied upon learning several techniques of fishing, hunting, gathering and transformation according to the seasonal rhythm of the climate (i.e., winter versus summer), hydrology (e.g., widening waters in the rainy season versus narrowing in the dry season), and biology (e.g., seasons of fruition, fattened animals, fish spawning, and fish schools). This labour is characterized by enjoyment and community cooperation, as *bosquecinos* are free to organize their activities as they please: as an exercise of solidarity according to social rights and obligations; as activity founded on reciprocity with family members, friends, and other community members; and that was grounded in the appreciation of a respected community authority who was believed to be able to influence the forces of nature through "visions," "conversations," and "healing." In this chapter, I use the terms Indigenous people and *bosquecinos* interchangeably as both of them live in the forest and from the forest.

The rainforest sociological unit of society and nature, described by Gasche and Vela, started to change in 1971,when fossil fuel extractive corporations backed by the military government of Juan Velazco Alvarado (1968-1976) granted permission for oil exploration on large tracks of Indigenous territories. However these changes accelerated under the extractivist model when states were given responsibility for sustainable development, understood as economic growth, by the United Nations Earth Summit in 1992 (Isla, *The "Greening"*).

Guided by globalized institutions, such as the World Bank, Alberto Fujimori's administration (1990-2001) came up with a number of laws and policies to enclose and privatize Indigenous land that violated the Peruvian Constitution. Fujimori's Constitution of 1993 eliminated the inalienable character of Indigenous communal land; removed the imprescriptibly character of perpetual right; and seized the land, arguing abandonment. At the same time, Indigenous women became victims of Fujimori's Reproductive Health and Family Planning

Program (Lizarzaburu). As part of the global states, the agenda of "democracy" centred on the voting process was incorporated into the Constitution. In 1993, Fujimori's administration also ratified the ILO Convention 169. The Convention requires that Indigenous peoples be consulted about activities that may take place in their territories and either approve or reject them. In 1995, the Convention become law and obtained constitutional status; however, it was never incorporated into national legislation. Furthermore, criminalization of the poor by the Peruvian state is recorded in *The Truth and Reconciliation Commission of 2003*.

The Johannesburg Earth Summit of 2002 transferred responsibility for sustainable development to corporations and their shareholders. As a result, in Peru, Alejandro Toledo's administration (2002-2006) ignored Convention 169 and furthered Indigenous land privatization to accommodate corporations, particularly Odebrecht ("Diez grandes proyectos"). By 2004, under The Initiative for the Integration of the Regional Infrastructure of South America (IIRSA), his administration created conditions for infrastructure investments to facilitate the transport of Brazil's merchandise en route to the world market. Along the "imaginary" railway lines, Toledo imposed the system of forest concessions to change the use of the land for developing monoculture and cattle ranching in the rainforest. This decision has made *bosquecinos'* logging activities illegal by requiring them to buy a permit, which they cannot afford. Thus, *bosquecinos* who do not have access to forest concessions or live far away from them have been criminalized (Alvarez).

Allan Garcia's two separate terms of office advanced land privatization and the criminalization of protest. His first administration (1982-1986) executed 300 inmates accused of "terrorism" while his second (2006-2011) exacerbated conflicts on Indigenous land such as the one in 2008, in Andoas (the case examined in this chapter), and in 2009 in Bagua (Isla, "Eco-Class-Race" 22). In an attempt to justify Indigenous land privatization required by Peru Trade Promotion Agreement (PTPA), on October 28, 2007, President Garcia wrote an article titled "El Perro del Hortelano" ["The Syndrome of the Gardener's Dog"]. Here he likened Indigenous people (and *bosquecinos*) to mad dogs who have resources they neither exploit nor allow anyone else to exploit. By 2008, Garcia nullified the consultation process in communal lands in order to sign oil and mining contracts. Further, he expanded the system of concessions and changed the use of land, from forest to agricultural projects to support the ethanol production of Dionisio Romero's corporation. But, forced by international concern for Indigenous peoples' rights, in May

2010, his administration approved the Law of Prior Consultation (LPC) of Convention 169.

Ollanta Humala's administration (20011-2016) signed and approved the regulations for LPC in 2011. The Regulations were going to clarify how the consultation with Indigenous peoples should take place, to what extent the result bound the state, and who were considered Indigenous peoples. However, these concerns have never been implemented. Instead, the regulation only defines who were not Indigenous people (Ruiz). Furthermore, his administration created Law No. 30230, which modifies the expropriation law, sets tax measures, and simplifies procedures and permissions for the promotion and encouragement of capital investment. Title III of this Law establishes the ownership of domain in favour of business interests, above the ancestral rights of Indigenous peoples (Tovar et al.). In addition, as Humala is a retired military officer, his government has produced new political instruments of repression, exclusion, and domination, such as Law 30151[1] that grants impunity to military and police personnel in cases of violation of human rights, and Law 1095[2] that authorizes the Armed Forces to intervene in conflicts without the declaration of a state of emergency.

Since 1990, the extractivist economic model of Sustainable Development, based on the exploitation of oil, minerals, and crops for biofuel production, has led to increased environmental conflicts in Peru. Table 1 shows the high percentage of active socio-environmental conflicts since 2009. The Ombudsman Office for the Prevention of Conflict and Governance argued that economic growth linked to resource extraction/ exploitation that takes place near or within community common land has led to such conflicts. It indicated that the majority of the country's conflicts were related to socio-environmental factors (Mendoza).

Table 1. *Active Socio-environmental Conflicts in Peru*

	Total conflicts	Active socio-environmental conflicts
2009	284 (November)	129 (46%)
2012	227 (December)	164 (72%)
2014	212 (November)	138 (65.1%)
2015	210 (June)	141 (67.1%)

Source: *Defensoría del Pueblo (2009, 2012, 2014, 2016).*[3]

Resistance comes from the Indigenous time-tested "art of the good life," which is grounded in an understanding that humans depend on functioning ecosystems to sustain themselves, and that fragile ecosystems require large territories and healthy forests in order to reproduce and maintain healthy populations of millions of mammals, reptiles, birds, fish and insects. Indigenous people know that a sick forest (fragmented and deforested) or a contaminated river, means hunger and misery for humans as well as for all animals and plants.

INDIGENOUS ACHUAR UPRISING:
THE BACKSTORY TO THE *DORISSA ACT*

In 1971, Peru's state oil corporation, Petro-Peru, together with the Occidental Petroleum Corporation of the United States operating in Peru (Oxy), signed the exploration concession in Block 1AB, in Achuar territory. By 1972, Oxy had drilled Blocks 1AB and 8. By the end of the 1970s, the North-Peruvian Oil Pipeline connected the Achuar territory with the Bayovar Refinery on the Pacific coast. Between 1999 and 2001, Blocks 1AB and 8 were sold to Pluspetrol Corporation S.A. (Martinez et al.), then sold to Pluspetrol Norte S.A., owned in partnership with the Korea National Oil Corporation, Daewoo International Corporation, and SK Energy.

The petroleum concessions granted in Blocks 1AB and 8 and the subsequent activities exploiting them affected four Indigenous ethnic groups. Table 2 presents these ethnic groups (Achuar, Quichua, Urarina, and Cocama) and the contaminated river basins they have occupied for hundreds of years (Corrientes, Pastaza, Tigre, and Marañon Rivers). However in this chapter, Block 8, Indigenous Cocama from Rio

Table 2. *Ethnic Groups and Rivers where Pluspetrol Norte S.A. Blocks 1AB and 8 are located*

Ethnic Group	Rivers	Population	Percentage
Block 1AB Achuar	Rio Corrientes	10,919	23.6
Quichua	Rio Pastaza	19,118	41.3
Urarina	Rio Tigre	4.854	10.5
Block 8 Cocama	Rio Maranon	11,307	24.4
Total		46,198	100

Source: *Instituto Nacional de Estadistica e Informatica (2009)*

Marañon, is only marginally included as the main actors in Andoas uprising were those in Block 1AB.

In 2003, the Achuar community in Rio Corrientes rose up against Pluspetrol after such heavy metals as lead and cadmium were found by the regional Ministry of Health in the drinking water, rivers, ecosystems and in the blood of the people. Contamination in Block 1AB was documented by Martinez et al. (5). Despite the fact that official documents had identified petroleum activities as the source of the contaminations, neither the governments nor the corporation took any action. As a result, in 2006 for two weeks, Indigenous Achuar stopped production of 40,000 barrels of oil per day. This uprising, led by Federación de Comunidades Nativas de Corrientes (FECONACO) [Federation of Native Communities of Corrientes River], pressured the central government to reach an agreement—the so-called *Dorissa Act*[4]—between FECONACO, the Ministry of Energy and Mines, the Ministry of Health, the Regional Government of Loreto, Pluspetrol, and the Ombudsman. In the *Dorissa Act,* one of the points approved was to initiate a process of remediation (involving the cleaning up of the exposed pools where waste oil and extremely toxic drilling mud were deposited) and reforestation.[5] This is the telling case examined in this chapter .

SUBSISTENCE CONFISCATION: INDIGENOUS PEOPLE IN POORLY PAID JOBS

Background

To fulfill the process of soil remediation and reforestation in oil-contaminated areas, Pluspetrol contracted the services of Graña & Montero (G&M), a Peruvian corporation. By 2007, 24 Indigenous communities living around the contaminated rivers were incorporated into discriminatory and exploitative salary work in soil remediation. Here is how it happens: Every month G&M employed between four and five young men from each of the twenty-four communities. They received less than USD$100.00 per month; and as G&M hired each community member for only one month, each worker and their family had 100 dollars to live for the entire year.

In March 2008, Indigenous workers, who used to live by hunting, fishing, gathering and slash/burn agriculture, stood up for their rights to live as temporary paid workers. The following account is based on my notes from an interview at the local jail in Iquitos with John Vega, Clever Cruz, and Marco Polo Ramirez (Isla "Eco-Class-Race"

37). They claimed that Indigenous workers complained about their situation to the 24 "Apus" or community leaders who, according to Indigenous custom, are consulted before any activity is carried out in the community's territory. After listening to the workers' complaints, the Apus set up an agenda to negotiate with the management of Graña & Montero (G&M) with the objective of securing better wages for their members. The Apus decided on the meeting location, and sent a request to meet with G&M administration.

At the first meeting, set in Titayacu-Andoas village, the G&M authorities did not show up. Instead, the area was policed as the government intervened to prevent the meeting. At the second meeting, this time in Nueva Jerusalen-Andoas village, three negotiators from G&M attended. The Apus proposed that the $100 paid to the Indigenous workers was not enough to live on and requested that each Indigenous worker should earn at least USD$800 as they have to live for an entire year without any other income. The company representatives rejected the proposal. At the third meeting, no G&M negotiators arrived. As these "meetings," initiatives for better wages failed, the workers, with the support of their communities, planned an occupation of the Pluspetrol Norte installations. The day of the occupation was decided and each community from the three river basins—Pastaza, Corrientes, and Tigre—planned the takeover of the Pluspetrol Norte installation in their respective jurisdictions.

THE EVENT

The following information is drawn from my notes from the proceedings of Court Iquitos, Peru, *Sentencia de* la Sala II, de la Corte de Iquitos, which I attended during my sabbatical year in 2009. The trial took seven months and two days, beginning in February and concluding in November (with breaks in the court proceedings).

Here is what happened: On March 20, 2008, more than 1,000 workers, community members, women (including the wives of the protesting workers) and children, and Apus congregated at an abandoned airport in Andoas town. They hoped to discuss their issues with any invited institution interested in helping to resolve the problem, such as Graña & Montero or Pluspetrol, the government, or the ombudsperson. Women and men participated together in the occupation, as all were upset by their deteriorating living conditions resulting from the oil economy. With oil spills and water contamination, they had lost their traditional economy based on common grounds. Several of their young children were dying from contamination while their teen-aged children were forced to sell their bodies to oil workers as prostitutes. Women's active participation was concentrated on organizing food and sleeping arrangements to meet survival needs, as they prepared themselves for a long occupation.

In Table 3, the first column provides the town name and the second column is the name of the Pluspetrol's installation occupied by the protestors.

Table 3. *Pluspetrol installations occupied on March 20, 2008*

Towns	Installations
Jerusalen	Pluspetrol Dorissa
Antioquia and Sauqui	Pluspetrol Jibarito
12 de Octubre	San Jacinto encampment
Andoas	Titayacu, and Los Jardines,
El Porvenir	Airport
Huararay	Airport

In addition, from Capihuari Sur, one of several Pluspetrol installations, the rebels demanded from the regular, full-time Pluspetrol oil workers the use of three pickup trucks. These trucks were used to transport community members from their towns to the abandoned airport that the Indigenous people and their families had occupied.

On March 21, 2008, the police arrested 20 men driving and riding in the three Pluspetrol pickups. Many of these riders were not directly involved in the protest but rather were hunting in the forest with single-shot weapons and ammunition. In the rainforest, single-shot weapons are viewed as instruments essential for the survival of hunter/gatherer peoples and therefore do not require legal permits. These individuals caught rides in the back of the pickups as they returned from hunting. The 20 arrested men were taken to Pluspetrol installations where a standard-sized metal shipping container was used as a detention center. In total, 53 men were arrested in different circumstances including an eleven-year-old boy and a priest, Jose Noa. They were held in the closed, airless and severely hot metal shipping container for three days. Some were beaten during this three-day ordeal.

At about 12:30 p.m. on March 22, 2008, armed police violently attacked the airport occupiers in Andoas town. In this encounter,

Indigenous men resisted, resulting in three protestors injured, one killed, and three "disappeared." There is no report on what happened to the Indigenous women and children. During the police raid and rampage, one police officer was also killed. On March 23, at about ten a.m., a police unit stepped into the home of Saulo Sanchez Rodriguez, later suspected of being responsible for the death of the police officer, and conducted him to a Pluspetrol shipping container. Afterwards, Pluspetrol "prisoners" were taken by air to jail in Iquitos (362 kilometres away) where local people and the Catholic Church intervened to rescue them or bail them out. Of the 53 people arrested, four young Indigenous men were accused of police assassination and terrorism and kept in jail.

Following the events described above, the state decided in 2009 to take 26 Indigenous people to Court in Iquitos on the charges described below. The state ignored the fact that the Indigenous defendants, living on subsistence or as temporary salaried labourers, were not familiar with money and cities. Further it ignored their travel and living arrangements in the city. As a result, the Indigenous defendants lived in grim conditions.

THE 2009 ANDOAS COURT CASE: SOCIAL CRIME OF THE STATE

Beginning in 1990, the Fujimori administration, in organizing the global neo-liberal Peruvian state, formulated a policy of criminalization of social protest, meaning that civil rights guaranteed in Article 1 and 2 (e) of the Political Constitution (wherein "Every person has the right to be presumed innocent until proved guilty") have been suspended, and security forces have responded to protestors more violently. Furthermore, during Garcia's administration, in 2007, the Congress approved Legislative Decree 982, where Article 20 declares the Armed Forces and the National Police as exempt from criminal liability during the carrying out of their duties and the use of their weapons in standard procedures causes injury or death.[6] It is a legally binding international instrument that deals specifically with the rights of Indigenous and tribal peoples (ILO). As a result, there is an open conflict between the central government and Indigenous communities.

This antagonism was expressed in the Andoas court case. In February 2009, Indigenous men were charged by the Prosecutor's office, representing the Public Ministry. Some of the defendants were the hunters who happened to catch a ride with the protestors. Others were active participants in the strike. Table 4 presents the seven charges against the defendants.

Table 4. *Defendants accused of seven violations of the law*

1	Disturbance against Pluspetrol
2	Violence
3	Aggravated robbery against Pluspetrol and Graña& Montero
4	Illegal arms possession against the Peruvian state
5	Resistance to authority
6	Aggravated assault
7	Aggravated homicide against Jaime Reyna Ruiz

At the beginning of the case, the Prosecutor's office affirmed that "the uprising was organized by four young men (John Vega, Jose Fachin, Clever Cruz, and Marco Polo Ramirez), who were born and raised in Iquitos city, and only 'pretending' to be native, while forwarding their own interests in the name of the native community." The prosecutor also accused "Saulo Sanchez, of killing a police officer on 22 March." The prosecutor requested between eight-and twenty-year jail sentences for the hunters and other participants for firearm possession, civil reparations of USD$1,736 each, and a 25-year jail sentence for Saulo Sanchez with civil reparations of USD$ 8,680.

Four lawyers, hired by *Organización de Pueblos Indigenas del Oriente* (ORPIO), defended the twenty-six accused. They argued that due to the criminalization of the protest, the prosecutor was unable to see that the goal of the occupation was to find solutions to the discriminatory wages imposed by the enterprise Graña & Montero (G&M), and to find grounds for an agreement with Pluspetrol, similar to the *Dorissa Act* in Rio Corrientes.

Defence lawyer Jorge Tacuri pointed out that the protest had followed normal procedures before his clients occupied the abandoned airport; that the occupation was organized, with a formal request to the enterprise G&M, Pluspetrol, and the Ombudsman; that the occupation was known by all community members and was not a chaotic action, as Pluspetrol's lawyer argued. He also argued that the Prosecutor could not prove the accusations, since four individuals could not persuade more than a thousand people from three different river basins to go to an abandoned airport. Further, lawyer Walter Cambero, defending the hunters, argued that according to the videos presented by the police—*Dirección de Operaciones Especiales* (DINOES)—none of the accused resisted the police authority; consequently, he requested the handing back of the firearms to the hunters, who urgently needed these tools in order to feed their families. Lawyer Victor Alva also called for Saulo Sanchez to be declared not guilty of killing the police via the application of the principle *indubio pro reo* (innocent until proven guilty), because there was no evidence to convict him for committing the crime. In summary, these advocates requested that the case be dismissed because there was no evidence to convict the defendants of any of the seven charges.

As the court case continued, all of the defendants described the daily activities of their lives, and linked their responses to the destruction of their provisions (fish, meat, fruits) and their subsistence economy due to oil contamination. During the proceedings, the lawyers from the defense requested an assessment by Iquitos-based anthropologist, Jorge Gasche of IIAP, on the motives of the accused rebels. The court accepted the request and here, I present portions of an unpublished transcript of the anthropological assessment provided to me by Jorge Gasch on December 15, 2009:

Daniel Dahua Mayna was born in Titayacu, Pastaza River, a Quichua community. He does not speak Spanish, never left his territory, and never went to school. Close to Titayacu, the Northwest oil canal was built within 300metres of his house. He protested against oil contamination as his closest family members died with stomach pain and bloody diarrhea. Further, he was a hunter, fisherman, and agriculturalist, but his hunting products had a bad taste, were rotten, and poisonous. In order to hunt and fish he has been walking for days to the border with Ecuador.

Jose Fachin Ruiz was born in Vista Alegre, Quebrada Santa Barbara, Tigre River. He lives in Andoas and speaks Quichua and Spanish, and went to high school in Iquitos. He wrote a Quichua dictionary and produced a video on oil contamination for the NGO *Red Ambiental*. He worked for Graña & Montero once, but decided not to work there anymore due to a hostile environment where Peruvian mestizos and foreigners humiliate local workers. He recognized his participation in the uprising and described the protest as a way to show their fatigue with abuse. He used to hunt, fish with a hook, and make canoes to sell to community members, but these activities have become difficult due to oil waste from around thirty abandoned and shutdown wells. Water has a bad odour, and it cannot be consumed. He stated: "Our watersheds [San Antonio, Montano and Yanayacu] have been contaminated as well as our wildlife. Many children have died vomiting and with diarrhea. Despite this tragedy, there is no water available to the community. There is only one artisan well, which is not enough for all community members; therefore people are forced to drink from contaminated rivers."

Martin Rober Cubas recognized that he went to the airport and had driven one of the pickups to distribute food among the demonstrators. He was aware that the occupation was due to inadequate and discriminatory wages and oil contamination. However, he said, he was not armed as the police indicated. Instead he pointed out that the police forced him and other "prisoners," as they were being held in Pluspetrol's container, to hold or carry their hunting rifles in order that the police could take pictures of the prisoners. These pictures were distributed to the media and used to accuse the Indigenous protestors of being terrorists. Martin recalled that after the pictures were taken, he was beaten.

Miguel Zuniga, 68, did not participate in the strike. He maintained that he did not know about the strike. He asserted that on March 20, 2008, he went hunting. At dawn on March 21, upon arrival from the jungle, he found two young men driving a pickup truck, and they offered him a ride to town. He jumped into the vehicle, without having any bad feeling because in the past the enterprise's drivers had invited him to ride with them. He said, "Nobody told me that the vehicle had been used for other reasons. I learned about the strike only when the police intercepted the vehicle, arrested the occupants, and conducted the car occupants to Pluspetrol installations. Once there, I and other community members were pictured carrying hunting rifles. But let me say," he added, "I was not assaulted physically by the police. However, after three days of detention, the police forced me to sign an incrimination letter in Spanish, despite the fact that I told them that I am illiterate."

Following the anthropological assessment, the court invited all regional organizations to show evidence of contamination by Pluspetrol. For instance, NGO *Red Ambiental* presented a video with serious evidence of oil contamination of the people and of the land. The court also invited Jose Noa, a Catholic priest who was jailed and tortured in a Pluspetrol container, to present his testimony.

The final remarks in the court proceedings started with Pluspetrol's lawyer who expressed disagreement with the anthropological assessment that identified habitat destruction as central in the uprising. He argued that the anthropologist did not explain how Indigenous traditions excused the event. Further, he maintained that, "the accused used Pluspetrol contamination in their favor, even though this was not the cause of the uprising." He recognized, however, that it was true that several denouncements of the contamination had been filed at the Prosecutor's office; but no one, he remarked, had been able to prove that Pluspetrol polluted, thus there was reasonable doubt with respect to the contamination. In addition, he challenged the audience, even if this were the truth, that those arguments were not justifications for the uprising.

Dr. Walter Cambero, defending the hunters, argued that the Prosecutor's Office violated the Peruvian Constitution, where people are declared innocent if there is no evidence of judicial responsibility; that on March 20, the hunters did not participate in the uprising; that they did not participate in taking the three pickups; and that the same police detained the hunters on the same day, in the same place, and at the same time. "Why," he asked, "would they steal a pickup since they do not know how to drive? In addition, where can a pickup go in Andoas? Nowhere, because there are no roads." Then he remarked, "They could not have participated in this protest organized on March 22, 2008 because they were imprisoned inside a Pluspetrol container. Treaties on international human rights recognize that corporations have a legal obligation to prevent harm to the fundamental right to life, and that governments have the legal obligations to defend its citizens' rights."

The final remarks of the defendants are also presented. Here I cite three statements from those who were included in the anthropological assessment:

Now, the enterprise Pluspetrol wants to condemn us to jail time, but 40 years ago oil production had already sentenced us to death. We cannot confront oil destruction by ourselves. This is a call for all people of the planet to maintain our Amazon as the lungs of the world.

—Defendant Martin Rober Cubas, November 27, 2009

If I am condemned to jail, I will be proud because it was as a result of defending my territory. But, I also demand sanction for the enterprise, and for the Fiscal [prosecutor's office], which has been used as an instrument of the enterprise.
—Defendant Daniel Dahua Mayna, November 27, 2009

The fact that I do not dress up as an Indigenous man, does not obliterate my indigenous identity. I am the son of a native woman and man from Andoas who sent me to Iquitos to get an education and a profession to avoid living from scraps, as my people are living now.
—Defendant Jose Fachin Ruiz, November 27, 2009

COURT VERDICT:
RECOGNITION OF INDIGENOUS PEOPLE'S COLLECTIVE RIGHTS AND CRITIQUES OF THE STATE BY THE JUDICIAL SYSTEM

On November 27, 2009, Judge Carlos del Pielago, in association with Judges Rosa Peláez and Roxana Carrión, charged the State prosecutor with incompetence. The judges' statement reads, "the State is not paying attention to the claim of the Indigenous people and *ribereños* (those living on river banks). Consequently it ignores their rights to land use and occupation where they have been living for hundreds of years, and disregards Indigenous peoples' rights to be heard, as mandated by the Constitution and international treaties. Moreover, the State Attorney and the local Prosecutor acted inadequately in reporting on what happened. During the oral examination the prosecutors released modifications to their complaint.[7] Therefore, *indubio pro reo* [when in doubt, decide in favour of the accused] is the reason to absolve the accused of the seven supposed crimes."

In delivering their verdict, the judges showed thoughtfulness and sensitivity. Judge Carlos Del Pielago, the Chair of the Court, offered an exemplary public process on how to proceed in cases where Indigenous people are involved in social conflicts. Consequently, the judges:

1. Relied in their deliberations on international agreements, such as, the International Labour Organization (ILO) Convention 169, Article 9, which states that on penal issues, authorities and tribunals must take into consideration the tradition of the communities, and Article 10, which maintains that a sanction must take into consideration the social, economic, and cultural

characteristics of the communities involved. And it accepted the defence's proposal that Indigenous people must be judged in their own languages.

2. Accepted the report produced by the Loreto Ombudsperson, which showed that Amazonia's Indigenous communities are judicially in a vulnerable position, and in need of protection of their fundamental human rights as well as their territorial rights. The report said "there is a need to defend one of the most fundamental rights, the right to land, in which indigenous people develop their livelihood, hunt, fish, and produce agriculture. All of these rights are based on relations of reciprocity, and linked essentially to conservation and development of their cultural identity." The Ombudsperson's report linked community and ecology as the reasons that underpinned the struggle.

3. Valued the Amazon Catholic Church's pronouncements during the strike, which stated that: "The Church values Indigenous peoples' respect of nature and love for mother earth as source of food, commons, and humanity."

4. Signaled globalization as causing the devastation of the forest and its biodiversity through depredation and greed that put in danger the livelihood and the habitat of millions who have been forcefully evicted onto useless land or to live in miserable conditions in big cities. Further, the decision indicated that it is public knowledge that rivers are contaminated with lead, heavy metals and toxic substances as a result of mining (formal and informal), and irresponsible oil extraction. Globalization has brought not just exploitation and oppression, but a new social exclusion of those living in the rainforest.

The verdict concluded:

We (three Judges) take on the view of our judicial authority, who said regarding the criminalization of the social protest that we draw attention to what we witness daily—the incapacity of the State to give solution to the claims formulated by low or no income social sectors and groups. That instead of solving the exclusion problems, the State has responded with criminalization and persecution of social activists, involving the Judicial System in issues outside of its competence—social conflicts. We do not accept becoming instruments of persecution of the poor. We are judges with sensibilities who evaluate the impact on the victim, the prosecuted, and society. We remark that social conflict forces us to be cautious in order to avoid the criminalization of poverty, and the protest of Indigenous communities and their membership. We must be careful to avoid trimming down rights that need the maximum protection.

Following the verdict, Pluspetrol's lawyer decided to bring an action for annulment of the verdict by the Supreme Court. In 2011, the Supreme Court upheld the judgment of Sala II, Iquitos Court.[8]

MY ASSESSMENT OF THE COURT CASE USING AN ECOFEMINIST PERSPECTIVE

From the ecofeminist perspective, Petro-Peru, the government oil corporation, and Pluspetrol, a private corporation, have enclosed for oil production, the territories of Achuar, Quichua, Urarina, and Cocama. Meanwhile, Graña & Montero's corporation has treated Indigenous people as housewives, as if there is no real value attached to their labour, so they do not have to be equitably remunerated, or remunerated at all. Consequently, the State prosectuor criminalized the Indigenous people.

First, the prosecutor overstated the case and generated a false hypothesis to dichotomize the relationship among the defendants, by separating Indigenous people who were born and raised in Iquitos and spoke Spanish from those who were born and raised in the town of Andoas. From this dichotomy, the prosecutor intended to legally dispossess the Indigenous identity of the four defendants, and requested them to be bound by Peruvian law; while, the "real Indians," who live in the jungle, should be bound by cultural tradition. By generating this false hypothesis, the prosecutor gave four individuals, schooled in the city, the role of key instigators and terrorists, to explain the labour uprising on March 20, 2008.

Second, the prosecutor separated the act of rebellion, as a result of the devaluation of their labour from the enclosure and destruction of their habitat. However, the prosecutor ignored the high cost to nature and society by the degradation of the ecosystem, the fact that oil contamination had been altering land use, and that extractivism had interfered with Indigenous people's survival by destroying their subsistence economy. Pluspetrol oil production has freely appropriated nature, diminished natural zones through contaminated soils, devoured forest, and increased emission of carbon dioxide. All of this was done at the expense of Indigenous people living in Corrientes, Pastaza, and Tigre. In addition, oil production begat oil spills along rivers, streams, lakes, wetlands, degraded pristine land and flooded healthy forests, and transformed the land into a massive, hellish oilfield with more than 150 wells, over 1,000 kilometres of road, and a pipeline network threatening the local communities. Moreover, the prosecutor discounted the fact that oil spills and so-called "down well water" laced with toxins had already sparked several rebellions in Rio Corrientes among the Achuar community, which resulted in the 2006 *Dorissa Act* discussed above. Further, the prosecutor disregarded the fact that toxic chemical pollution had already killed several Indigenous people and their children and that at the time of the uprising, bereaved communities were in mourning.

Third, the prosecutor replicated historic colonial cultural stigmatization by stating that the Indigenous people were manipulated, that they lacked understanding of what they had done. This ideology of the ignorant native being led into mayhem failed utterly to acknowledge that Pluspetrol's destructive oil production methods were central to the uprising, and a key contributor to the violation of the rights of present and future generations of Indigenous peoples, as well as countless non-human residents of the river basins. Moreover, the prosecutor ignored the fact that Pluspetrol participated directly in the violation of Indigenous human rights when it provided the container with no lavatories, no food, and no water, where the police of *Dirección de Operaciones Especiales* (DINOES) beat and tortured some of the prisoners during three days of illegal internment. Further, the prosecutor ignored Article 46 of the Peruvian Political Constitution, which states that, "No one owes obedience to a usurper government or to anyone who assumes public office in violation of the Constitution and the law. The civil population has the right to insurrection in defense of the constitutional order. Acts of those who usurp public office are null and void."

Fourth, the prosecutor claimed compensation from the defendants, whose self-sustaining and autonomous economy was destroyed, and whose work, like that of the housewife, was not seen as legitimate labour. Absurdly, the prosecutor tried to penalize Indigenous community members when requesting money as reparation, knowing that these community members earned USD$100 annually. By proposing monetary reparation, the prosecutor effectively passed a death sentence on the culture of Indigenous people living in a quasi-non-monetized economy.

Indigenous communities were never going to have the ability to raise money to pay reparations.

ECOCIDAL (IN)ACTION OF THE PERUVIAN STATE

Despite of enclosure and territorial destruction substantiated during the legal proceedings, violence against Indigenous territory continues. The ecocidal (in) action of the state results from many elements of the extractive industry and the absence of sanctions against corporate crimes. A Peruvian NGO, Convoca, documented more than 1,000 environmental reports shelved during the four administrations discussed here. It revealed hidden reports of systematic oil contamination in soils and open areas, inadequate management of organic and solid waste, and lack of environmental management permits obtained by oil corporations. In Loreto from 1999 to 2011, for example, Blocks 1AB and 8 operated by Pluspetrol Norte S.A. and Pluspetrol Peru Corporation S.A. have incurred nineteen instances of oil contaminations that governments overlooked. Most of the archived reports, 686 out of 832, occurred during the government of Allan Garcia.

In September 2011, as the area experienced a water emergency and as incidents of oil contamination continued to pile up, a letter was sent to President Humala by the four Indigenous groups living in the area where Pluspetrol operates. The letter states that for four decades, the Pastaza, Tigre and Corrientes Rivers have been polluted by toxic water from oil production while the Marañon River was contaminated by a series of oil accidents. It also stated, a deficient and decaying pipeline had created havoc among those communities. As Pluspetrol's contract was ending in August 2015, the Humala's government accepted the Indigenous people's request for "the establishment of a parliamentary commission to investigate the environmental and social impacts in the affected rivers due to oil activities in North Amazonia" (Congreso de la República 6).

A Parliamentary Commission (PC) of four members was nominated. The PC was accompanied by a member of *Organismo de Evaluación y FiscalizaciónAmbiental* (OEFA) [Organism on Environmental Evaluation and Fiscalizatio], and a journalist from *La Republica* newspaper, Jonathan Castro. In July 2013, the commission reported that in Andoas, Rio Pastaza, there were several water bodies destroyed by oil contamination, specifically:

1. Shanshococha, a lake, was abandoned and left in 1995

by the U.S. based Occidental Petroleum Corporation (OPC). Supposedly, Pluspetrol, which bought the assets and liabilities of OPC between 1999 and 2001, was to clean up such toxic lakes but never complied with the required remediation. As this problem became public in 2015, *The Guardian* reported that, "Members of the Indigenous Achuar tribe from [Corrientes River] the Peruvian Amazon have won an undisclosed sum from Occidental Petroleum in an out-of-court settlement after a long-running legal battle in the U.S. courts" (Collyuns).
2. Ushpayacu, a lake, apparently badly "remediated" by Pluspetrol. The oil spill originated from layered barrels that were stacked and overflowed close to swamps and lakes.
3. San Jacinto town, Rio Tigre, experienced another oil spill due to corroded barrels and pipelines.
4. Marsella, a lake, an oil-spill site badly covered with soil, leaves, and declared remediated by the State, despite an inadequate clean-up.

The Commission concluded that the impact of hydrocarbon activities in Amazonia not only degrades biomass and biodiversity, but as the territory is an integral part of the universe of Indigenous peoples, it delivers a body blow to culture, social organization and spirituality, thus precipitating multidimensional negative effects that threatens their existence (Congreso de la República 12-13).

On July 4, 2012, journalist Jonathan Castro summarized his examination as follows:

1. Corrientes River oil contamination continues in Choroyacu, where an oil carrier sank and is now covered with leaves and fallen trees. Part of this oil spill was recovered and then abandoned in 50 corrosive barrels, only covered with plastic. The rain washes the oil spill toward river basins and becomes a permanent cause of contamination; and
2. Marañon River had "bad faith reparations" (blocked out the impacted area, and/or use of toxic dispersants designed to sink the crude oil below the surface of the river) where five thousand five hundred (5,500) barrels of crude oil spilled in 2000, while in 2010 another 500 barrels was added. There is not an account of spills before 2007.

Humala's government disregarded the Parliamentary Commission's

Final Report. Instead it advanced the cause of the extractive industries by eliminating the requirement that oil companies conduct and publish detailed Environmental Impact Studies (EIS) of the seismic exploration in oil and gas concessions, and introduced the possibility that companies could submit a more expeditious EIS. The new procedure takes approximately 35 days for approval instead of 150 days. As a result, this new regulation regime allows corporations to submit a less rigorous study of protected areas, fragile ecosystems, Indigenous reserves and conservation areas (Garcia 2). In addition the "fast-track" approval process eliminates even a faint possibility of meaningful consultation with affected communities, amongst which Indigenous peoples are most represented.

To add insult to injury, Pluspetrol Norte S.A. was hoping to get a contract renewal in August 2015, giving the corporation access, yet again, to its expired concession. Strong Indigenous opposition against Pluspetrol's contract renewal was expressed. They called the renewal illegal, outrageous, and an infringement of their rights ("The misleading art"; "Piden no renovar"). They also insisted that before any oil contract restoration occurred, prior consultation with the entire community must take place because these areas have been declared in states of emergency due to oil contamination. In 2015, despite all Indigenous opposition to oil extraction, Block 192 (the former Block 1AB) was granted to Pacific Stratus Energy, a subsidiary company of Canadian Pacific Exploration & Production Corporation. In 2016, there were ten oil spills from PetroPeru's pipelines (Pipeline NorPeruano) which directly affected the people of the departments of Loreto and Amazonas who ended up drowning their cultural wealth in a sea of impotence, frustration and resignation. As a result, the Peruvian indigenous communities and bosquecinos were left without water, fishing, agriculture, cassava, banana, cacao, maize, land or dreams. Furthermore, these communities were left with bronchitis, diarrhea, warts, headaches, vomiting, crying, hunger and a great deal of hopelessness (Fermin).

CONCLUSION

This chapter argued that capital accumulation comes not only from waged labour but from the bodies of the un-waged and from nature. It has shown that the Peruvian states' articulation with the global state, led by the United Nations Earth Summits, has exacerbated the political practice of exclusion and domination of the rainforest Indigenous communities in Peru. Here I have shown how the state has the land

and hunting grounds, and deprived of self-sufficiency and autonomy the Achuar, Quichua, Urarina and Cocama Indigenous people living near Blocks 1AB (now 192) and Block 8. As a result, they were obliged to sell their labour for almost nothing in service of hydrocarbon spill soil remediation on Pluspetrol oil field production sites organized by Graña & Montero Corporation. Indigenous people's labour on these projects has been "housewifized," meaning indentured servitude, near enough to slavery. However, in 2008, these communities actively resisted the exploitation of their labour and nature and unveiled the state's crimes in the Judicial Court in 2009.

Indigenous people in the rainforest of Peru are at the forefront of preserving the health of the forest in order to continue a centuries-old subsistence economy that depends on a healthy ecosystem (forest, rivers, watershed, and air) that maintains basic nature (flora, fauna, and fishery). Their struggle to protect their own forest-based economies against the extractivism of sustainable development with all its intense destruction is a critical element of the global fight against climate change and for the elaboration of post-oil economies. Ecofeminists have long argued for wider recognition of the unwaged work of women household workers, peasants, Indigenous people and nature itself. This invisible "housewifized" labour is essential to the process of capital accumulation and thus has enormous power to simultaneously interrupt the destructive path of industrialization and re-route humanity towards an alternative civilization, one based on the Commons.

Update: Currently, the Peruvian ex-presidents named in this chapter—Alejandro Toledo (2001-2006), Alan Garcia (2006-2011) and Ollanta Humala (2011-2016)—are under investigation for money laundering, as Odebrecht, a construction corporation from Brazil, paid them bribes in offshore accounts. Meanwhile, Alberto Fujimori was sentenced to jail for twenty-five years for crimes committed during his mandate.

Note: The methods used to research this chapter were a literature review, information from an article published by Isla ("Eco-Class-Race") in Capitalism Nature Socialism, *interviews, and observation of courtroom proceedings in 2009. The study's limitation is a lack of physical inspection of the Indigenous territory in Blocks 1AB and 8.*

ENDNOTES

[1]Decreto Legislativo 30151: Establishes the use of weapons by the

police and army (online).
[2]Decreto Legislativo 1095: Sets rules for employment and use of force by the armed forces in the national territory (online).
[3]Defensoría del Pueblo, Reporte de Conflictos Sociales N° 69, 2009; Reporte de Conflictos Sociales N°106, 2012; Reporte de Conflictos Sociales N°129, 2014; Reporte de Conflictos Sociales No. 171, 2016. Lima, Peru (online).
[4]*Dorissa Act*, 2006. Acta que complementa y precisa los acuerdos subscritos entre las comunidades indígenas del Rio Corrientes–FECONACO, el Ministerio de Energia y Minas, el Ministerio de Salud, el Gobierno Regional de Loreto, la Empresa Pluspetrol Norte S.A., y la Defensoria del Pueblo. Signed at Dorissa, Loreto, Peru (online, accessed October 2009).
[5]For more information on reforestation, see Congreso della República, *Informe Final*.
[6]Decreto Legislativo 982: Amends the Criminal Code approved by Legislative Decree No. 635 (online).
[7]Dr. Victor Alva, lawyer for the man accused of aggravated homicide, asserted:
•The police officer that supposedly recognized Saulo, as someone who shot the police, had contradicted himself. The police was unable to explain the colour of Saulo's pants, as well as unable to bring forth any witness.
•Ballistics did not find remains of blood or any organic element of the victim in the clothing of the accused. The atomic absorption spectroscopy proved that Saulo had lead in his clothing, but not *antimony and barium*, two elements needed to prove that he was the shooter.
•Another man apprehended at the same time as Saulo said that they were held around 10:00 a.m. (and the crime occurred at 12:30 p.m.) but the police registered their arrest at 1:00 p.m., without reason.
•During the investigation, ten new suspects surfaced as possible shooters.
•The Act of Reconstruction or review of evidence from the crime scene, involving three prosecutors and a colonel, confirmed that the now dead police officer was wounded in the backyard of Maria Dahua Betuna's house and not at the abandoned airport as the police stated.
[8]"Sentencia del Tribunal Constitutional," 2011. EXP. No. 0022-2009-PI/TC (online, accessed July 2012).

REFERENCES

Alvarez, Jose. "Perú: Comunidades, territorio y concesiones forestales."

Servindi.org, 23 September 2011. Web.

Castro, Jonathan. "Pasivos de explotación petrolera en Rios Tigre and Pastaza aun no se solucionan." *La República* 3 July 2012. Web.

Collyns, Don. "Indigenous Peruvians win Amazon pollution payout from U.S. oil giant." *The Guardian* 5 March 2015. Web.

Congreso de la República de Peru. *Informe Final sobre la Situación Indígena de la Cuencas de los ríos Tigre, Pastaza, Corrientes y Marañón.* Comisión de Pueblos Andinos, Amazónicos y Afroperuanos, Ambiente y Ecología. Grupo de Trabajo, 2013. Web.

Constitution of Peru, 1993. Lima: Congress of Peru. Web,

Convoca. "Mapa de Informes Encarpetados. Lima, Peru." Archivos Ocultos Convoca, 2016. Web. 29 April 2016.

"Diez grandes proyectos que tiene Oodebrecht en el Perú." *El Comercio,* 08 March 2016. Web.

Federici, Silvia. *Caliban and the Witch: Women, the Body and Primitive Accumuation.* New York: Autonomedia, 2009.

Fermin, Carlos. "Los 10 ecocidios del 2016 en Latinoamérica." Servindi. org, 27 Dec. 2016. Web.

García, Francesca D. "Reglamento flexibiliza medidas ambientales para favorecer inversión." *La República* 20 Dec. 2014: 2. Web.

Gasche, J. "Anthropological assessment of Andoas defendants." Unpublished report presented to the court during the legal proceedingsin the Andoas court case, 2009.

Gasche, J. and N. Vela. *Sociedad bosquecina.* Tomo 1. Loreto, Peru: Instituto de Investigaciones de la Amazonia Peruana, 2011.

International Labour Organization (ILO). *C169–Indigenous and Tribal Peoples Conention, 1989.* Geneva, Switzerland: ILO, 27 June 1989. Web.

Instituto Nacional de Estadistica e Informatica, 2009. Web.

Isla, Ana. "The Eco-Class-Race struggles in the Peruvian Amazon Basin: An Ecofeminist Perspective." *Capitalism Nature Socialism* 20.3 (2009): 21-51.

Isla, Ana. *The "Greening" of Costa Rica: Women, Peasants, Indigenous People and the Remaking of Nature.* Toronto: University of Toronto Press, 2015.

Lizarzaburu, Javier. "Forced sterilization haunts Peruvian women decades on." *BBC Online News,* 2 December 2015. Web.

Martinez, Martí Orta, Dora A. Napolitano, Gregor J. MacLennan, Cristina O'Callaghan, Sylvia Ciborowski, and Xavier Fabregas. "Impacts of Petroleum Activities for the Achuar People of the Peruvian Amazon: Summary of Existing Evidence and Research

Gaps." *Envionmental Research Letters* 2.4 (2007): (n.p.). Web.

Mendoza, Raúl. 2009, July 12. "Cazador de conflictos." *La República*. 12 July 2009. Web.

Mies, Maria. *Patriarchy and Accumulation on a World Scale: Women in the International Division of Labour.* London: Zed Books, 1999.

Mies, Maria and Vandana Shiva. *Ecofeminism* . London: Zed Books, 1993.

"Piden no renovar concesion del lote 192 a Pluspetrol." *La República* 5 March 2014. Web.

Ruiz, J. C. "Ministerio de Cultura bloquea la consulta previa de las concesiones mineras e invisibiliza a los PPII (Pueblos Indigenas) en Espinar." Servindi.org, 22 April 2015. Web. July 2015.

"The misleading art of commercial deception; or how to transform a major polluter into an environmentally sustainable and socially responsible company." Blog post, Alianza Arkana, 9 Dec. 2013. Web.

Tovar, J. G., I. Monterroso, C. Mora, S. Baldovino, I. Calle, P. Peña and J. L. Capella. "Ley 30230: Efectospara la institucionalidadambiental y la tenencia de la tierra en Perú." Center for International Forestry Research (CIFOR) Brief no. 102, 2014. Web. July 2015.

Truth and Reconciliation Commission. *Final Report, Peru,* 2003. Web.

On Sustainable Development

15.
The "Greening" of Costa Rica

A War Against Subsistence

ANA ISA

THE 1987 BRUNDTLAND REPORT, *Our Common Future,* was
instrumental in forging links between development and the
environment through the concept of sustainable development (World
Commission). In this chapter, I evaluate sustainable development as
discussed at the 1992 Rio Earth Summit, and extend the assessment
to the "green economy" as developed in Costa Rica, the first green
neoliberal project. In the "green economy," goods and services provided
by nature depend on the stock exchange. A key concept in this latest
iteration of sustainable development is "natural capital," which is used
to refer to the goods and services that the planet's stock of water, land,
air, and renewable and non-renewable resources provides (Fenech et
al.). As a result of nature being defined as natural capital, new areas of
global intervention were opened up and nature entered the domain of
neoliberal politics. The argument is that sustainable development is good
and desirable for the entire world, including and most particularly for
the so-called underdeveloped world. Costa Rica has been an important
"laboratory" for experimentation in innovative environmental
governance mechanisms using debt-for-nature exchanges. I have called
this political ecology "greening" to indicate how the ecosystems of an
indebted country are increasingly becoming destabilized.

In this chapter I draw on an ecofeminist perspective to cast doubt
on the overarching message that sustainable development creates social
and gender equality, reduces poverty, confronts ecological destruction
and combats climate change, and to show the need for a new approach
grounded in critical feminism. In the first part, I present ecofeminist
critiques of capitalism and sustainable development. Critics have
underlined "enclosure" (Federici) and "housewifization" (Mies) as
part of the working of patriarchal capitalism, which is a system that
maintains relationships of domination and subordination in general

311

and men's domination of women in particular. I then apply these insights to analyze a case study of the debt-for-nature exchanges between Canada and Costa Rica that took place from 1995 to 1999 by two environmental non-government organizations (ENGOs) operating in the Arenal-Tilaran Conservation Area (ACA-Tilaran), now Arenal Tempisque Conservation Area. I document a war against subsistence, understood as livable economy, by examining the triple crises brought about by the solutions provided by sustainable development using debt-for-nature exchanges: crises with extremely harmful effects on the lives of women and children, peasants and Indigenous peoples and nature itself. I conclude by introducing "green" capitalism as another phase of capital accumulation and arguing that its real war is not against poverty and unsustainability but against subsistence.

METHODOLOGY

Data was collected from 1997 to 2009 in a number of sites in the Arenal-Tilaran Conservation Area (now the Arenal-Tempisque Conservation Area) located in northwest Costa Rica where I interviewed seventy community members. In Tilaran, I collected information on the structure and functioning of the conservation area and interviewed several officials from the Costa Rican Ministry of Natural Resources, Energy and Mining (MINAE). In San Jose, I interviewed one official from the World Wildlife Fund-Canada and two officials from INBio.

ECOFEMINIST CRITIQUES OF CAPITALISM AND SUSTAINABLE DEVELOPMENT

I have chosen the ecofeminist subsistence perspective to discuss Costa Rica's experimentation in innovative environmental governance because this perspective is critical of sustainable development within the framework of economic growth. It argues that it has been possible to sustain the illusion that economic growth is a positive and benign process because the costs have been borne primarily by what has been devalued: women, peasants, Indigenous people, the so-called underdeveloped world, and nature. Ecofeminist theorists in particular have contributed to the analysis of domination of nature. This has a special significance for women because, in patriarchal thought, women are believed to be closer to nature than men. Ariel Salleh maintains that, "By introducing the nature-woman-labour nexus as a fundamental contradiction, ecofeminism affirms the primacy of an exploitative, gender-based division of labour, and simultaneously shifts the analysis of all oppressions toward an ecological problematic" (110).

Programs associated with economic growth tend to result in enclosure and housewifization. The enclosure movement began in England in the sixteenth century and entailed both the closing off of common land and appropriation of the common wealth of workers through the elimination of customary rights (Hobsbawm). Silvia Federici (notes that primitive accumulation was not just a precondition for, but is an essential and lasting quality of capitalism. She observes that the privatization of the commons happened in the same historical period as the European witch trials that led to the devaluation of women. This produced a new gendered division of labour, in which women's position in society as providers was redefined in relation to men, to become wives, daughters, mothers, and widows, all of which "hid their status as workers, while giving men free access to women's bodies, their labour, and the bodies and labour of their children" (97). Federici draws a long historical line from the enclosure of the commons movement and the witch trials to modern day prostitution and the monetization of nature in her theorizing of the systematic subjugation and appropriation of women and nature, bodies and labour, under patriarchal capitalism. She revalorizes the work of reproduction and reconnects our relation with nature, with others, and with our bodies to regain a sense of wholeness in our lives.

Maria Mies is critical of the Eurocentric assumption that women's work has no value and that all unpaid work has no value. She locates the origin of women's oppression in the interconnected systems of patriarchy, capitalism and colonialism. She argues that when women, peasants, and Indigenous people are described as 'closer to nature', they are made exploitable. Mies uses the term "housewifization" (106) to capture the process whereby hitherto life-sustaining work is captured, confined, devalued, and put to use in support of monetized production. This division of labour into reproductive labour (housewife, subsistence) and productive labour (breadwinning, salaried) is necessary for capitalism to function. Housewifization is a devaluation of women's work but it also applies to those who are not biologically women. Similar to families and households, this division exists on a worldwide scale through colonialism. These feminized, socially marginal and externalized economic sectors and actors are Indigenous people, peasants, and the peoples of the so-called underdeveloped world in the North and the South when their land and products are taken from them through structural violence with little or no compensation.

Ecofeminists propose a subsistence perspective to transform the nature of our economy and to create social and political frontiers before capitalist accumulation reaches the last ecological limit. Some ecofeminist scholars see subsistence economies as an alternative model of social, economic, gender, and environmental justice. For example, Veronika Bennholdt-Thomsen and Maria Mies propose what they call "the subsistence perspective," which is based on production that includes all work that is expended in the creation and maintenance of immediate life and which has no other purpose. For these authors, there is no insatiable appetite to devour nature, "subsistence production ... stands in contrast to commodity and surplus value production for profit" (20). They present it as form of resistance to the effects of global capitalism and colonialism by acknowledging that "life comes from women and food comes from the land" (80). Subsistence producers all over the world—the majority of whom are women and/or feminized—should be seen as experts who can lead the way to recovering autonomous ways of living, starting from territorial and bodily autonomy, food security in small farms and energy efficiency.

In this chapter, I link Federici's "enclosure" by witch hunts and Mies' "housefization" by enclosure to explore the gendered character of the so-called green economy, in which nature is selective monetized and turned into goods and services, devalued and traded in global financial markets.

THE "GREEN" ECONOMY OF NEOLIBERALISM: THE WAR AGAINST SUBSISTENCE

In this section, I describe the "green" economy of neoliberalism and argue that the market-based solution for ecological and social crises is the real war against subsistence, as it has advanced a new capital accumulation strategy that shifts the burden onto the indebted periphery and its inhabitants. To support my argument, I present the war against subsistence that has been waged by USAID and ENGOs in Costa Rica.

At the 1992 Rio Earth Summit, participating governments negotiated Agenda 21, which officially linked development and environment within the concept of sustainable development. Neoliberal economists of the World Bank articulated a set of prescriptions designed to manage global capital. Sustainable development thereby became equated with economic growth that would rescue poor countries from poverty (Pearce and Warford). Within this frame, the ecosystem must be embedded in the economic system through the price system to ensure a steady supply to capital. In other words, the environment requires a fully monetized world in order to be protected. This means that ecological commons, such as the atmosphere, oceans, seas, land, forests, mountains, biological diversity, ecosystems, fresh water and so on, need to be priced as natural capital (Hamilton), thus legitimizing the privatization of the ecological commons.

The prescriptions formulated by the World Bank in Agenda 21 were first applied in Costa Rica, due to its foreign debt that was used by the United States to impose USAID as a direct agent to take control of Costa Rica's economy (Petch). This was in large part due to the fact that the government of Costa Rica sold its political and military location in conflicted Central America to USAID that became a parallel government and pumped a great deal of money in development aid to soften the blow of privatization while simultaneously funding the war against Nicaragua (Petch). During USAID's tenure the neoliberal political ecology of Costa Rica was put together by the World Wildlife Fund United States (WWF-US) at the Estrategia de Conservación para el Desarrollo Sostenible (ECODES) conference in 1988 (Quesada). Biologist Christopher Vaughan, head of the WWF-US, emphasized the role of human population growth and peasants in hastening environmental degradation (Vaughan). He defended the case that the local biodiversity would be seriously threatened if the ecological and social landscapes were not transformed into conservation areas. Participants of ECODES recognized that the pursuit of development as economic growth, capital accumulation, and technological innovation applied to agriculture, was the cause of the debt crisis, poverty, dispossession, and deforestation in Costa Rica. Consequently, a new form of economic growth was promoted as a form of sustainable development (Quesada). At ECODES the ideology was promoted that debt-for-nature exchange was an instrument to develop a model of environmental management in conservation areas that could be applied to all indebted countries in the world. Participants also agreed that, "an effective way to achieve national unification and consolidation of existing protected areas is the creation of The National System of Conservation Areas (SINAC) under the supervision of MINAE. SINAC divided the country into eleven conservation areas and was designated to manage the country's wildlife and biodiversity for sustainable development. Meanwhile, MINAE w was set up to manage the conservation areas through regional offices and to set policies for the use of energy, natural resources, mines, and water. One of these areas is the Arenal Conservation Area (ACA-Tilaran), now the Arenal-Tempisque Conservation Area. ACA-Tilaran is a telling case that highlights several key aspects of 'greening' in Costa Rica. As Agenda 21 established that sustainable development was a government responsibility, in 1992, Costa Rica received an award for being an example of sustainable development in the periphery. In 1999, more than ten years after ECODES, I interviewed its former president Carlos Quesada, then director of the Sustainable Development Research Centre based in San Jose. He criticized ECODES' framework and said "This discourse [sustainable development] was formulated outside of Costa Rica with the goal of reducing the state's capacity to organize itself as a country" (Quesada, personal communication, August 1999).

This section has shown the global participants in Costa Rica's articulation of the concept of conservation areas using debt-for-nature exchanges as a way to newly connect nature and labour into the international power relations of creditors. In the next section, I discuss the case of the Canada-Costa Rica debt-for-nature exchange, because it is an example of the core mechanism used within the sustainable development framework that aims to develop a model of environmental management "to reduce poverty and environmental degradation." My analysis uncovers what is behind the structural changes in the nature of land governance and gendered class formations to facilitate conditions for capital accumulation. Debt-for-nature opens new areas for environmental management and is one way of "getting the price right" where the bulk of the resources come from people's commons.

THE CANADA-COSTA RICA DEBT-FOR-NATURE EXCHANGE: GENDER AND NATURE IN THE WAR AGAINST SUBSISTENCE

To reshape nature as a means of turning local commons into globally-significant conservation areas, several debt-for-nature swaps were signed (Sheikh 2007, 2010). Among them is the Canada-Costa Rica debt-for-nature exchange which was a bilateral initiative that implemented so-called sustainable development programs in ACA-Tilaran. This exchange is the result of Costa Rica using the Official Development Aid Loan Agreement with Canada for three lines of credit to purchase fertilizers between 1983 and 1985. Costa Rican economic development during the 1980s debt crisis focused on increasing revenue from agriculture and cattle ranching. This agreement, signed in 1995, forgave half of Costa Rica's outstanding debt to the Canadian government (CAD$11.36 million), while the other half of the funds were donated. As part of international rules created by the creditors, the Canadian and Costa Rican governments

316

were not allowed to receive debt titles directly—these they gave to the World Wild Life Fund Canada (WWF-C), which co-directed the Arenal-Tilaran Conservation Area with MINAE and the Instituto Nacional de Biodiversidad (INBio), and at the same time, became the government's creditors (Government of Canada and Government of Costa Rica 1995). The aim of this debt-for-nature swap, according to WWF-C, was to protect the area from further environmental degradation, to help stabilize land use by strengthening natural resource management capacities, and to improve the quality of life of the Costa Rican people (Tremblay and Malefant 9,10, 11).

In ACA-Tilaran, the Ministry of Natural Resources, Energy and Mining (MINAE), the Canadian International Development Agency (CIDA) and the WWF-C developed the Plan General de Uso de la Tierra (or Land Plan) (MINAE *Plan General*). The Land Plan allowed managers to strategically remove the right to land from small-and-medium-sized farms, and to place the farmland into the hands of NGOs and MINAE. It disassembled the natural commons and enclosed 250,000 hectares of land on which people depended for their livelihood. In the ACA-Tilaran conservation area, I discuss three of several types of enclosures: the enclosure of biodiversity to promote competition for inventory and prospecting of the local folkloric knowledge of plants and animals; scenery for ecotourism; and use of medicinal plants through microenterprise. I will discuss each type of enclosure before explaining how these have resulted in a war against nature, peasants, Indigenous people, and women in Costa Rica.

i) Enclosure of Biodiversity for Biotechnology
The Convention on Biological Diversity of the 1992 Earth Summit opened Costa Rica's gene resources, turning nature and community knowledge into areas of secrecy and paranoia. In 1998, Costa Rica created Biodiversity Law No. 7788, in which Article 6 determines that biochemical and genetic properties of wildlife and domestic biodiversity are in the public domain; therefore, the state authorizes exploration, investigation, bioprospecting, and use of biodiversity elements (Costa Rica Biodiversity Law 7788). The government also created Comision Nacional para la Gestion de la Biodiversidad (CONAGEBIO), to develop and coordinate policies on biodiversity. It established four types of permits, one of which was for bioprospecting, which CONAGEBIO placed at the centre of its activities (Organization for Tropical Studies). The collection of highly selective genes from plants and animals was initiated in the conservation areas by parataxonomists working for

international NGOs, and further developed through experiments by the pharmaceutical, medical, and agricultural industries of the developed world (Mateo).

The protection of natural resources through central management is very costly. To obtain international money, environmental NGOs managed international donations and were central to most commercial debt-swap transactions (Hitz) that have been opened up for private investors to construct a multi-billion dollar extraction machine. For instance, in 2009, the Costa Rica Forever Association (CRFA), organized by the Nature Conservancy, the Linden Trust for Conservation, the Gordon and Betty Moore Foundation, and the Walton Family Foundation, set up a debt-for-nature fund. In 2010, SINAC and CRFA signed a five-year cooperation agreement to facilitate delivery to CRFA of the intellectual property rights on research and other studies to be carried out in the protected wild areas. By 2010, the growing foreign debt had become the best way to entrap forested countries into debt-for-nature transactions under a new name—the *Tropical Forest Conservation Act* (Sheikh 2010:14). "Best known for enabling "debt-for-nature" swaps, the *Tropical Forest Conservation Act* (TFCA) of 1998 offer eligible developing countries options to relieve certain official debt owed the U.S. Government while at the same time generating funds in local currency to support tropical forest conservation activities" (USAID).

Bioprospecting often becomes biopiracy. Biopiracy amounts to the appropriation of the traditional knowledge and biogenetic resources of Indigenous peoples and peasants to feed the knowledge systems of colonialist corporations.

ii) Enclosure of Scenery for Ecotourism

By the early 1990s, under pressure from the International Monetary Fund (IMF) and the World Bank, indebted Costa Rica had become the most important ecotourism destination in Latin America, adopting travel and ecotourism as a strategy of sustainable development and as an employment priority in the hope that it would bring foreign exchange and investment to repay its foreign debt (Honey).

In 1996, President Jose Maria Figueres signed Forestry Law No. 7575 and put into effect Article No.2, on land expropriation affecting small-and-medium sized landholders, in most cases without compensation to the owners (Vizcaino). New areas of intervention for ENGOs became "wild areas," which they proposed to "sell" to mostly Northern consumers for recreation. Ecotourism is promoted as an activity that contributes toward economic growth and generates income for local communities while purportedly protecting the environment. Ecotourism promoters promise aesthetic and recreational benefits that will restore a visitor's physical, emotional and spiritual health—a world of leisure, freedom and good taste, risk-free for those with money to spend. Simultaneously it is supposed to be politically-empowering and economically advantageous for some of the most disadvantaged groups of society, namely poor peasants, rural women, and Indigenous peoples (Honey; Stronza and Durham).

iii) Enclosure of medicinal plants through microenterprise
Since the 1992 Earth Summit, in Costa Rica every project of international cooperation has included both sustainable development and women in development (WID) components to ensure gender equity. But in Costa Rica, women's programs have very little government support, as a result of which NGOs have used debt-for-nature swaps as a credit source to initiate income-generating activities.

In the microenterprise model, credit is given to a group, which manages the money and within which individuals can borrow and repay small amounts that they use to set up a small business or microenterprise. Under neoliberalism, microenterprises firstly strengthen NGOs with access to international donors, and secondly, the preference of northern consumers for organically produced plants and vegetables is considered a growing market expected to benefit rural areas economically (Rodriguez).

SUSTAINABLE DEVELOPMENT'S TRIPLE CRISIS

From an ecofeminist perspective, the past practices of colonialism and the current day practices of economic development and sustainable development projects have been tools used by patriarchal capitalism to commit violence against feminized bodies and nature. If nature is violated for capitalist gain, then peasant and Indigenous women and men, who have been feminized and defined as closer to nature than others, can be equally violated. In this section, I explain how the three types of enclosure that have come with the "greening" agenda of sustainable development have resulted in a war against nature, peasants, Indigenous people, and women in Costa Rica. I thereby paint a stark picture of sustainable development's triple crisis.
i) The "Greening" of Biodiversity for Biotechnology and Intellectual Property Rights

Several anti-democratic characteristics exemplify the weakness of Public-Private Partnerships of the Conservation Areas. Here I will highlight only one: a privileged relationship to state power played a central role in controlling large segments of public assets, particularly nature. These assets freely passed into the private domain of NGOs, securing their tremendous economic power.

The WWF-C established itself in the Arenal-Tilaran Conservation Area (ACA-Tilaran), considered one of the richest biodiversity areas in Costa Rica (MINAE, *Plan General*). The PROACA project of WWF-C in partnership with the Monteverde Conservation Association collects material and researches flora and fauna in national parks, biologic reserves, protected zones, the national wildlife sanctuaries and forestry reserves (Asociacion Conservacionista Monteverde and WWF-C). The project consisted of two research components, each of them limited to five years. The first phase aimed to "help regenerate the tropical forest" and to carry out an inventory of flora. In the second phase, the production of commodities from products originating in the areas of biodiversity was to be attempted.

The National Biodiversity Institute (INBio) was established in 1991 by fifteen shareholders, most of them high-level functionaries in the Costa Rican government. INBio's main objective was to provide the country with research and development in the areas of biotechnology and chemistry of the rainforest by prospecting for industrial products of high value (Mateo). Under the neoliberal framework, these functionaries were able to transfer to INBio the right to sell the biological resources that belonged to the nation and to the local communities. In 1994, it established a partnership with MINAE to collect samples of plants and animals from the conservation areas for interested industries. To manage biodiversity, it created four divisions where the Biodiversity Prospecting Division received the Canada/Costa Rica debt-for-nature funds. This division did systematic research on new sources of chemical compounds, genes, and proteins produced by plants, insects and micro-organisms that may be of use to pharmaceutical, medical, and agricultural industries, worldwide (Mateo).

The new management transformed zoning in ACA-Tilaran; for instance, the Arenal Volcano category of "Forestry Reserve" with five protected hectares was changed to Arenal Volcano National Park with 12,010 hectares, which subsequently became the centre for bioprospecting (or biopiracy) (MINAE, "Area de Conservación"). Once research centres were organized, these areas became off-limits to rural communities. Local community members' intimate knowledge of and connections with the land were broken, unless they agreed to become part of the taxonomic research program. This transformation of nature into a commodity robbed communities of their livelihood. Living organisms became raw material and living knowledge is eliminated. Communities' work over centuries as the keepers of nature has no value, while scientific labour was perceived as adding value to the "free" resources of nature.

The newly declared nucleus area became private land controlled by park rangers organized into a Police Control Unit, trained and designed to counter land invasions. When park rangers find community members in designated areas without permission or without paying the necessary fee, they confiscate any fish or game these individuals might have obtained and whatever tools they used to do so. They then report the offence to the Office of Public Prosecutor.

As the eco-laboratory model developed in Costa Rica disassembled the natural commons on which rural people depended for their livelihoods, hunters initiated a campaign against national symbols. For instance, in Guanacaste, on May 9, 2001, "illegal hunters" as the Costa Rican government calls them, allegedly burned La Casona de Santa Rosa. This attack on a symbolic building revealed the existence of these traditional local hunters and their status as a marginal segment within society. Then head of MINAE, Elizabeth Odio, acknowledged that the hunters were acting in retaliation to being harassed by park guards when hunting deer and other species (Loaiza and Zeledon). MINAE also reported that in 2014, 699 hectares were burned; and in 2015, 2999 hectares were torched, including close to 1500 hectares in Diriá National Park in Guanacaste (cited in Arguedas). Luis Diego Roman, from MINAE, said that those who originated these fires were presumed to be deer hunters (qtd. in Arguedas).

In sum, the Land Plan undermined the rights of local communities to use their surrounding environment because conservation areas are used as collection centres of single samples for potential profits by interested industries.

ii) The "Greening" of Women and Children

Ecotourism creates waged employment, often at the expense of subsistence activity. For peasant women and men, the disappearance of forests is an issue of survival, forcing many to migrate to the large cities in the hope of earning an income for themselves and their dispossessed families. Many women migrate to ecotourism areas in search of jobs. In ACA-Tilaran, for example, the building of hotels, cabins, bed and breakfasts, and eco-tourist lodges has meant that not only have volcanoes, mountains, rivers, and forests been marketed and sold as recreation products, but the resident community has also been turned into a branded product for sale to customers (tourists) in a variety of forms.

Women and children are affected in an especially acute way by the processes of enclosure that come with "greening." When forest dwellers are evicted from their lands and move to urban areas, dispossessed women and children are usually the most vulnerable, becoming victims of predatory industries and individuals. Hidden beneath the veneer of ecotourism, sex tourism offers women and children's feminized bodies as commodities that are pure, exotic, and erotic.

As Costa Rica slid into a subordinated position internationally, the country became a paradise for sexual trafficking, paedophilia, and child pornography. In this way, just as nature has been commodified, the bodies of women and children are turned into another form of human capital so they can to contribute to the global tourism industry, to the wealth of businesses, and to state coffers to pay its debt. This image of Costa Rica entangles two aspects of capitalist patriarchal economics: the domination of creditors (core countries) over debtors (the periphery); and the domination of men over women. The two are related: as Costa Rica is increasingly impoverished by foreign debt and the enclosure of the commons, the mark of international gender power relations is stamped on the bodies of Costa Rican women and children.

Jacobo Schifter estimates that there are between 10,000 and 20,000 sex workers in Costa Rica, and between 25,000 and 50,000 sex tourists. He calls them "whoremongers," meaning regular clients who visit each year. The vast majority—eighty percent—are U.S. citizens (43). Tim Rogers reports that the U.S. has become Costa Rica's pimp, as crack cocaine and sex with prostitutes helps male tourists and old retirees affirm their masculinity and "escape reality" from their dissatisfied financial and social decline back home.

Since 2001, international human rights groups have put the Costa Rican government under intense scrutiny for lack of action against sexual abusers of children, most of them tourists. In 2001, the ex-president of Costa Rica, Miguel Angel Rodriguez, said on ABC's *20/20* news program that there were only "twenty or thirty" children being sexually exploited in Costa Rica, even though the U.S. Department of State had estimated three thousand children to be victims of commercial sexual exploitation in Costa Rica ("Costa Rica policeman").

Most of the pimps who profit from the organization of sex tourism are men from countries of the Global North, such as the U.S., Canada,

THE "GREENING" OF COSTA RICA

Spain, Germany, and Italy. At *Tico Times*, an English language weekly newspaper in Costa Rica, agencies advertise various types of prostitution, such as mail order brides, and dating and escort agencies, aimed specifically to tourists and U.S. migrants. On the internet there are hundreds of websites selling Costa Rican women and children.

These problems affecting women and children in Costa Rica are also the concern of foreign NGOs. Since 2004, World Vision has been launching campaigns in Costa Rica to deter potential child-sex tourists. In 2006, the NGO AYUDA, of the Inter-American Development Bank and the Ricky Martin Foundation, established an anti-trafficking campaign, entitled "Llama y Vive" ("Ricky Martin Fights"). The United Nations Population Fund/Costa Rica revealed that every ninety minutes a baby is born whose mother is between twelve and seventeen years old. Its project *Iniciativa Mesoamerica* intends to reduce the number of pregnant children; however, in 2016, due to poverty and economic inequality, the phenomenon of fathers "renting" their daughters as child prostitutes that was still occurring throughout the country, particularly in Limon Province (Mata).

iii) The "Greening" of Women's Knowledge

Until the end of the 1980s, medicinal plants had no market value, but they were used by rural populations whose heath depended on the ecosystems where they lived. Through centuries of growing medicinal plants, women in rural areas acquired the necessary skills and the knowledge of seeds, soil preparation, and optimum growing seasons. They also have the know-how to make combinations of plants used in healing practices. This knowledge was central in developing organic medicinal plant microenterprises.

In microenterprise development, medicinal plants become commodities, thus belonging to the market place. There, medicinal plants lose their cultural, and even their biological power, because the removal from their natural surroundings and different exposure to sunlight change their chemical properties (Biologist Celso Alvarado, personal communication, August 1999). Consequently, the women who grow them become commodity producers. As such, exploitation and surplus value are located in the bioproductive aspect of this knowledge—that is, in the privatization and monopolization of the results.

The results showed that although they had experienced some increased status and sense of agency in the community, the Abanico Project members had suffered serious negative effects. Since the 1990s, neoliberal policies have dictated that credits must be sustainable, that is, credits to microenterprises must cover operational costs. As a result, women became indebted to NGOs. For instance, each stage of the Abanico project was built on loans from NGOs acting as banks. Funding from FUNDACA provided loans at an annual interest rate of 20 percent, while ANDAR lent the money at an annual interest rate of 33 percent.

The women's working time thus expanded, decreasing the time they could spend on activities important to the community and their families. Their work was time-intensive and labour-intensive as they use machetes to cultivate and natural pesticides to reduce pests. Despite working nine hours in the medicinal plant plots, the women also worked many more hours at home, engaged in cleaning, cooking, washing, ironing, caring for their elders and rearing children, and doing community work.

To develop a microenterprise, families had to convert a substantial part of their land from food production to the production of medicinal plants. Yet the financial return to the women was not sufficient to buy food for the family's subsistence needs, because they no longer owned plots of land for personal use. In this way, women's autonomous subsistence work was eliminated in favour of poorly paid work that exploited their labour and Indigenous knowledge. In the face of daily inflation and devaluation imposed by the IMF's and the World Bank's, stabilization and structural adjustment policies, wages were inadequate. Working for less than the minimum wage and exhausted by a never-ending competitive spiral of reduced real wages, women's labour power was housewifized.

In a conversation regarding the market value of medicinal plants, one woman explained that most members were not talking to each other:

[D]eciding how many kilos each member could sell has produced a crisis in the group. Many of us are not talking to each other, although we are sharing the same misfortune, because we have in common the need to improve the economic situation of our families, we need to find support. We accepted the Consejo Nacional de Producción (CNP) initiative to discuss our problems with a psychologist. Our betterment will help our children and our community. (Yerbabuena, personal communication, August 2001)

In a word of one CNP official, "The group was not tough enough to survive the globalization project." The message was that the women needed to alter their psyches to be able to participate in and profit from the globalized market. Compelling women into visits with a psychologist displayed the tendency to see women as pathologized, instead of recognizing that the women's lagging economic performance resulted from the reality of their working conditions and global market mechanisms.

When the conversation turned to the use value of medicinal plants, it became clear that the plants embodied relationships. Naming the benefits that medicinal plants brought to their lives connected the women, generating a lively exchange and great communication. It brought memories of family members, grandmothers and mothers, times of sickness and happiness, scents and pleasures. The women recognized how their individual lives had been enriched because of their regular use of the plants and their ever-increasing knowledge. Speaking about the medicinal plant knowledge of their mothers and elders, they realized that they possessed an entire knowledge system that had been made invisible and undesirable by a world-view that sees as backward rural women's work to protect and conserve nature, work that sustains human life and ensures self-provision.

CONCLUSION

In this chapter, I have showed how sustainable development discourse, developed by the neoliberal "green" economy in Costa Rica, enforced the ideology that environmental commons are international, part of the global economic rationale, which means that individual states lost their decision-making power over use of their own territories in order to serve a global market machine. I have exposed processes by which mountains of debt and interest on loans undermined sustainable and livable economies, and inaugurated "greening" as a new phase of capital accumulation that entails four aspects. First, it entails the expansion of credit instruments, such as debt-for-nature exchanges by financial capital to create economic growth. In this frame, the debtor's obligation is to allocate domestic resources for financing ecological projects in exchange for extinguishing a limited portion of the country's foreign debt. Countries must be considered close to default to sell foreign debts at a fraction of their value in the secondary markets where one investor purchases a debt title from another investor rather than from the issuer country.

Second, it involves the World Bank licensing big ENGOs to broker the indebted countries' resources with large corporations involved in economic restructuring and globalization. The role of ENGOs is to establish the monetary values of the "'global commons" of the indebted periphery, such as biodiversity, forest, scenery, and mountains, and to export these values through stock exchanges. I gave the example of the Canada-Costa Rica debt-for-nature agreement that channelled funds to WWF-Canada and the Costa Rican INBio. These new experts, most of them biologists grouped in ENGOs, have emerged as new models of modernization and environmental protection by using the discourse of "protecting" the global commons in ecologically sensitive areas.

Third, new types of markets were organized—biodiversity for biotechnology and Intellectual Property Rights, forests for carbon credits, scenery for ecotourism, mountains for open-pit mining—in conservation areas. Fourth, the "greening" process results in peasants and Indigenous people acquiring new roles as service providers in new industries such as ecotourism. These roles are gendered in that they reproduce existing gender roles: men work as tour guides while women work in hotels and resorts as maids or as prostitutes. Viewed through the lens of ecofeminism, "greening," enclosure, and "housewifization" come together to wage war on subsistence.

Based on this analysis, my conclusion is that UN sustainable development promotes unsustainability and poverty. In Costa Rica, assigning monetary value to the commons required the devaluing of other forms of social existence, such as transforming agriculture skills into deficiencies; commons into resources; knowledge of biodiversity into ignorance; peasants and Indigenous peoples' autonomy into dependency; and self-sufficiency of men and women into loss of dignity for women's and children's bodies.

This chapter is excerpted from my book, The "Greening" of Costa Rica: Women, Peasants, Indigenous People, and the Remaking of Nature.

REFERENCES

Arguedas, Carlos. "Mano criminal provoca cuatro incendios en Parque Diriá." *La Nación* 21 April 2015. Web.
Asociacion Conservacionista de Monteverde and World Wildlife Fund-Canada (WWF-C). *Programa de Investigación para el uso racional de la biodiversidad en el Area de Conservación Arenal: Proposal Presented to MINAE,* 1996.
Bennholdt-Thomsen, Veronika and Maria Mies. *The Subsistence Perspective: Beyond the Globalized Economy.* London: Zed Books, 1999.
Boserup, Ester. *Women's Role in Economic Development.* London:

Allen & Unwin, 1970.

Costa Rica Biodiversity Law 7788. WIPO (n.d.). Web.

"Costa Rica policeman convicted for helping child pimp escape." Online forum post, Casa Alianza. 31 January 2001. Web.

"Costa Rica–Unemployment. Unemployment, Youth Female (% of female labour force) (national estimate)." Index mundi (n.d.) Web. Online. 15 October 2015.

Federici, Silvia. *Caliban and the Witch: Women, the Body and Primitive Accumulation*. New York: Autonomedia, 2004.

Fenech, A., R. Hansell, A. Isla, and S. Thompson, eds. *Report of an April 23, 1999 Workshop on Natural Capital: Views from Many Perspectives*. Toronto: University of Toronto, Institute for Environmental Studies, 1999.

Government of Canada and Government of Costa Rica. *Memorandum of Understanding between the Government of Canada and the Government of the Republic of Costa Rica concerning the Canadian Debt Conversion Initiative for the Environment*, 1995.

Hamilton, K. "Genuine Savings, Population Growth and Sustaining Economic Welfare." Paper presented at the Conference on Natural Capital, Poverty, and Development, Toronto, Ontario, Canada, September 2001.

Hitz, W. "The 'Debt for Nature Swap': Meeting Costa Rica's Conservation Needs?" Master's thesis. University of California, Los Angeles, California, 1989.

Hobsbawm, Eric J. *The Age of Revolution, 1789-1848*. New York: Vintage Books, 1996.

Honey, Martha. *Ecotourism and Sustainable Development: Who Owns Paradise?* Washington, DC: Island Press, 1998.

Loaiza V. and I. Zeledon. "Golpe a la Historia: Ardío la Casona." *La Nación* 10 May 2001. Web.

Mata, Esteban. "Niñas son 'alquiladas' con fines sexuales por sus padres." *La Nación* 26 July 2016. Web.

Mateo, N. "Wild Biodiversity: The Last Frontier? The Case of Costa Rica." *Globalization of Science: The Place of Agriculture Research*. Eds. C. Bonte-Friedheim and K. Sheridan The Hague, Netherlands: International Service for National Agricultural Research, 1997. 113-122.

Mies, Mies. *Patriarchy and Accumulation on a World Scale: Women in the International Division of Labour*. London: Zed Books, 1986.

Ministry of Natural Resources, Energy and Mining (MINAE). *Plan General de Uso de la Tierra*. Vols. I-IV. San José: MINAE, Agencia

Canadiense de Desarrollo Internacional, and Fondo Mundial para la Naturaleza de Canadá, 1993.

Ministry of Natural Resources, Energy and Mining (MINAE). "Area de Conservación Arenal." ACA-Tilaran Nucleos Areas, 1997. Organization for Tropical Studies. Web.

Pearce, David W. and Jeremy J. Warford. *World Without End: Economics, Environment and Sustainable Development*. New York: Oxford University Press, 1993.

Programa Estado de la Nación (PEN). *Estado de la Nación en Desarrollo Humano Sostenible*. San Jose, Costa Rica, 1996.

Programa Estado de la Nación (PEN). *Estado de la Nación en Desarrollo Humano Sostenible*. San Jose, Costa Rica, 2010.

Petch, T. "Costa Rica." *The Dance of the Millions: Latin America and the Debt Crisis*. Ed. J. Roddick. London: Latin America Bureau, 1988. 191-215.

Quesada, C. *Estrategia de Conservación para el Desarrollo Sostenible de Costa Rica*. [San José, Costa Rica]: Ministerio de Recursos Naturales, Energía y Minas, República de Costa Rica, 1990.

"Ricky Martin Fights Human Trafficking." Dalje.com, 30 April 2008. Web.

Rodriguez. S. "Los determinismos mercantiles y tecnocraticos en el 'modelo' de funcionamiento del INBio." *Ambien-Tico* 32 (1995): 11-16.

Rogers, Tim. *Costa Rica's Sex-Tourism Is Growing. Tico Times* 16 October 2009. Web. 15 October 2015.

Salleh, Ariel. "Nature, Woman, Labor, Capital: Living the Deepest Contradiction." *Is Capitalism Sustainable? Political Economy and the Politics of Ecology*. Ed. M. O'Connor. New York, London: The Guilford Press, 1994. 106-124.

Schifter, Jacobo. *Viejos Verdes en el Paraiso: Turismo Sexual en Costa Rica*. San José, Costa Rica: Editorial Universidad Estatal a Distancia, 2007.

Sheikh, Pervavez A. *Debt-for-Nature Initiatives and the Tropical Forest Conservation Act: Status and implementation (CRS report for Congress RL31286)*. Washington, DC: Congressional Research Service, Library of Congress, 11 October 2006. Web.

Sheikh, Pervavez A. *Debt-for-Nature Initiatives and the Tropical Forest Conservation Act: Status and Implementation*. Research Gate, April 2010. Web. 15 October 2015.

Shiva, Vandana. *Staying Alive: Women, Ecology and Development*. London: Zed Books, 1989.

Stronza, Amanda and William H. Durham, eds. *Ecotourism and Conservation in the Americas.* Ecotourism Series No. 7. Cambridge, MA: CABI, 2008.

The Costa Rica Forever Association. Web.

Tremblay, C. and D. Malenfant. "Estrategias locales para favorecer la sostenibilidad de acciones de desarrollo El Caso del Proyecto de Conservacion y Desarrollo Arenal, Costa Rica." Paper presented at Congreso Mundial para la Conservacion. Montreal, Quebec, 1996.

Ulloa, C. *Diagnostico socioambiental de la unidad territorial priorizada.* La Fortuna, San Carlos: Proyecto de Conservación y Desarrollo Arenal, Tilaran, Costa Rica, 1996.

United Nations Environment Programme. *What is the Green Economy?* (n.d.). Web. 15 October 2015.

USAID. "Financing Forest Conservation: An Overview of the *Tropical Forest Conservation Act.*" (n.d.). Web. 30 March 2018.

Vaughan, C. "Ponencia Sectorial: Biodiversidad." Paper presented at the Memoria-Primer Congreso Estrategia de Conservacion para el Desarrollo Sostenible (ECODES). San Jose, Costa Rica, October 1988.

Vizcaino. I. "Deuda millonaria por las expropiaciones." *La Nación* 29 August 1999. Web.

Waring, Marilyn. *If Women Counted: A New Feminist Economics.* San Francisco, CA: Harper and Row, 1988.

World Commission on Environment and Development, ed. *Our Common Future.* Oxford, UK: Oxford University Press, 1987.

World Vision. "World Vision's work to prevent child sex tourism." 2006. Web.

16.
Earth Love

Finding Our Way Back Home

RONNIE JOY LEAH

HOW DO WE FIND OUR WAY BACK HOME to the ancient Goddess cultures that embrace the sacredness of all life? How do we move beyond the centuries long disconnection of humans from nature fostered by patriarchal cultures and religions? How do we remember our kinship with the Earth community?

These are questions posed to students in "Goddess Mythology, Women's Spirituality and Ecofeminism" the course I teach at Athabasca University. Earth Love, a concept introduced to me by June Watts, my teacher of sacred circle dance, expresses this paradigm shift. Earth Love celebrates the power and beauty of the Earth, indeed of all life. It carries us forward to wholeness, balance, and reconnection in today's fractured world. Earth Love reminds us: *"The Earth is our Mother, we must take care of her"* (chant by Libana).

This tribute to Earth Love expresses my own journey with Goddess. I am dancing with the Goddess, embodying Her timeless stories, sharing the teachings of ancient cultures which honour the sacredness of women and the Earth.

This new / ancient paradigm of ecological awareness is being carried forward by a growing river of understanding. It is fed by many streams: Goddess spirituality, ecofeminism, deep ecology, earth-based spirituality, Indigenous wisdom, and engaged Buddhism. They speak to us and we listen.... The Earth is sacred. She is alive. She is our Mother.

> We have a beautiful / Mother
> Her green lap / Immense
> Her brown embrace / eternal
> Her blue body / everything / we know.
> (Walker, cited in Bolen 39-40)

Hearing these words of African American writer Alice Walker inspires me to take better care of this planet I love as my Mother, this planet I love as myself. What is this paradigm shift, the shift in consciousness that brings us closer to Earth Love? It is "awareness of Earth as a living system... we belong like the cells in a living body" (*Joanna Macy and The Great Turning*). We are embedded in the sacred Earthbody (Spretnak). "The world is a being, a part of our own body" (Seed 6). We are nature, nature is us, and we are all sacred manifestations of the Goddess (Wells and Leah 123).

PARADIGM SHIFT: THE SACRED EARTH

Cultural ecofeminism, "the hands and feet of the Goddess in today's world," draws much of its inspiration from goddess mythology (Wells and Leah 113). Unlike the transcendent god of patriarchal religions who is separate from creation, the goddess is immanent in creation; she *is* creation. She is the source of life, death and regeneration. Ecofeminism, which links the domination of Earth and the exploitation of women, presents a vision of life free from all forms of oppression, including "naturism," the oppression of nature by humans (Wells and Leah 114). Ecofeminism, the union of feminism with deep ecology, looks to transform the destructive relations of humans and nature to a life-affirming culture that respects the web of life (Reuther 13). Deep ecologist Joanna Macy calls on us to move from "the industrial growth society to a life affirming society" (*Joanna Macy and the Great Turning*).

Feminist ecological responses reconnect spirituality with the material world, challenging patriarchy's false separation of spirit and matter. The subsistence ecofeminist perspective acknowledges the material "connection and continuity between the human and the natural" and recognizes the sacredness of the living Earth (Mies and Shiva 20; Shiva 4). Nature, "the complex web of processes and relationships that provide the conditions for life," is not separate from or external to our being (Shiva 8). Spirit is the life-force in everything: "we ourselves with our bodies cannot separate the material from the spiritual" (Mies and Shiva 17).

As an activist for the world's rainforests, deep ecologist John Seed calls on us to embody these understandings. He recalls a moment of "intense, profound realization" while defending the trees: "I knew then that I was no longer acting on behalf of myself or my human ideas, but on behalf of the Earth... on behalf of my larger self, that I was literally part of the rainforest defending herself" (Seed 5). Feminist-pacifist writer Barbara Deming beckons us to remember this "Spirit of Love" which connects us with the earth.

We are earth of this earth, and we are bone of its bone.
This is a prayer I sing, for we have forgotten this and so
The earth is perishing.
(Deming, reprinted in Seed i)

PRACTICES FOR SACRED ECOLOGY: COUNCIL OF ALL BEINGS

How do we open to these understandings? How do we come to "hear within ourselves the sound of the Earth crying?" This phrase borrowed from Buddhist teacher Thich Nhat Hanh underlies "The Council of All Beings," a practice developed by Joanna Macy and John Seed in 1985, which "opens us to experiencing our fundamental interconnectedness with all life" (Seed 5). I was fortunate to train with Joanna Macy in 2009, as a facilitator for the Council and "The Work That Reconnects." My recent experience facilitating "Peace with the Earth," a sacred ecology workshop in Calgary, has convinced me that, at a visceral level, we are indeed learning how to listen....

I played a heartbeat on my hand drum as we walked ceremonially out into the garden. We gathered in a semi-circle under the branches of a large tree. I spoke as the Guide:

This Council is called to order, on behalf of the future generations.

One by one, the beings introduced themselves through us...

I am a mama grizzly bear protecting her cubs. I am the krill in the ocean. I am the dirt under your feet. I am a tree dying from pollution...

After each being spoke, the circle answered:

We hear you.

As I guided this exercise, I no longer spoke in my human voice: I was the grizzly bear protecting her cubs. I spoke in her voice

... and I was angry with the humans! Other beings spoke:

Hear us, humans. This is our world too. Our days are numbered because of what you are doing. Listen to us.

We took turns listening as humans:

We hear you.

When all the beings had a chance to address the humans and call them to account, I spoke again as the Guide / Grizzly:

The humans are now frightened. Our life is in their hands. If they can awaken to their place in the web of life they will change their ways. What wisdom do you have to offer to the humans?

The beings offered us their insights, their powers, to stop the destruction of the world. We listened and we accepted these gifts with thanks, on behalf of all humans. I picked up my drum again to announce the closing of the Council, as I shifted back into human form.

The Council of All Beings was a transformative experience, a journey into the imaginal realm. It was a communal ritual experience, where we allowed the Earth and other life forms to speak through us. We expanded our human identities into our larger ecological selves. We spoke on behalf of the earth; this is an important step in experiencing the shift in consciousness that Joanna Macy describes in her film (*Joanna Macy and The Great Turning*). Buddhist teacher Thich Nhat Hanh reminds us: "Only when we recognize our connectedness to the earth, can real change begin."

The "Law of Dependent Co-Arising" expresses the fundamental Buddhist concept of "interbeing": our nonseparateness from a world where "all events and beings are interdependent and interrelated" (Kaza 57; "Joanna Macy").

We are seeing a convergence of three streams of thought and practice: deep ecology, Buddhism and ecofeminism. Most of all, we are remembering our Earth Love.

When we can truly see and understand the earth, love is born in our hearts. We feel connected. That is the meaning of love: to be at one. Only when we've fallen back in love with the earth will our actions spring from reverence and the insight of our interconnectedness. (Thich Nhat Hanh)

This is the revolution, the shift in consciousness that needs to happen: "We need to wake up and fall in love with the earth" (Thich Nhat Hanh).

CREATIVE AND SPIRITUAL PRACTICES TO EMBODY EARTH LOVE

The creative arts play a crucial role in this expression of Earth Love. "Ecofeminst arts... [are] essential catalysts of change" that (re)connect us with nature and spirit (Orenstein 279). In my own journey with the Goddess, it is through sacred circle dancing that I feel most connected to the earth and all life. Dancing creates a shift in consciousness, it provides a way to embody the peace, wholeness and unity I envision for the world (Leah 74). Our circle dances in Calgary often incorporate rituals to honour the sacred earth and the turning of the year. Ritual practices help to awaken and deepen our spiritual connections with the earth and her continuing cycles of birth, growth, decay, death and regeneration (Starhawk; Spretnak). Rituals help us to remember, to recognize and celebrate the sacredness of everyday life, to show gratitude and respect for our larger family, our nonhuman relations (Sanchez 222). We are all children of the Goddess. "In ritual we can feel our interconnections with all levels of being" (Starhawk 184). Rituals help to restore these sacred relations, allowing the Goddess to come alive in our bodies, minds and spirits.

Themes of interconnectedness with the living earth echo the ancient earth wisdom of First Nations. "The land, and all it has to teach, to give, and all it demands, is what it means to be Indigenous" (Alfred 10). Many Aboriginal cultures express values similar to the ancient goddess cultures. "All females are the human manifestations of the Earth Mother, who is the first and ultimate giver of life. Our instructions are *"Minobimaatisiiwin*—we are to care for her" (Williams and Johnson 252). Nishnaabeg women speak out about their sacred relationship with water, their responsibilities as caretakers of the land and keepers of the water:

We call the Earth "our real Mother," the land as our "Mother's lap" and water the blood of this Mother the Earth ... the

Nishnaabeg people view the land, water, plants, animals and sky world as one unified and interdependent living system that works to sustain us all. (Bedard 96)

Through movements such as Idle No More (which began in December 2012), Indigenous women in Canada have taken the lead in protecting the sacred earth and waters of our land. In an interview with Naomi Klein, Anishinaabeg writer and academic Leanne Simpson articulates a clear alternative vision to destruction of the environment: "Our systems are designed to promote more life ... (through) resisting, renewing and regeneration." The concept of *mino bimaadiziwin*—continuous rebirth—is a guiding cultural principle of Anishinaabeg society: it's about the fertility of ideas, bringing forward and acting on your dreams and visions, it's the principle of regeneration (Klein and Simpson 5). Leanne Simpson describes the Round Dances which became a feature of Idle No More events. The dances reflect the joy of building "authentic relationships with the land and the people around you.... Let's make this fun. It was the women who brought that joy" (Klein and Simpson 9).

I was fortunate to be involved in Round Dances organized by Idle No More activists in Calgary in 2013 and 2014. These were expressions of community celebration and renewal, rooted in Indigenous traditions and spirituality, drawing in the youth and elders, native and non-native participants. Simpson goes on to describe how the dances embodied joyful transformation: "Watching the transformative nature of those acts (the Round Dances), made me realize that it's the embodiment, we have to embody the transformation." What were the emotions generated by the dances? Simpson affirms that it was love, on an emotional, physical and spiritual level. "Like the love I have for my children or the love that I have for the land.... It was a grounded love" (Klein and Simpson 11).

EARTH LOVE: DANCING THE WORLD INTO BEING

In the words of Naomi Klein and Leanne Simpson, we are "dancing the world into being." This is the world of Earth Love: the paradigm shift to a life affirming culture; reconnected with our ecological selves, we are one with the sacred earth. Through ritual and dance we are remembering who we are. This is embodied transformation. "We dance to know ourselves and our place in the great scheme of things" (June Watts). "No longer separate from the web of life, *we are Gaia*" (Leah, "Foreword" iii). We have found our way back home.

REFERENCES

Adams, Carol J. *Ecofeminism and the Sacred*. New York: Continuum, 1995. Print.

Alfred, Taiaiake. Opening Words. *Lighting the Eighth Fire*. Ed. Leanne Simpson. Winnipeg: Arbeiter Ring, 2008: 9-12. Print.

Bedard, Renee Elizabeth Mzinegiizhigo-kwe. "Keepers of the Water." *Lighting the Eighth Fire*. Ed. Leanne Simpson. Winnipeg: Arbeiter Ring, 2008: 89-110. Print.

Berry, Thomas. *The Dream of the Earth*. San Francisco: Sierra Club Books, 1988. Print.

Berry, Thomas. *The Great Work: Our Way into the Future*. New York: Three Rivers Press, 1999. Print.

Bolen, Jean Shinoda. *Urgent Message from Mother: Gather the Women, Save the World*. Boston: Conari Presds, 2005. Print.

Deming, Barbara. "Spirit of Love." *Thinking Like a Mountain*. Gabriola, BC: New Society Publishers, 1988: 1. Print.

Joanna Macy and the Great Turning. Film by Christopher Landry. Video Project, 2015.

Kaza, Stephanie. "Acting with Compassion." *Ecofeminism and the Sacred*. Ed. Carol Adams. New York: Continuum, 1995. 50-69. Print.

Klein, Naomi and Leanne Simpson. "Dancing the World into Being: A Conversation with Idle No More's Leanne Simpson." *Yes Magazine*. Web. March 05, 2013.

Leah, Ronnie Joy. "Foreword to the Revised Edition." *Study Guide: Women's and Gender Studies 333*." Rosalie Wells and Ronnie Joy Leah. Athabasca University, Rev. 2010. iii-iv. Print.

Leah, Ronnie Joy. "Dancing for Peace and Healing: Spirituality in Action." *Canadian Woman Studies/les cahiers de la femme* 29.1,2 (2011): 72-76. Print.

Libana, "The Earth is Our Mother." *A Circle is Cast*. Chant.

Macy, Joanna and Molly Young Brown. *Coming Back to Life: Practices to Reconnect Our Lives, Our World*. Gabriola, BC: New Society Publishers, 1998. Print.

Macy, Joanna and Chris Johnstone, *Active Hope*. Novato, CA: New World Library, 2012. Print.

Nicholson, Shirley. Ed. *The Goddess Re-Awakening*. Wheaton, IL: Theosophical Publishing House, 1989. Print.

Macy, Joanna. "Dependent Co-Arising." Web. 2012.

Mies, Maria and Vandana Shiva. *Ecofeminism*. Halifax: Fernwood, 1993. Print.

"Peace with the Earth: A Sacred Ecology Workshop." Ronnie Joy Leah and Karen Huggins. Calgary: Project Ploughshares, April 18-19, 2015. Web.

Orenstein, G. F. "Artists as Healers: Envisioning Life Giving Culture." *Reweaving the World: The Emergence of Ecofeminism.* Eds. Diamond and Orenstein. San Francisco: Sierra Club, 1990: 279-309. Print.

Pollack, Rachel. "The Body Alive" (chapter 9). *The Body of the Goddess.* London: Vega, 2003. 223-238. Print.

Reuther, Rosemary. "Ecofeminism: Symbolic and Social Connections of the Oppression of Women and the Domination of Nature." *Ecofeminism and the Sacred.* Ed. Carol Adams. New York: Continuum, 1995. 13-23. Print.

Sanchez, Carol Lee, "Animal, Vegetable, and Mineral: The Sacred Connection." *Ecofeminism and the Sacred.* Ed. Carol Adams. New York: Continuum, 1995. 207-228. Print.

Seed, John. "Introduction." *Thinking Like a Mountain.* Gabriola, BC: New Society Publishers, 1988: 5-18. Print.

Shiva, Vandana. "Introduction: Women, Ecology and Health: Rebuilding Connections." *Close to Home.* Ed. Vandana Shiva. Gabriola Island, BC: New Society, 1994: 1-9. Print.

Simpson, Leanne, ed. *Lighting the Eighth Fire.* Winnipeg: Arbeiter Ring, 2008. Print.

Spretnak, Charlene. "Earthbody and Personal Body as Sacred." *Ecofeminism and the Sacred.* Ed. Carol Adams. New York: Continuum, 1995: 261-280. Print.

Starhawk. "Feminist, Earth-based Spirituality and Ecofeminism." *Healing the Wounds.* Ed. Judith Plant. Toronto: Between the Lines, 1989. 174-185. Print.

"The Work That Reconnects." Web.

Thich Nhat Hanh. "Wake Up to the Revolution." Web. April 8, 2015.

Watts, June. "Earth Love" *Sacred Circle Dance Workshop.* Calgary, August, 2015.

Wells, Rosalie and Ronnie Joy Leah. *Study Guide: Women's and Gender Studies 333:* "Goddess Mythology, Women's Spirituality and Ecofeminism." Athabasca University, Rev. 2010. Print.

Williams, Marjorie Johnson and Nadjiwon Johnson. "*Minobimaatisiiwin* – We Are to Care for Her." *Sweeping the Earth: Women Taking Action for a Healthy Planet.* Ed. Miriam Wyman. Charlottetown: Gynergy Books, 1999. 251-257.

Contributor Notes

Dorothy Attakora-Gyan studies mainstream feminist solidarity as an assemblage. She is interested in the role of shame and fear as it pertains to feminist solidarity, how not understanding and devaluing these multiplicities leads to a weaponization of them, thus, posing a threat to solidarity.

Jennifer Bonato is a feminist educator in the Sociology and Women's and Gender Studies programs at Brock University, Canada. Her activist work has involved positions as the Board President of YWCA Niagara Region, and community and national advocate for women and girls. Jennifer recently co-founded a small business that aims to embody many of her eco-feminist values.

Leigh Brownhill teaches Sociology and Communication Studies at Athabasca University in Alberta, Canada, with a focus on environment, social movements, food sovereignty and eco-feminism. She is an editor of the journal *Capitalism Nature Socialism,* as well as an artist, activist, and water protector. She is the author of the 2009 book, *Land, Food, Freedom: Struggles for the Gendered Commons in Kenya, 1870-2007,* and lead editor of the 2016 book, *Food Security, Gender and Resilience: Improving Smallholder and Subsistence Farming.*

Born in Saskatchewan, Margaret Kress, a woman of Métis, French, English, and German ancestry, is guided by the words of Elders in her quest of a transformative education and a conscious society. As teacher, advisor, and learner, Margaret works to explore and present discourses encompassing inclusivity, gentleness, traditional land-based knowledges, and justice frameworks to help others see in new ways. She has worked closely with Elders and knowledge keepers throughout Canada in the area of Indigenous wellness, matriculture, and environmental justice. Currently, she supports students and faculty at the University of New Brunswick in teaching, research, and critical issues associated with Aboriginal education, Indigenous research methodologies, environmentalism, storywork, and decolonizing and self-determining practices.

Veronika Bennholdt-Thomsen, professor, ethnologist and sociologist, has lived and researched in Mexico for many years. Her topics include feminist research and rural and regional economy in Latin America and Europe. She has lectured and researched at universities in Germany (Bielefeld, Berlin), The Netherlands (The Hague) and Austria (Vienna, Klagenfurt). She currently works in the Institute for Theory and Practice of Subsistence (Institut für Theorie und Praxis der Subsistenz, ITPS e. V.) in Bielefeld, Germany, and teaches "The Culture of Subsistence" at the University of Natural Resources.

Irene Friesen Wolfstone holds an MA in Integrated Studies from Athabasca University (2016) and is currently engaged in doctoral studies at the University of Alberta where she studies Indigenous matricultures and Indigenous food sovereignty as models for cultural continuity needed for climate change adaptation. She lives on sacred land in Pinawa, Manitoba, located in Treaty 1 territory. Living in a round home is a catalyst for thinking outside the box.

Klaire Gain is settler-ally on Turtle Island and is an academic and activist dedicated to working in solidarity with mining impacted communities. Klaire completed her Masters at Brock University in Social Justice and Equity studies and is pursuing a PhD in Health and Rehabilitation Sciences at Western University. Klaire is passionate about ecofeminism and currently works at a feminist agency for women impacted by violence. Klaire believes in using her voice of privilege to share the narratives of women impacted by mining and utilizes scholarship as a platform to do so.

Ana Isla is a professor at the Department of Sociology and the Centre for Women's and Gender Studies at Brock University. She has taught Environmental Justice in the Sociology Department for several years. In 2000, Ana's doctoral thesis was nominated for the Governor General Gold Medal Award, and in 2001, it was selected as one of three best theses at the 25th Annual World-System Conference, Virginia

Polytechnic Institute and State University. In 2002, she was awarded the Rockefeller Fellowship at the University of Kentucky. She is the author of *The "Greening" of Costa Rica: Women, Peasants, Indigenous People and the Remaking of Nature* (2015).

Wahu M. Kaara is a global social justice activist from Kenya. She has been deeply involved in political and ecofeminist activism in her own country as well as through leadership participation in the World Social Forum and the Africa Social Forum, among other eminent organizations.

Ronnie Joy Leah is an educator, dancer, and feminist activist. She has a Ph.D. in Education (OISE/University of Toronto, 1986) and is a Certified Expressive Arts Practitioner (2009). She teaches Women's and Gender Studies at Athabasca University and is instructor/co-author for the online course "Goddess Mythology, Women's Spirituality, and Ecofeminism." She is currently writing a book about the life-affirming teachings of matriarchal goddess cultures. She trained with deep ecologist Joanna Macy in "The Work That Reconnects" and she facilitates sacred ecology workshops. She leads Sacred Circle Dances and is being mentored in the Dances of Universal Peace. She realized her vision of "Dancing for Earth Love" with a workshop at the Parliament of the World's Religions (2018, Toronto, Ontario). Ronnie Joy is a mother and grandmother living in Treaty 7 land in Mohkinstsis (the Blackfoot name for Calgary, Alberta) where she co-creates community dances and rituals for peace and healing.

Rachel O'Donnell is an Assistant Professor in the Writing, Speaking, and Argument Program at the University of Rochester and also teaches in Gender, Sexuality and Women's Studies. Her ongoing work is on feminist critiques of science, colonialism, and biotechnology. She has lived and worked in Latin America, and has previously published on Sor Juana de La Cruz, revolutionary movements, and migration.

Patricia E. (Ellie) Perkins is an ecofeminist Professor in the Faculty of Environmental Studies, York University, Toronto, where she teaches ecological economics, community economic development, and critical interdisciplinary research design. Her research and community projects with civil society and university partners address environmental and climate injustice, economic inequities, and the transition to sustainable provisioning.

Reena Shadaan is a PhD Candidate in the Faculty of Environmental Studies at York University. Currently, Shadaan is a member of the International Campaign for Justice in Bhopal-North America's (ICJB-NA) Coordinating Committee, which, under the leadership of survivors and grassroots activists in Bhopal, seeks justice for survivors of the 1984 Bhopal gas disaster. In addition, Shadaan is part of the Endocrine Disruptors Action Group, which is concerned with the widespread prevalence of endocrine disrupting chemicals in Canada.

Teresa E. Turner is an independent scholar working mainly on petroleum issues. She is a Marxist feminist who has taught at the University of Guelph, the University of Massachusetts, Rutgers University and elsewhere. She is the author of *Arise Ye Mighty People! Gender, Class and Race in Popular Struggles* (1994) among many other publications. She is also a member of the ecofeminist editorial collective of the journal *Capitalism Nature Socialism* and founding member of the International Oil Working Group, an NGO registered at the United Nations Department of Public Information.

Photo: Mario Rossini

Ana Isla is a professor at the Department of Sociology and the Centre for Women's and Gender Studies at Brock University. She has taught Environmental Justice in the Sociology Department for several years. She is the author of *The "Greening" of Costa Rica: Women, Peasants, Indigenous People and the Remaking of Nature* (2015).